Introduction to the History of
Plant Pathology

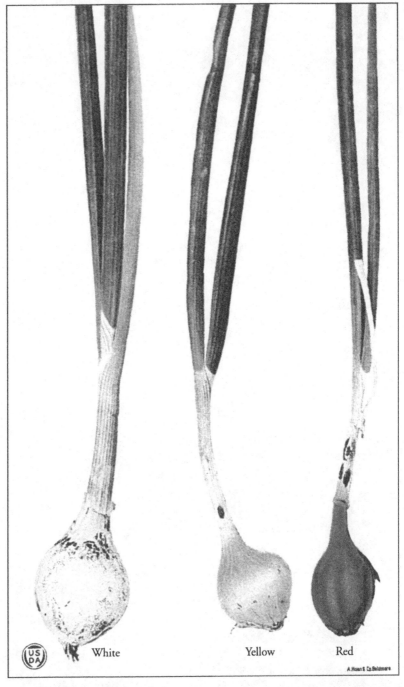

White   Yellow   Red

G. C. AINSWORTH

Introduction to the history of plant pathology

The author was formerly Director of the Commonwealth Mycological Institute, Kew

CAMBRIDGE UNIVERSITY PRESS

Cambridge

London   New York   New Rochelle

Melbourne   Sydney

CAMBRIDGE UNIVERSITY PRESS
Cambridge, New York, Melbourne, Madrid, Cape Town, Singapore, São Paulo, Delhi

Cambridge University Press
The Edinburgh Building, Cambridge CB2 8RU, UK

Published in the United States of America by Cambridge University Press, New York

www.cambridge.org
Information on this title: www.cambridge.org/9780521112833

© Cambridge University Press 1981

First published 1981
This digitally printed version 2009

A catalogue record for this publication is available from the British Library

ISBN 978-0-521-23032-2 hardback
ISBN 978-0-521-11283-3 paperback

To F.H.A.
who introduced me to plant pathology
fifty years ago

CONTENTS

PREFACE

The losses caused to staple crops by diseases and disorders and by depredations by pests are a world-wide problem at a time of population increase and food shortage. Plant pathology is thus an increasingly important branch of applied science. The exact scope of plant pathology is, as discussed in the Introduction, somewhat indefinite. Here emphasis is on the essential core: the study of infectious diseases of plants; the growth of which is traced by considering three basic aspects – diagnosis, treatment, and prevention – of the practice of the discipline. If this pattern is not inappropriate to an exposition of the principles of plant pathology, it is hoped that the orientation to the past rather than the present will give both plant pathologists and others an introduction to the history of the subject.

In general, the approach is similar to that of my *Introduction to the history of mycology* in which pathogenicity of fungi to plants was touched on and from which a few paragraphs have been adapted. An attempt has again been made to provide a concise and straightforward account of the historical development of the diverse and interwoven themes of which the subject is composed. This may be read without reference to the documentation which gives supplementary information and additional clues. The concluding bibliography cites the primary and secondary sources on which this study is based and also includes a number of representative plant pathological publications not specifically mentioned in the text. Many of the biographical references will be found to yield additional period detail.

G. C. A.

Topsham, Devon
November 1979

ACKNOWLEDGEMENTS

Much material, sometimes difficult to locate, had to be consulted during the preparation of this history and I am particularly grateful to the librarians of the British Museum (Natural History), Commonwealth Mycological Institute, Exeter University, Harpenden Laboratory of the Ministry of Agriculture, Rothamsted Experimental Station, and Royal Botanic Gardens, Kew for allowing me access to the collections in their charge and for other help.

I have to thank a number of friends – including Colin Booth, J. F. Bradbury, R. J. W. Byrde, D. L. Ebbels, S. D. Garrett, P. H. Gregory, Paul Holliday, P. L. Lentz, R. W. Marsh, Sheila M. Robb, and John Webster – for lending me books and offering me advice, either at my request or after reading sections of the manuscript, while my wife has read successive drafts and helped me with the proofs.

Photographs for many of the illustrations were taken by Mr J. Saunders of Exeter University Library's photographic department and Mrs M. G. Samuels prepared the final typescript.

For permission to use quotations I am indebted to the American Phytopathological Society, Professor R. H. Estey and Pflanzenschutz–Nachrichten Bayer and to reproduce illustrations to the following copyright holders or owners of the originals copied: Academic Press Inc., Fig. 57; American Phytopathological Society, Figs. 20, 23, 30, 33, 37, 47, 69, 70; Association of Applied Biologists, Figs. 32, 51, 52, 54, 68; Lady Bawden, Fig. 29; *Botanical Gazette* (University of Chicago Press), Fig. 31; The British Library, Fig. 9; Trustees of the British Museum (Natural History), Fig. 8; British Mycological Society, Figs. 2, 56, 59, 80; Cambridge University Press, Fig. 33; Christie's Contemporary Art, Fig. 27; Commonwealth Mycological Institute, Figs. 15, 25, 79; Controller of HM Stationery Office (Crown Copyright), Figs. 53, 61; Department of Agriculture, Victoria, Fig. 71; DSIR (Plant Disease Division) Auckland, Fig. 72;

Exeter University, Fig. 5; Imperial Chemical Industries Ltd (Plant Protection Division), Fig. 41; *Mycologia*, Figs. 1, 76; *Pflanzenschutz–Nachrichten Bayer*, Fig. 58; Rothamsted Experimental Station, Figs. 11, 28; Royal Botanic Gardens, Kew (Crown Copyright; reproduced with the permission of the Controller of Her Majesty's Stationery Office and of the Director, Royal Botanic Gardens, Kew), Figs. 4, 6, 12, 19, 23, 24, 65, 66, 78; School of American Research and the University of Utah, Fig. 3; John Wiley & Sons Inc., Fig. 38.

Introduction

I

Historical patterns of plant pathology

Introduction

The first point to clarify is one of circumscription. The terms medicine and veterinary medicine have, in English, no generally accepted equivalent for plants.[1]* A human sufferer from an infectious disease, a nutritional disorder, genetic abnormality or mental disturbance, internal or external infestation by animals, or accidental injury consults a practitioner of medicine who may refer the case to a colleague specialising in one of the numerous subdivisions of medicine. For diseases and disorders of plants the position is very different. Growth defects resulting from lack or imbalance of macronutrients (that is, manurial problems) and adverse effects on growth of other environmental factors (such as weather) typically fall into the province of the agriculturist or horticulturist, pest infestations into that of the applied entomologist (because of the large number of insect pests and the major economic importance of the depredations and injuries they cause to staple crops) while infectious diseases are the concern of the plant pathologist (or phytopathologist). But this last delimitation is far from clear cut. In some continental European countries plant pathology (phytopathology) covers both diseases and pests and elsewhere it is not uncommon for nematode infestations to be included in plant pathology although these, in the absence of a nematologist, are often referred to the 'entomologist' because of his zoological training and because he usually deals with a wide range of pests (e.g. slugs, snails, mites, etc.) which are not insects, while deficiency diseases (disorders induced by lack of essential trace elements), basically chemical problems, tend to be a plant pathological concern. The plant pathologist, who is usually crop orientated, must, in order to ensure the confidence of his grower clients, always have a working knowledge of the cultural requirements necessary to ensure well-grown plants and must be able to recognise the common pests.

* See Notes on the Text, pp. 245-51.

To define disease is not easy. *The Oxford English Dictionary* notes that the term is applied to 'a disordered condition in plants'. Plant pathologists, both as individuals and in committee, have made frequent attempts at an acceptable definition.[2] Currently, the Federation of British Plant Pathologists' *A Guide to the use of terms in plant pathology*, 1973 (p. 10) offers two – 'Deviation from normal functioning of physiological processes, of sufficient duration to cause disturbance or cessation of vital activity' (American Phytopathological Society, 1940); 'harmful deviation from normal functioning of physiological processes' (Plant Pathology Committee of the British Mycological Society, 1950) – and notes that these definitions 'are broad enough to allow malfunctions caused by nutritional deficiencies or excesses, toxic chemicals, adverse environmental factors, genetic anomalies etc. to be classed as diseases in addition to those caused by *infective* agents'. In conclusion, the *Guide* recommends that the use of the term 'disease' should be restricted to malfunctions caused by pathogenic organisms or viruses and that those caused by other factors (the so-called 'non-infectious diseases') should be termed 'disorders', which has long been the practice in the *Review of plant pathology* and is the one adopted in this book.

Plant pathology is concerned not only with determining the cause (*aetiology*) of diseases and disorders (an activity now acceptably covered by the term *diagnosis* which in its original medical usage was limited to the observation and interpretation of signs and symptoms) but also with their *treatment* (i.e. therapy, cure, amelioration) and *prevention* – two terms for each and both of which *control* is widely used. A plant pathologist's activities are thus wider than those of a medical pathologist whose interests are mainly confined to questions of aetiology. A very fundamental difference between the practice of medicine and that of plant pathology is that the former typically deals with individuals, the latter with populations (growing crops). Further, a human patient is willing to pay for treatment, either directly or through taxation, and thus medical treatment can be both elaborate and costly. Owners are frequently willing to pay for the treatment of their sick animals so that veterinary practice also employs sophisticated techniques. This is in marked contrast to the simpler skills used for the treatment of diseased plants where the economic value of the individual is usually small, its structure less elaborate than that of an animal, and the emotional involvement of the owner low.

Like other applied sciences, plant pathology is difficult to delimit from the branches of pure science on which it is based. There is a central core of investigatory work in which the scientific method is applied in both the laboratory and the field to practical problems of diseases and disorders of plants. This is overlaid (or, as some would have it, underpinned) by descriptive and experimental studies on the nature of the pathogens, their relationship to the host, and the effects of the environment on host–parasite interaction. Such studies fan out into a multitude of research projects in mycology, bacteriology, virology, entomology, nematology, general botany, plant physiology, chemistry, biochemistry, statistics, and other branches of science undertaken by workers who may be only remotely concerned with practical problems of plant disease. These studies, which may be considered to be a reflexion of the debt that 'pure science' owes to the 'applied science' of plant pathology,[3] do much to deepen knowledge of disease in plants but this increased understanding does not necessarily have a feedback which enhances the control of plant disease in the field. For example, the many investigations on the nature of the tobacco mosaic virus, though they have given major insights into the question 'What is life?', have, so far, done little or nothing to save tobacco and tomato crops from a common and widespread disease. J. G. Horsfall,[4] director of the Connecticut Agricultural Experiment Station for 24 years up to 1972, makes a distinction between the 'science of plant pathology' (that is, research designed to deepen knowledge of plant disease) and the 'art of plant pathology' by which the findings obtained from the 'science' are applied to the amelioration of disease in plants. He even suggested that those engaged in these two activities belong to different professions. It is often possible to detect a certain shrillness in the not uncommon insistence during recent years, usually by plant pathologists, that plant pathology is a science, and this leads to the suspicion that such claims originate from a feeling of inferior status – the remuneration of those who practise medicine, veterinary medicine, and plant pathology is a descending series. It is forgotten that many of those engaged in research on plant disease would (as already noted) repudiate the designation plant pathologist for biochemist, mycologist, or other label. Here again there is a parallel with medicine where a distinction can be made between 'scientific' and 'clinical' medicine in both of which research work is undertaken to deepen knowledge of human disease but the first category includes workers from numerous branches of science who do not legally belong to the medical profession. The standpoint from which this historical survey is written is that, as for medicine, the right end of plant pathological practice is the diagnosis, treatment, and prevention of diseases and disorders. Emphasis throughout is, therefore, on the more practical aspects of plant pathological endeavour.

Historical patterns

Losses from disease must have occurred since man first cultivated plants and although there are references to such depredations in early written records (some of which are noted in Chapter 2) these records are largely of antiquarian or anthropological rather than phytopathological interest. This, again, is in marked contrast to medicine which has been practised by professionals in all past civilisations – and even today primitive tribes have their 'medicine men' and 'witch doctors'. Thus for several millennia there has been a considerable body of knowledge (if of varied reliability) on human diseases, knowledge which has over the centuries gradually become more accurate and more extensive and given rise to modern medicine. Phytopathology is, on the other hand, a recent development. It began with the acceptance of the concept of pathogenicity, between 1750 and 1850, and the pattern of its general development shows a multiplicity of strands which are identified according to the interests and attitudes of any particular observer.

Major advances in science, like outstanding works of art and literature, depend on 'the man' and 'the power of the moment'. Historians commonly tend to stress one or other component of this duality. Professor H. H. Whetzel of Cornell University (Fig. 1) in his attractive little history of phytopathology (Whetzel, 1918) emphasised the first. As he wrote in the Preface, he 'endeavored ... to set forth in outline ... the most outstanding features in the evolution of the science, and to indicate the proper relation thereto of the men who have chiefly shaped its progress'. Whetzel treated phytopathology very broadly and two of the subdivisions of his 'Ancient Era' were the 'Theophrastian Period' and the 'Plinian Period' after Theophrastus and Pliny, the chief Greek and Roman classical authors to make reference to disease in plants. Divisions of the 'Premodern Era' included the 'Zallingerian' (eighteenth century) and the 'Ungerian' (early nineteenth century) periods while the first two subdivisions of the 'Modern Era', considered to have begun about 1850, were the 'Kühnian' and 'Millardetian' periods,[5] a nomenclature which again emphasised the outstanding men of the time as do the 22 illustrations which accompany the text which are all portraits of phytopathologists. By contrast E. C. Large (Fig. 2) in *The Advance of the fungi*, 1940 (a much appreciated popularisation of historical aspects of plant pathology) tended to emphasise 'the moment'. He reviewed 'the successive epidemics of fungal and other plant disease which have swept the crops of the modern world' to quote from the dust jacket. That is to say, he used famous outbreaks of plant disease such as potato blight in the 1840s in Ireland, vine mildew a few years later in France, coffee rust in Sri

Lanka in the 1870s, and fire blight in North America in the following decade, etc. to landmark advances in phytopathology because such epidemics, like epidemics of disease in man or other crises in human affairs, tend to stimulate the advance of knowledge and precipitate change.

A number of other trends are discernible. One is fragmentation. At first plant pathology was treated comprehensively as, for example, by the Rev. Miles Joseph Berkeley in his famous series of 173 articles on 'vegetable pathology' which appeared in the *Gardeners' Chronicle* between 7 January 1854 and 3 October 1857. After four introductory articles the next thirty-eight dealt with plants in a state of health and succeeding articles covered every aspect of plant abnormality whether genetical in origin, the result of insect damage, or caused by infectious disease. Such an approach has rarely been attempted since,[6] and as already discussed plant pathology currently limits its concern mainly to infectious disease. Another common distinction is to categorise the second half of the nineteenth century as the 'descriptive'

Fig. 1. Herbert Hice Whetzel (1877–1944).

Fig. 2. Ernest Charles Large (1902–76).

phase in contrast to the succeeding 'experimental' approach but this division like so many is not clear cut for Prévost, among others, was experimenting ably before 1810.

The 75 years following the general acceptance of the concept of pathogenicity by 1850 was the age of the pathogen when first fungi, then bacteria, and at the turn of the century viruses (but not until 1967 mycoplasmas) were recognised as major causal agents of disease in plants. Much effort was devoted to the characterisation of pathogens, particularly fungi,[7] and studies of their life histories and if at times undue emphasis was given to this aspect of disease much valuable data potentially applicable to the control of plant disease were accumulated. During this period, while the majority of plant pathologists – the 'pathogeneticists' in Whetzel's terminology – emphasised the role of the pathogen, a minority – the 'predispositionists'; the ideological descendants of Franz Unger, including Paul Sorauer and

Marshall Ward – emphasised the environmental factors which predisposed the plant to infection.

Since plants were first cultivated there must have been conscious (or unconscious) selection for desirable characters or uses including, presumably, disease resistance. The discovery in the opening years of the twentieth century that rust resistance in cereals was inherited as a Mendelian character stimulated the initiation and expansion of the large breeding programmes which have done so much, and still do, to mitigate crop losses from disease.

Another pattern can be delimited by considering the major chemicals used for disease control. Sulphur, employed in medicine since classical times, was the first fungicide to be used against plant disease. Then, from the introduction of Bordeaux mixture in 1885 followed 50 years during which copper was the dominant fungicide until the deployment of organic fungicides and later, though to a lesser extent, antibiotics.

The first legislative measures against plant diseases were *ad hoc* responses to emergencies. From the last quarter of the nineteenth century more comprehensive mandatory legislative approaches to the problem were developed in many countries. It was, however, early apparent that plant disease is no respecter of political boundaries but there were several abortive attempts before international agreement could be reached. Even today such international organisations as have been established usually lack mandatory powers, they are only able to recommend the ratification and implementation of desirable measures to ensure uniformity by international collaboration.

Currently the approach to disease in plants is better integrated than ever before. Disease is recognised as being dynamic rather than static and the total situation is considered – that is, the host–pathogen–environment interactions are taken into account. Intervention by chemicals or other agents is used in the most efficient way with the help of forecasting outbreaks of disease supported by legislation regarding the movement of plants, quarantine procedures, and the notification of disease outbreaks. Alternatively, the economic situation may indicate that non-intervention is in the best financial interest of the grower.

Yet another pattern is the geographical. Plant pathology like modern developments of science in general was of European origin and much of the early-nineteenth-century work is associated with German-speaking investigators who were at first in the van of progress. The French, Italians, and Dutch all made major contributions to phytopathology before the end of the century when the lead passed to North America. The European colonial powers were, however, responsible, particularly before and between the two World Wars, for initiating plant pathological services in their dependent territories, especially India and tropical Africa, while the United States did a similar service for the Philippines and elsewhere. This geographical development was paralleled by a transition of phytopathological concern from the national to the international arena (see Chapter 11*).

As for other branches of biology, students of plant disease have during the past two centuries become more and more professional. Lively-minded clergymen, school teachers, lawyers, and members of other professions whose hobbies[8] contributed so much to the foundations of phytopathology up to the mid-nineteenth century and beyond have been replaced by specially trained workers who have adopted phytopathology as a career, a trend that has been encouraged by the development in most countries of the world of plant pathology as a branch of the public services. They also have tended to specialise; inevitably on the diseases of one or more related crops, but also as mycologists, bacteriologists, virologists, geneticists, physiologists, fungicide chemists, etc. whose researches not infrequently lead them on divergent paths. Phytopathologists also share characteristics in common with research workers in general. Up to the beginning of the twentieth century most workers published as individuals while during recent years there has been a marked tendency for plant pathologists to work in teams (which supplement the traditional teacher–student collaboration in research) – a tendency encouraged by the necessity to enlist the help of other specialists such as biochemists, meteorologists, and statisticians in the solution of complex problems when the collaborators are credited with joint authorship of an eventual publication. This trend is well brought out by a random sampling of journal articles abstracted in the *Review of plant pathology* (see Table 10, Chapter 12) from which it appears that there has been an increase in the number of authors per paper from 1.2 to 2.0 over a 50-year period. The sample also shows that there has been a decrease in the length of journal articles since 1950. This is in part due to editorial directives resulting from escalating costs of publication and in part because with the large increase in the number of those engaged in research the average investigator tends to make smaller additions to new knowledge. Other factors which help to shorten papers are that research workers, particularly in fashionable fields, minimise delay in publication in order to establish priority by submitting their findings to the press in what are essentially progress reports and that it is usually easier to find a publisher (or pub-

* Cross-references not specified by page may be elucidated via the Subject Index.

lishers) for long papers when offered as a series of shorter papers, a procedure which, as a bonus, increases the number of an author's publications. The widely held belief that it is the number of an author's publications that accelerates preferment undoubtedly has an element of truth.

2

Beginnings: problems of aetiology up to 1858

Early references to plant disease[1]

Greeks and Romans

Historical studies on plant disease, as on many other aspects of plants, lead back to Theophrastus of Lesbos [*c*.371–287 BC], a pupil of Plato and Aristotle, whose *Historia plantarum* and *De causis plantarum* have earned him the appellation 'father of botany'. Both these works include references to disease in plants and, although it is uncertain to what extent the detail (which is frequently difficult to interpret in modern terms) is a compilation from other writings and hearsay or based on observations by himself or his contemporaries, it is clear that Theophrastus had a better and more objective grasp of plant disease than many of his successors for the next two thousand years.

Both *Historia plantarum* and *De causis plantarum* have special sections devoted to plant diseases (including depredations by insects) which Theophrastus distinguished from

certain affections due to season or situation which are likely to destroy the plant, but which one would not call diseases: I mean such affections as freezing and what some call 'scorching'. Also there are winds which blow in particular districts that are likely to destroy or scorch. (*HP*, Book IV, chap. 14)[2]

He noted a variety of diseases, for example:

The olive, in addition to having worms . . ., produces also a 'knot' (which some call a fungus, others a bark blister), and it resembles the effect of sun-scorch. (*HP*, IV, 14)[3]
 The fig is also liable to scab, and . . . is also often a victim to rot and to *krados*. It is called rot when the roots turn black, it is called *krados* when the branches do so; for some call the branches *kradoi* (instead of *kladoi*), whence the name is transferred to the disease. (*HP*, IV, 14)[4]

and he included a number of references to differences of disease incidence:

they say that wild trees are not liable to diseases which destroy them . . . Cultivated kinds however, they say, are subject to various diseases, some of which are, one may say, common to all or most, while others are special to particular kinds. (*HP*, IV, 14)[5]

Generally speaking, cereals are more liable to rust than pulses, and among these barley is more liable to it than wheat; while of barleys some kinds are more liable than others, and most of all, it may be said, the kind called 'Achillean'. (*HP*, VIII, 10)[6]

As to diseases of seeds – some are common to all, as rust, some are peculiar to certain kinds; thus chick-pea is alone subject to rot ... Some again are subject to canker and mildew, as cummin. (*HP*, IV, 14)[7]

Attention was also drawn to seasonal and predisposing factors which favour disease:

Scab [of figs] chiefly occurs when there is not much rain after the rising of the Pleiad [the beginning of summer]; if rain is abundant the scab is washed off. (*HP*, IV, 14)[8]

... lands which are exposed to the wind and elevated are not liable to rust [of cereals], or less so, while those that lie low and are not exposed to the wind more so. And rust occurs chiefly at the full moon. (*HP*, VIII, 10)[9]

The fig ... becomes diseased if there is heavy rain; for then the parts towards the roots and the root itself become, as it were, sodden, and this they call 'bark-shedding'. (*HP*, IV, 14)[10]

Moreover the wounds and blows inflicted by men who dig about the vines render them less able to bear the alterations of heat and cold ... Indeed, as some think, most diseases may be said to be due to a blow. (*HP*, IV, 14)[11]

In *De causis plantarum*, after repeating the assertion that rust of cereals occurs in hollow and windless places, Theophrastus writes (Book III, chap. 22):

Grain plants not blighted stand with bent head and not upright. So it is advantageous that the head be inclined to allow water and dew to run off. Large heads are more inclined to bending. Flat and short ones are more rigid and so they become 'rusted' by being saturated. It is also advantageous if the head is at a distance from the leaves, for moisture remains more in the leaves. If the head is near it (the leaf), it (the head) is touched at once; if it is farther away (it is) less (affected). Achillean barley both black and white are liable to 'rust' because their heads are upright. But genuine barley is safe on account of its bending heads. In areas exposed to the winds there is less 'rust', for the movements (created by the wind) shake off the moisture and droplets. When wind follows on rain and then night sets in, you have less 'rust'. This is because the wind shakes the head and the sun does not come out at once causing putrefaction, but the head is dried. 'Rust' is a kind of rot and nothing rots without heat from another source. Grain 'rusts' most at the full moon, for the moon gives heat that causes putrefaction during the night.[12]

and in Book IV, chap. 14 (see Orlob (1973):105–6) these observations are supplemented by the statement that 'Barley is more susceptible than wheat because the latter has no awns. Moisture can adhere to leaves putrefying these parts and when the head emerges, it comes into contact with the diseased part.'

Some three hundred years later diverse references to plant diseases were included by Caius Plinius Secundus [AD 23–79] in his *Historia naturalis*. [13] In this massive work (comprising 37 'Books') Pliny uncritically compiled (with or without acknowledgement) views culled from Theophrastus and other Greek and Roman authors. Among the causes of tree disease he gave prominence to 'sideration' – the evil influence of the heavenly bodies [14] – and with reference to garden plants he states that 'All plants, indeed, will turn of a yellow complexion on the approach of a woman who has the menstrual discharge upon her.' [15]

Although Pliny was more concerned with control measures than was Theophrastus, many of the remedies offered perpetuated superstitions, e.g.

As for mildew, that greatest curse of all to corn, if branches of laurel are fixed in the ground, it will pass away from the field into the leaves of the laurel. (Book xvii, chap. 46) [16]

and

Many persons, for the more effectual protection of millet, recommend that a bramble-frog should be carried at night round the field before the hoeing is done, and then buried in an earthen vessel in the middle of it. (xvii, 46) [17]

Pliny also drew attention to the Robigalia – festival days to propitiate the rust god Robigus –

which were established by Numa in the fortieth year of his reign, and are still celebrated on the seventh day before the calends of May [25 April], as it is at this period that mildew mostly makes its first attack on the growing corn. Varro fixes this crisis at the moment the sun enters the tenth degree of Taurus. (xviii, 69) [18]

In addition to these major accounts of plant disease by Theophrastus and Pliny, a number of other Greek and Roman authors (including Xenophon, Strabo, Cato, Varro, Virgil, Ovid, Palladius, and Columella) made incidental phytopathological references, many of which are noted by Orlob (1973). Of particular interest is the *Geoponica*, a compilation on Byzantine agriculture made by Cassianus Bassus during the seventh century AD, it is believed, and rewritten, at the command of the Emperor Constantine VII, in the middle of the tenth century. According to Orlob this work includes extracts from 33 writers, 31 of whom lived before AD 350. Like Pliny's *Historia naturalis*, the *Geoponica* has a practical bias so that when plant diseases are touched on control is emphasised but the remedies advocated are, almost all, irrational and unreliable. The *Geoponica* was, however, a popular and much consulted work. Printed editions appeared in Basle in 1539, London, 1704, and Leipzig, 1895.

Other civilisations

There are a number of pre-Hellenistic records of plant disease. A Sumerian clay tablet, dating from about 1700 BC, bears a Babylonian farm almanac in cuneiform script (see Orlob (1973), fig. 3). This includes a reference to barley turned red as suffering from 'Samana' disease. (Possibly the same disease was referred to somewhat earlier (*c.*2000 BC) in one of the very few references to plant disease found in Egyptian writings.) There are several citations in the authorised version of the Old Testament (*c.* eighth–fifth centuries BC of crops affected by 'blasting and mildew'[19] terms translated as 'black blight and red' in the New English Bible, 1970) and a few oblique references to plant disease in the post-biblical Talmud.

In the Far East, also, plant disease was recognised from early times. Charms against harmful influences on crops are included in the Vedic literature (1500–500 BC) and during the next millennium there are more specific references to 'mildew' and 'blight' in India. Marco Polo [1254–1324] in the account of his famous travels noted Chinese rites (including the sacrifice of a black-faced sheep) to ensure the health of farm crops and animals and Han Yen chih writing on oranges *c.*1170 attributed disease to 'hsien or fungus' and 'tu or boring insect'.

The three major pre-Columbian civilisations of the Americas, the Aztec, Maya, and Inca, were all firmly based on agriculture. The staple cereal of the first two was maize in which the Aztecs, at least, recognised disease (see Fig. 3) while the illiterate Incas of Peru cultivated the potato and, as it is from the Andes that potato blight (*Phytophthora infestans*) is now believed to have spread, they too almost certainly had plant pathological problems.

Finally mention must be made of the Arabs who during the Middle Ages provided one of the bridges between Greek and Roman culture and the Renaissance in Western Europe. The best known Arab writer in this field was Ibn-al-Awam who lived in Seville during the twelfth century where he wrote his comprehensive agricultural treatise *Kitab al-Felahah* of which there are nineteenth-century translations into Spanish and French. This compilation, which includes a chapter on disorders, diseases, and pests of the fig, olive, and other fruit trees, the grape vine, and a few herbaceous plants, is comprehensive if well spiced with irrationality.

Interpretation of early records

The interpretation of early records of plant disease, when not impossible, is usually difficult and the conclusions uncertain. The inclination to strain hindsight, when not resisted, frequently leads to more precise interpreta-

Fig. 3. Healthy (left) and smut (Ustilago maydis) *infected (right) maize cobs after Paso y Troncoso from the sixteenth-century manuscript on Aztec culture prepared by Barnardino de Sahagún (see Sahagún 1963). Cf. Fig. 19.*

tions than the evidence would appear to justify. One reason for this is that while some disease names such as 'blight' and 'blast' ('blasting') have survived as general and nonspecific terms, others including 'rust', 'smut', and 'mildew' at first used loosely (and in overlapping senses) are now circumscribed and relatively precise designations. This confusion was noted as early as 1707 by J. Mortimer in *The whole art of husbandry* where he wrote: 'Blights and Mill-dews have generally been taken to be the same thing which has begotten much error.' The Greeks used 'erysibe' (èrysíbh) for rust which the Romans called 'robigo' (or 'rubigo'). Linnaeus (*Philosophia botanica*, 1751) further complicated the situation. Following as he thought the usage of Theophrastus, he wrongly applied Erysiphe to both downy and powdery mildews and *Erysiphe* was subsequently taken up as a generic name for powdery mildews by Hedwig *fil.* and accepted by Fries (*Systema mycologicum* **3**: 234, 1829); an attempt by Walroth in 1819 to restore the use of the term for rusts having failed. In England 'mildew' was the usual term for rust from the time of Shakespeare[20] to the middle of the nineteenth century. Even as late as 1884 Worthington G. Smith in his textbook dealt with *Puccinia graminis* infection of wheat as 'corn mildew' (which he still believed to be unrelated to 'barberry blight'). Similarly, the German Rost (rust) and Brand (brand), now generally applied to rust (Uredinales) and smut (Ustilaginales) fungi (and the diseases they cause), respectively, have a history of less restricted applications.

Rusts and smuts must have attracted attention throughout historic times. One interesting record is the identification of the spores of *Ustilago hordei* (loose smut of barley) and also sclerotia of ergot (*Claviceps purpurea*) in the stomach contents of the corpses of two Iron Age (500 BC–AD 500) men found preserved in a Danish peat bog.[21] Other, indirect but convincing, evidence of early outbreaks of ergot of rye and other cereals is provided by

Fig. 4. Loose smut of wheat (Ustilago tritici) *from Jerome Bock (1551). Cf. Fig. 11.*

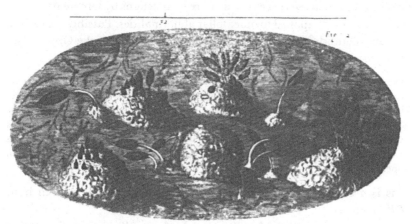

Fig. 5. Rose rust (Phragmidium mucronatum) *from Robert Hooke's* Micrographia, *1665 (part of pl. 12). The first illustration of a plant pathogenic microfungus. Note that the scale (in inches) is indicated.*

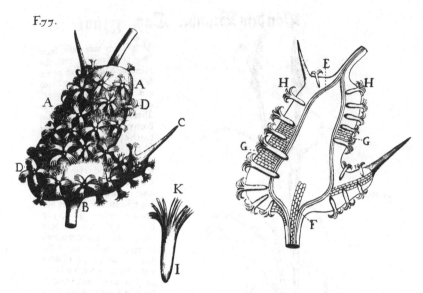

Fig. 6. Aecial Gymnosporangium clavariiforme *on hawthorn. (Malpighi (1679), Tab. 22, fig. 77.)*

medical records of ergotism in man. According to Barger (1931) in his classical monograph, a 'plague of fire' is first mentioned in and around Paris in AD 945, although holy (or St Antony's) fire was not identified with gangrenous ergotism until the eighteenth century. The fungal nature of Claviceps was established even later. It is highly probable that some of the Greek and Roman references to rust were to infections by Uredinales even if the term was also applied to other foliar abnormalities. Possibly the earliest account of an identifiable cereal disease is that by Basil the Great (*c.* AD 330–370), bishop of Caesarea, who wrote:

Does the earth allow each seed to grow according to its kind, if it happens that we harvest black kernels from clean seed? This, however, is not a change into another kind, but instead a sort of disease and weakness of the seed for it does not cease to be wheat. It merely blackens like being burned as indicated by its name (Brand). It is burned because of excessive heat and thereby changes its colour and taste. If it is sown again in suitable soil and the weather is suitable, it is said to regain its normal appearance.[22]

This is a convincing account of a smut that may well have been bunt (*Tilletia caries*).

What appears to be the first illustration of a specific fungus disease that can be identified with complete confidence is in the *General history of the things of New Spain*, a manuscript compiled during the sixteenth century by a

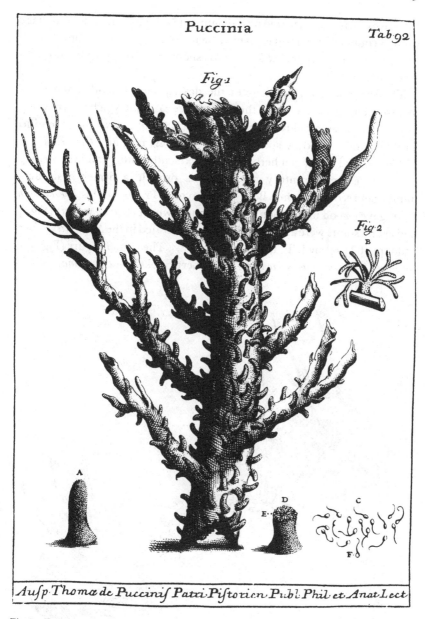

Fig. 7. *Telial* Gymnosporangium clavariiforme *on juniper.* (*Micheli (1729), Tab. 92.*) *Note the teliospores (c).*

Franciscan friar, Barnardino de Sahagún, in Nahuatl (the Aztec language) from accounts of the natives, which includes an illustration of a head of maize attacked by smut (*Ustilago maydis*) showing the diagnostic spore-filled galls (Fig. 3).

The same source also includes the description of another maize cob disease: 'The green maize ear, the dried maize ear: that which rots, develops fungus, becomes mouldy at harvest'; which Orlob (1973:189) identified with 'ear rot' caused by species of *Diplodia*, *Fusarium*, and *Gibberella*.

The printed European herbals of the sixteenth century and the writings of the seventeenth-century microscopists provide a firmer basis for a number of records. Jerome Bock (Hieronymus Tragus) in his *Kreutterbuch*, 1551, gave a good illustration of loose smut of wheat (*Ustilago tritici*) (Fig. 4) and cereal smuts were also described and illustrated in the *Neuw Kreuterbuch*, 1588 (pp. 683–4) by J. T. Tabernaemontanus. The famous figure (Fig. 5) of rose rust (*Phragmidium mucronatum*) in Robert Hooke's *Micrographia*, 1665, in

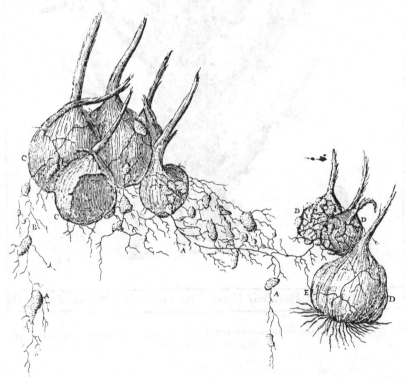

Fig. 8. Helicobasidium purpureum *on saffron crocus corms showing sclerotia. (Duhamel de Monceau (1728) pl. 2.)*

which the teliospores are recognisable, is the first illustration of a plant pathogenic microfungus. Later both Malpighi (*Anatome plantarum*, 1679) and P. A. Micheli (*Nova plantarum genera*, 1729) illustrated the rust *Gymnosporangium clavariiforme*. Malpighi, under 'Tumours and Excrescences', described the aecia on hawthorn as 'flowers' (Fig. 6) while Micheli recorded the telia (including teliospores) on juniper as *Puccinia non ramosa* (Fig. 7). Two other interesting records are those of hop mildew (*Pseudoperonospora humuli*) in England by Stephen Hales (1727) (see Chapter 9) and the sclerotia of *Helicobasidium purpureum* on Saffron crocus in France by Duhamel de Monceau (1728) (Fig. 8) who recognised the contagiousness of the disease.

Disease classification

The first attempts to systematise knowledge of plant diseases were to classify the diseases. As already noted, Theophrastus distinguished diseases from damage by such agents as frost or wind. Joseph Tournefort, the famous Parisian botanist, in 1705, classified diseases into those caused by internal and external factors, respectively (mouldiness being included in the latter) and Adanson (1763) employed the same two categories. C. S. Eysfarth in his doctoral thesis *De morbis plantarum*, 1723, grouped plant diseases according to the time of occurrence: at germination, during growth, or at maturity. Others attempted aetiological classifications. For example, the German physician J. S. Elsholz (1684)[23] attributed plant diseases to meteors, animals, and pests while Linnaeus in *Philosophia botanica*, 1751, listed *Erysiphe* [mildew], *Rubigo* [rust], *Ustilago* [smut], *Nidus insectorum* [galls and other deformations caused by insects], and *Insecta* as agents of disease in plants. The most popular approach, however, during the eighteenth and early nineteenth centuries was to classify diseases according to symptoms and the part of the plant affected, an approach much influenced by medical practice which frequently led to the adoption of terms applied to human diseases. The most influential example of this approach was that of J. J. Plenck in *Physiologia et pathologia plantarum*, 1794 (written in Latin) of which three German, as well as French and Italian, versions appeared during the next 20 years. Plenck recognised eight classes of disease (see Table 1). Similar schemes were devised by Johan Baptista Zallinger in *De morbis plantarum*, 1773, in which five classes of disease were recognised and the Danish naturalist Johan Christian Fabricius in 1774 (six classes). In the closing years of the century Erasmus Darwin's appeal to the

Table 1 *Classification of plant diseases used by J. Plenck in* Physiologia et pathologia plantarum, *1794, pp. 123–5*

I	*Laesiones externae*
	Vulnus – Wunden
	Fissura – Spalt
	Fractura – Bruch
	Exulceratio – Geschwür
	Defoliatio – Entblätterung
II	*Profluvia*
	Haemorragia – Blutsturz
	Lachrymatio gemmarum – Tränen
	Albigo – Mehlthau
	Melligo – Honigtau
III	*Debilitates*
	Deliquium – Niedersinken
	Suffocatio incrementi – Misswachs
IV	*Cachexia*
	Chlorosis – Bleichusucht
	Icterus – Gelbsucht
	Anasarca – Wassersucht
	Maculae – Flecke
	Pythiriasis – Laussucht
	Verminatio – Wurmsucht
	Tabes – Abzehrung
V	*Putrefactiones*
	Teredo pinorum – Wurmtrocknis
	Rubigo cerealium – Kornrost
	Ustilago cerealium – Kornbrand
	Clavus secalinus – Kornzapfel
	Necrosis – trotner Brand
	Gangraena – feuchter Brand
VI	*Excrescentiae*
	Gallae quercuum – Galläpfel
	Bedeguar rosarum – Rosenbedeguar
	Squamatio gemmarum – Zapfenrosen
	Verrucositas foliorum – Warzen
	Foliculi carnosi foliorum – Fleischstacheln
	Carcinoma arborum – Baumkrebs
	Lepra arborum – Baumaussatz
VII	*Monstrositates*
	Plenitudo florum – Füllung
	Mutilatio florum – Verstümmelung
	Deformitas – Ungestaltetheit
VIII	*Sterilitates*
	Polysarcia – Vollsaftigkeit
	Sterilitas – Unfruchtbarkeit
	Abortus – Missfall

'Sylphs' (in *The botanic garden* (Part 1, The economy of vegetation): 202–3, 1791) reflected his clear idea of the range of disease symptoms:

> Shield the young Harvest from devouring blight,
> The Smut's dark poison, and the Mildew white,
> Deep-rooted Mould, the Ergot's horn uncouth,
> And break the Canker's desolating tooth.
> First in one point the festering wound confin'd
> Mines unperceived beneath the shrivel'd rind;
> Then climbs the branches with increasing strength,
> Spreads as they spread, and lengthens with their length...

Nine years later Erasmus Darwin returned to the topic in more detail, and in prose, in Sect. IV of his *Phytologia*, 1800 where his four main divisions were diseases resulting from internal and external causes, insects, and destruction by vermin. His treatment of fungal infections was mainly derived from Linnaeus (loc. cit.).

In 1807 the Italian Philipo Ré (professor of botany and agriculture at Modena) proposed a comprehensive disease classification in his *Saggio teorico-practico sulle malattie delle piante*. An English translation of the second (1817) edition of Ré's book appeared in the *Gardeners' chronicle* during 1849–50 and Ré's scheme was adapted by M. J. Berkeley as the framework for his famous series of articles on 'Vegetable pathology' (1854–7) in the same periodical. Ré accepted five classes of disease comprising 67 genera each of which included one to several species. It was not until the acceptance of the concept of pathogenicity during the second half of the nineteenth century that the vernacular names for infectious diseases could be given precision by their association with the scientific names of the pathogens and with increasing frequency diseases were classified, for the convenience of plant pathologists, either according to the pathogens or the host plants.

A notable recent departure from these last patterns is that by Horsfall & Dimond in *Plant pathology, an advanced treatise*, 1959–60, where the treatment of plant diseases and disorders is approached from the standpoint of host physiology – effects of disease on growth, reproduction, nutrition, water relationships, respiration – and it is interesting to note that B. E. J. Wheeler (1969) in an introductory student text is exceptional among modern authors in following the earlier tradition by chapters on different types of symptom expression – damping-off and seedling blights, wilts, rusts, smuts, blight, anthracnose, leaf spots, mosaics and yellows, etc.

Aetiology up to 1858

The emergence of phytopathology was impossible until the principle of infection had been elucidated and thus for two millennia after Theophrastus there was little change in the approach to plant disease. Then, between the mid eighteenth and mid nineteenth centuries, in less than a hundred years, experimental evidence was offered for the pathogenicity of fungi to plants, insects, and man. A decade or so later the reality of bacterial diseases of animals, man and finally plants was accepted and by the turn of the century virus diseases, too, had been distinguished. A secure basis for the modern phytopathological offensive was established.

Early views
The early explanations of disease frequently reflected the prevailing world picture. The Hebrew prophets accepted 'blasting and mildew' as punishments by God for wrong doing. The Romans attributed cereal rust to a particular member of their hierarchy of gods whom they attempted to propitiate by ceremonies and sacrifices. Similar views prevailed among the Aztecs of the New World who instituted ceremonial precautions designed to ensure successful crops – procedures which sometimes culminated in human sacrifice.

The weather was frequently implicated as a cause of disease and disorders of plants and as that too was widely considered to be under divine control petitionary prayer was resorted to, and still is – in the Anglican *Book of common prayer* the first two prayers for use 'Upon several occasions' are 'For rain' and 'For fair weather'. Movements of the moon and other heavenly bodies were also widely believed to influence the weather and many human activities. These latter beliefs were systematised in the practice of astrology which offered horoscopes or other guidance on the times most favourable for, among other things, the planting of crops and their treatment. Many such beliefs have survived in the folklore and agricultural traditions of today.

The Pythagorean philosopher Empedocles [504–443 BC] of Argentum in Sicily believed the universe to be composed of four elements – fire, air, earth, and water. Corresponding to the elements were four qualities – hot, cold, dry, moist – and it was from these notions that the human body came to be regarded as composed of four 'humours': blood (hot+moist; air), phlegm (cold+moist; water), yellow bile (hot+dry; fire), and black bile (cold+dry; earth) which in mediaeval times became the four 'temperaments': sanguine, phlegmatic, choleric, and melancholic. According to the 'doctrine of humours' as expounded by Hippocrates, health depended on the correct balance of the four components, disease resulted from their imbalance and drugs used to correct the imbalance were classified on a numerical scale according to the proportions of the qualities present. Over many centuries these views infiltrated beliefs on the aetiology and treatment of plant diseases. In ancient India, where the doctrine of humours had become well established in medical practice by contact with Greek physicians, Surapala (who lived during the ninth century AD) in his manuscript Vrksayurveda wrote that tree diseases:

may briefly be classified in two groups, one of these arising from the body, i.e. internal and the other of those attacking from outside, i.e. extraneous. The bodily or internal diseases arise from the disorders of wind, phlegm and bile, and the extraneous disease and ailments are caused by vermin, frost, etc.

A tree afflicted with the wind affection is gaunt, too tall or too stunted, dry, sleepless, less sensitive and does not bear flowers or fruit.

A tree which cannot stand the sun, which is pale, shorn of branches and of which the fruit ripens out of time, is said to be suffering from bilious disorder.

That which has glossy leaves and branches and is glorious with flowers and fruit but is rendered pale by the strangulating creepers, is said to be suffering from the disorders of phlegm.

The disorders of wind are removed by medicines of bitter, pungent and astringent taste, bilious affections are cured by bitter, hot, salt and acid medicines and the phlegmatic troubled by oily, sweet, acid and salt medicines.[24]

In eighteenth-century Europe Tournefort (1705) attributed plant diseases caused by internal factors to too much, too little, unequally distributed, or defective sap while Stephen Hales (1727) held that in wet weather 'the stagnating sap corrupts, and breeds moldy fen' – mildew of the hop – explanations showing clear overtones of humoral thinking. It was at this time that Jethro Tull in *The horse-hoing husbandry*, 1733, attributed wheat bunt to cold wet summers, an explanation he considered he had confirmed by inducing the disease by overwatering.

Another popular and persistent explanation of many diseases was the intervention of insects. 'All Blights proceed from insects' was in 1725[25] the rather extreme opinion of Richard Bradley, professor of botany at the University of Cambridge, but this notion was widespread and favoured as an explanation of rusts and smuts of cereals, especially in Italy (for example, by Ginanni, 1759).

Fungal pathogenicity

The concept of contagious disease is of ancient origin as is clear from the Biblical account of leprosy (Lev. 13:14). Similarly for plants, in ancient India Misra Chakrapani in his *Visva-Vallabha* wrote:

A tree diseased by contagion recovers quickly if the affected tree standing by its side

(?) is cut down and the earth under it dug out and replaced by healthy soil and then fed with water.[26]

The nature of the contagious principle and the concept of infectious disease caused by living pathogenic agents proved intractable problems. Girolamo Fracastoro (Latinised as Hieronymus Fracastorius) of Verona, a Renaissance polymath – he was physician, poet, physicist, geologist, astronomer, and pathologist – in his famous *De contagione* of 1547 distinguished contagion by contact, by 'fomites' (clothes or other contaminated articles), and at a distance. He also suggested that contagion was spread by 'seminaria contagiorum' which he appears to have envisaged as similar to the atoms of Lucretius rather than living 'seeds' or 'germs'. Although Fracastorius had followers both in Italy and elsewhere who elaborated on his views little direct progress towards an explanation of contagion was forthcoming for more than a century when in Tuscany the physician Giovan Cosimo Bonomo and the druggist Giaconto Cestoni in 1687 proved that scabies was caused by a parasitic mite and introduced effective local therapy in place of general therapy based on a supposed humoral concept of the disease. This discovery, which carried parasitism to the limit of visibility with the naked eye, combined with William Harvey's principle of *omne animal ex ovo* (1651) and Leeuwenhoek's observations on bacteria and other micro-organisms (1657) contributed to the formulation of a thoroughgoing and consistent doctrine of *contagium vivum* by Carlo Francesco Cogrossi, 'philosopher and physician of the city of Cremona' in Lombardy, to explain an epizootic of plague in cattle in the Venice region during 1711–14 following the introduction of an infected Hungarian ox. Cogrossi set out his hypothesis in a letter to Antonio Vallisnieri, professor of medicine at the University of Padua, which was published, together with Vallisnieri's reply, as *Nuova idea del male contagioso de' buoi*, 1714. Towards the end of this letter Cogrossi wrote:

If some day, my most illustrious sir, you could employ your very erudite pen to describe the nature of corn rust, as you gave us hope in one of your esteemed dialogues [*Dialog* 2:262], perhaps then on the example of those insects on which you think it depends the idea which I have set forth to this point will not seem so strange.

Vallisnieri is still remembered for his entomological writings. His views on the nature of cereal rust remained unsound but this would probably not have unduly disturbed Cogrossi who held 'that sciences, arts, and even more their discoveries never come into being completely and suddenly'. It was however in Italy, where scientific investigation was then in the ascendant, that some fifty years later the nature of wheat rust was correctly ascertained by two independent observers.

This discovery was precipitated by an epidemic outbreak of wheat rust in 1766 when 'the Rust was universal over the whole of Italy, and in all the different levels and exposures of its territory' (Targioni-Tozzetti, 1767:1 (Eng. trans.)). Among the many who investigated or speculated on the nature of the disease was the physicist and naturalist Felice Fontana who in 1767 published his findings in a pamphlet entitled *Osservazioni sopra la ruggine del grano*, the opening words of which (in Pirone's translation) read:

On the 10th June last year, I discovered that the rust, which had devastated the lands of Tuscany, is a grove of plant parasites that nourish themselves at the expense of the grain.

By careful observation Fontana accurately described both the macroscopic and microscopic features of both teliospores and urediniospores which he took to be the fruits of little plantlets of two distinct rusts. He was, however, unable to demonstrate within these bodies 'seeds' such as Micheli had found in fruiting structures of moulds. Fontana also had reservations as to their parasitic nature and concluded that unlike true parasites which attack undamaged plants rusts were semi-parasites which 'attach themselves to the grain only when there is a rupture of the vessels and a diffusing of humours'.

Also in 1767 the same discovery was announced by Giovanni Targioni-Tozzetti, a public-spirited physician, who had studied botany under Micheli in Florence and after Micheli's death in 1737 had acquired Micheli's manuscripts and specimens. In an attempt to improve agricultural practice Targioni-Tozzetti planned a comprehensive work, the *Alimurgia* or 'Means of rendering less serious the dearths. Proposed for the relief of the poor'. The final chapter of the first and only volume of this project to be published was devoted to a comprehensive account of diseases of plants, especially cultivated plants of importance in agriculture, and included Targioni-Tozzetti's elucidation of the nature of black stem rust (*Puccinia graminis*) of wheat.

Targioni-Tozzetti writes that he began his observation on wheat rust on 11 June 1766 with his father's microscope of three lenses made by the famous Guiseppi Campano with a 'very sharp lens' when 'Quickly I came to know that Rust was an aggregation of very minute little bodies of a form almost oval' but 'not being able to satisfy myself as much as I wished, for the lack of the necessary equipment' he persuaded owners of better microscopes to examine his material or borrowed their instruments and 'in the middle of the morning of the 19th July with the microscope of Sig. Doctor Guadagni ... it appeared without equivocation, that every single knot of Rust on the Stem and the Leaves of the Wheat, is an internal, very tiny, parasitic plant, which does not arise except between skin and skin, so to speak, of the Wheat'. Targioni-Tozzetti also described the subcuticular mycelium ('hairs' which 'steal and suck up for themselves the nutriment prepared, and destined for the Wheat') and the increase in size of the sorus which 'forces up the Cuticle of the Wheat outwards, and makes it swell, and finally crack in a very delicate spangle'. He also described and illustrated both teliospores and urediniospores (which he deduced were different stages of one rust), accepted the rust as a parasite, and concluded that the rust was dispersed by wind and that under humid conditions infection occurred via the stomata. Like Fontana, he took each spore to be an individual plantlet derived from its own seed of 'scarcely credible smallness, such as could not be seen by even the most acute microscope'. These observations on wheat rust were supplemented by illustrated accounts of rusts of rose (*Phragmidium mucronatum*), broad bean (*Uromyces fabae*), and mint (*Puccinia menthae*), wheat bunt (*Tilletia caries*), maize smut (*Ustilago maydis*), powdery and downy mildews of various plants and a number of other fungal infection so that this chapter, as the first monograph on fungal diseases of plants, is a landmark.

These two publications, by Fontana and Targioni-Tozzetti, constitute a climax in the descriptive approach to the pathogenicity of fungi. The complementary experimental approach which clinched the matter of fungal pathogenicity is also represented by two now classic publications reporting the results of studies on wheat bunt (*Tilletia caries*), undertaken in France some fifty years apart, the first of which (by M. Tillet) initiated phytopathological field experimentation, the second (by I. B. Prévost) laboratory research. Both authors embarked on their investigations at the instigation of their regional scientific societies.

Mathieu Tillet, Director of the Mint at Troyes, who was born about 1730, devoted his spare time to the improvement of agricultural practice. He earned a permanent place in the history of plant pathology by his 150-page *Dissertation sur la cause qui corrupt et noircit les grains de bled dans les épis; et sur les moyens de prevenir ces accidens*, 1755 (Fig. 9), which gained for him the prize, offered in 1750 by the Académie Royale des Belles-Lettres, Science & Arts de Bordeaux, for the best dissertation on the cause and cure of blackening of wheat. After preliminary experiments by which Tillet refuted Jethro Tull's explanation that bunt resulted from too much moisture, the first of his two main experiments on winter wheat was set up in the autumn of 1751 when a plot of land suitable for wheat, 24 by 540 feet, was divided into 120 subplots on which he compared the effects of unmanured soil with four manurial treatments (pigeon manure, sheep manure, fecal matter, horse and mule manure) on clean seed, seed naturally contaminated or

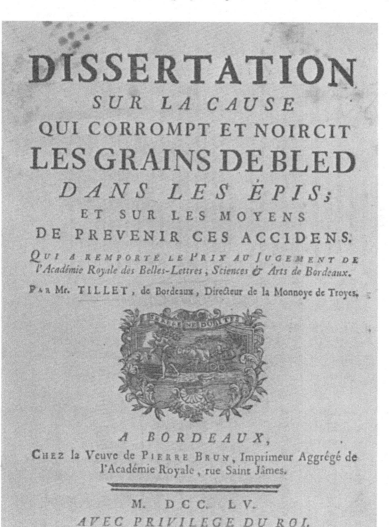

DISSERTATION
SUR LA CAUSE
QUI CORROMPT ET NOIRCIT
LES GRAINS DE BLED
DANS LES ÉPIS;
ET SUR LES MOYENS
DE PREVENIR CES ACCIDENS.
QUI A REMPORTÉ LE PRIX AU JUGEMENT DE
l'Académie Royale des Belles-Lettres, Sciences & Arts de Bordeaux.

PAR Mr. TILLET, de Bordeaux, Directeur de la Monnoye de Troyes.

A BORDEAUX,
CHEZ la Veuve de PIERRE BRUN, Imprimeur Aggrégé de
l'Académie Royale, rue Saint James.

M. DCC. LV.
AVEC PRIVILEGE DU ROI.

Fig. 9. M. Tillet (1755) title page.

experimentally inoculated with bunt spores, and seed left untreated or after treatment with several formulations of salt, lime, and saltpetre. In addition, comparison was made of six dates of sowing on each of which seed was sown on all five manurial treatments.[27] During the second season (1752–3) on a similar but slightly smaller plot, again divided into 120 subplots, the effects of various methods of inoculation, different types of seed treatment, and the effects on clean seed of the addition to the soil of contaminated wheat straw,

MÉMOIRE

SUR LA CAUSE IMMÉDIATE

DE LA CARIE OU CHARBON DES BLÉS,

ET DE PLUSIEURS AUTRES MALADIES DES PLANTES,

ET SUR LES PRÉSERVATIFS DE LA CARIE.

PAR M. BÉNÉDICT PRÉVOST,

Membre de la Société de Physique et d'Hist. Nat. de Genève, de celle des Naturalistes de la même ville, et de la Société des Sciences et des Arts du département du Lot, séante à Montauban: Correspondant de la Société Galvanique et d'Électricité de Paris, des Sociétés Médicale et de Méd. Prat. de Montpellier, de celle des Amateurs des Sciences de Lille, et d'Émulation de Lausanne.

Cet une botanique à faire, que celle des plantes microscopiques.
SENEBIER, *Physiol. Vég.* T. II. p. 11.

A PARIS,

CHEZ BERNARD, IMPRIMEUR-LIBRAIRE. QUAI DES AUGUSTINS, No. 25.

1807.

TABLE DES ARTICLES.

Fig. 10. I.-B. Prévost (1807) title page and contents.

manure from a Spanish horse which had eaten straw heavily contaminated
with bunt spores, and various other types of manure before or after admix-
ture with contaminated straw were compared. Further seed treatments
were also included as well as experiments with darnel, rye, and barley.
Additional trials were made during 1753–4.

These investigations created something of a stir and at the request of the
King, Louis XV, Tillet repeated his experiments at the Trianon in Paris,
the Jardin du Roi, founded by Louis XIII in 1626. The results were equally
successful. They were published as a pamphlet in 1756 (Tillet, 1756) and
reprinted under a Government order in 1785 because 30 years later bunt
was still prevalent and few farmers made a practice of seed treatment.

Tillet's results made it clear 'that' – in his own words (in H. B.
Humphrey's translation) – 'the common cause, the abounding source of
bunted wheat plants resides in the dust of the bunt balls of diseased wheat';
and further 'that the treatments I employed have protected the most
heavily infected seed against the effects of the contagion'.

Tillet was convinced that insects played no part in the aetiology of bunt.
He distinguished loose smut (*Ustilago tritici*) from bunt but although he
likened bunt spores to the spores of *Lycoperdon* he was not able to identify
them as being of fungal origin. He believed that the clean healthy seed,
inoculated with bunt dust 'receives through rapid contagion and a very
intimate communication the poison peculiar to it; that it transmits the
poison to the kernels of which it is the origin; that these kernels, once
infected become converted into a black dust and become for others a cause
of disease'.

Isaac-Bénédict Prévost, born at Geneva of Swiss parents in 1755 (the
year of publication of Tillet's monograph), who at the age of 22 became a
private tutor at Montauban, Département du Lot, France, where he spent
the rest of his life, was from 1810 professor of philosophy at the Faculté de
Théologie Protestante. His one mycological investigation originated from
an invitation in 1797 by the scientific section of the Society of Montauban to
its members to occupy themselves with the problem of bunt in wheat. Ten
years later saw the publication of *Mémoire sur la cause immédiate de la carie ou
charbon des blés* ... (Fig. 10) in which Prévost gave an account of his historic
experiments which demonstrated that 'la cause immédiate de la carie est un
plante du genre des urédo ou d'un genre très-voisin', a conclusion he had
reached by 1804. Prévost also made observations on various rusts and
mildews and he generalised from his observations by suggesting that plants
of this class should be designated 'plantes parasites intestines' or simply
'plantes intestines'. He established the nature of the 'grains' filling the

bunted kernels by a series of rigorous experiments on the external factors determining their germination (and illustrated sporidial production), including the effects of toxic agents (see Chapter 7), as well as age, previous treatment, and concentration of the spores. He undertook extensive inoculation experiments and defined factors determining infection. He observed germinating spores in soil and on the surface of young wheat plants and described the development of spores in the infected ovary, and correctly believed, though he could not prove it, that infection of the seedling spread through the developing plant to sporulate in the ovaries. In conclusion he claimed the parasite as the 'immediate' or 'direct' cause of bunt because he realised that infection occurs only when the environmental conditions are favourable. These results constitute the first experimental evidence for the pathogenicity of a micro-organism.

Although the combined publications of Tillet, Fontana, Targioni-Tozzetti, and Prévost established the role of fungi in the aetiology of bunt and stem rust of wheat the conclusions and the generalisation they suggested were not very widely accepted. It must be remembered, as already recalled, that these monographs appeared over a period of more than fifty years; and in small editions. Fontana's little book became famous (one gathers the impression from the introductory pages that he was a good publicist, and his writing is concise and straightforward) while the weightier and higher quality contribution by Targioni-Tozzetti (with its rather muddled presentation) was virtually forgotten until the appearance of the 1943 reprint of the *Alimurgia* by Professor Goidàñich (and the publication of an English translation of the final chapter in the *Phytopathological classics* series nine years later) made it more generally available. Likewise, although Tillet's work was remembered the significance of Prévost's results, if praised at the time, was not widely recognised in spite of the fact that the Montauban Academy submitted the published memoir to the Institute in Paris when the Abbé Tessier, himself the author of a notable, well-illustrated book (*Traité des maladies des grains*, 1783) on the rusts and smuts of cereals (Fig. 11), was the reporter of a commission appointed to examine it. Also, Prévost's monograph became inaccessible. In 1853 even de Bary had to cite it from secondary sources. Older beliefs were revived, plant disease was often attributed to the effects of weather or other environmental factors, or to insects. More perceptive opinions were, however, expressed. In Denmark Fabricius in 1774 concluded that 'the symptoms of smut can never be better explained than by assuming something organized to be the cause'. J. Robertson (1824) in England reported successful experimental transmission of peach mildew but the next year J. C. Loudon was writing of wheat bunt, in *An encyclopaedia of agriculture*:

Some have attributed it to the soil in which the grain is sown, and others have attributed it to the seed itself, alleging that the smutted seed will produce a smutted crop. But in all this there seems to be a great deal of doubt. Wildenow regards it as

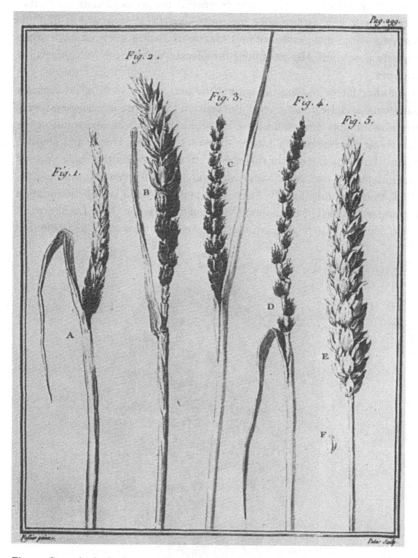

Fig. *11*. *Stages in the development of loose smut of wheat* (Ustilago tritici). *(Tessier (1783): 299.) Healthy ear on right. Cf. Fig. 4.*

originating in a small fungus which multiplies and extends till it occupies the whole of the ear (*Princip. of Bot.* p. 356). But F. Bauer of Kew,[28] seems to have ascertained it to be merely a morbid swelling of the ear, and not at all connected with the growth of a fungus. (Smith's *Introd.* p. 348)

Even as late as 1852 M. J. Schleiden considered that the attribution of potato blight either to vegetable or animal parasites 'merits no further notice, for it can escape only the most superficial observation that the diseased symptoms occur, without exception, before the least trace of the parasite is visible'. He attributed the disease to an excess of phosphatic manure.

Another factor mitigating against the acceptance of fungi as primary pathogens was the persistence of the belief in the spontaneous generation of microbes or that they could arise by heterogenesis from abnormal sap or tissue – an idea stemming back to the views of Robert Hooke and Stephen Hales. Towards the end of the eighteenth century and the early years of the nineteenth the latter view was widely held, even by mycologists. C. H. Persoon and Elias M. Fries (who between them laid the foundation for modern mycological taxonomy) were of this opinion. John Lindley in his *Ladies botany*, 1834 (p. 285) popularised the same view when he wrote that

Fig. 12. Anton de Bary (1831–88), aet. 46.

Fig. 13. Anton de Bary's first phytopathological monograph.

some fungi 'are generated by the living substance of the leaves and stems, which they afflict under the names of mildew or blight'. Heterogenesis was most fully developed by Franz Unger in *Die Exantheme der Pflanzen*, 1833, which was written under the influence of *Naturphilosophie*, then in high fashion. Fungi were what he called 'entophytes', 'after organisms' (or *Afterpilze*), but he did treat them as though they were independent organisms as regards their taxonomy and nomenclature.

The devastating outbreak of potato blight in Europe during the 1840s which resulted in the catastrophic Irish famine focussed attention on this and other plant diseases and was certainly influential in convincing the Rev. M. J. Berkeley (1846) and many others in both Europe and North America of the reality of fungal pathogenicity. This view which has now prevailed for more than a hundred years may be considered to have become current orthodoxy since the publication of *Untersuchungen über die Brandpilze*, 1853 (Fig. 13) by the 22-year-old Anton de Bary (Fig. 12) (who was to become the outstanding mycologist of the nineteenth century) reporting his

Fig. 14. Julius Kühn (1825–1910).

own observations on rusts and smuts, made while a medical student, and summarising and evaluating earlier work on the topic. Although most of de Bary's conclusions had already been reached by others, never before had the evidence been marshalled so thoroughly and assessed so critically against the current knowledge of fungi (including his own observations on the relationship of parasite to host) and subsequently no case could be made against de Bary's opinion that all reliable observations are in harmony with the view that the Brand fungi are true fungi and to be regarded as independent parasitic organisms.

Although de Bary subsequently made notable contributions to the knowledge of disease in plants he was essentially an academic, a research worker, and the beginning of the modern era of plant pathology is more appropriately associated with the name of Julius Kühn, a fellow countryman, his senior by six years, and the author of *Die Krankheiten der Kulturgewächse* published in 1858. Julius Kühn (Fig. 14), who has been claimed by Whetzel and others as the first plant pathologist, was the son of a Saxon landowner. He was born in 1825 and on leaving school spent 14 years gaining experience on his father's and other estates and it was not until the age of 30 that he had his first formal training, at Bonn. Subsequently he was awarded his

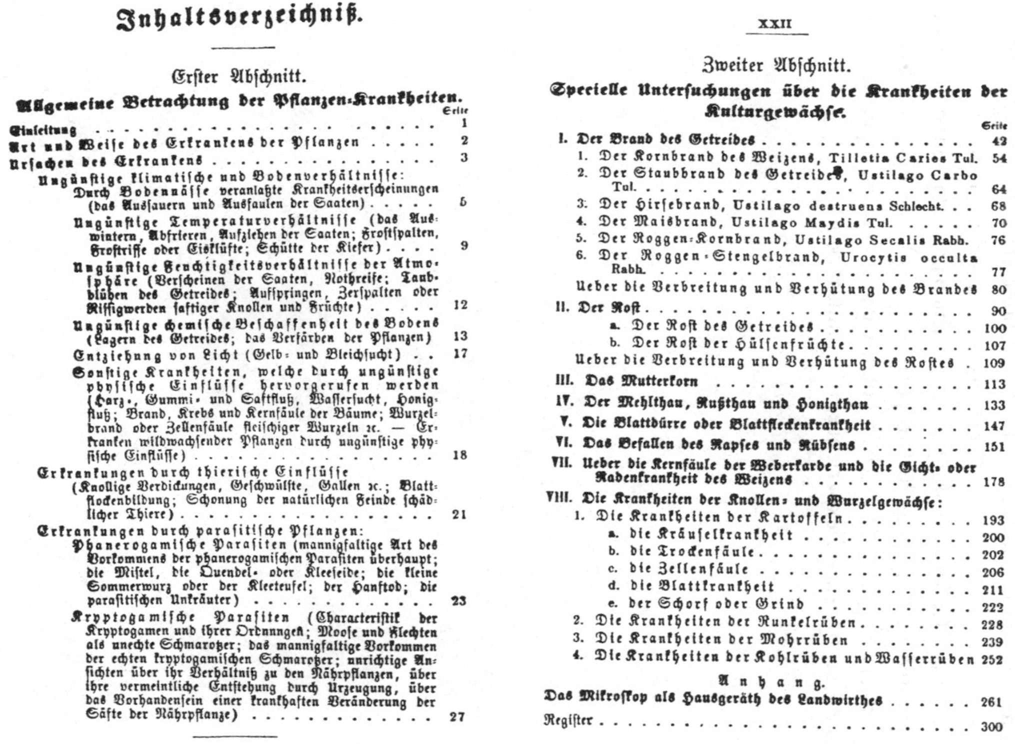

Inhaltsverzeichniß.

XXII

Erster Abschnitt.

Allgemeine Betrachtung der Pflanzen-Krankheiten.

Zweiter Abschnitt.

Specielle Untersuchungen über die Krankheiten der Kulturgewächse.

Fig. 15. Contents of Die Krankheiten der Kulturgewächse *by Julius Kühn, 1858.*

doctorate at Leipzig and after a further period of estate management became professor of agriculture at the University of Halle in 1862. Kühn early developed an interest in plant disease; first, about 1854, in diseases of beet and rape (which were the subject of his doctoral thesis). During the next four decades he published more than 70 papers on these and other diseases and on plant pathogenic fungi. In later years this interest took second place to the use of fertilisers (as advocated by Liebig) and the feeding of cattle. His book on plant diseases, which proved popular and had to be reprinted in 1859, was in two parts (Fig. 15), the first devoted to general aspects, the second to accounts of diseases of cereals, rape, and root crops based on first-hand experience, is a landmark.

Diagnosis: the pathogenic agents

The first step in the rational appraisal of a disease whether of man, animals, or plants is the correct diagnosis of the cause although it is sometimes possible to gain much knowledge of a disease of unknown aetiology, or one for which the causal agent has been wrongly identified, and to design appropriate control measures which are not invalidated when the true nature of the disease is elucidated. For example, the recent discovery that aster yellows, accepted as a virus disease for more than fifty years, is caused by a mycoplasma-like organism does not affect the large body of information on leaf hopper transmission and host range of the agent of aster yellows and its elimination by heat treatment. It does, however, now allow aster yellows to be considered in relation to the knowledge of mycoplasmas in general (see Chapter 5).

There are three distinct categories of deviations from the normal state shown by plants: (1) genetical abnormalities, (2) disorders induced by environmental factors, and (3) infectious diseases. Here, as already stated, emphasis is on infectious diseases while the first two categories are briefly reviewed in Chapter 6.

Although mistletoe (*Viscum album*) was apparently the first parasite of plants to be recognised (by Albertus Magnus *c*. AD 1200)[1] and although dwarf mistletoes (*Arceuthobium*) are of economic importance as pathogens of Pinaceae and Cupressaceae in North America[2] (as also is dodder (*Cuscuta*), world wide on diverse plants) phanerogams are of minor importance as plant pathogens. This is also true of algae.[3] The one notorious plant-pathogenic alga is *Cephaleuros virescens* which is common on tea (causing 'red-rust') and other tropical plants and was first described by D. D. Cunningham from India in 1879. These pathogenic plants will not be considered further.

The important plant pathogens, which were first recognised in order of their decreasing size, are fungi (Chapter 3), bacteria (Chapter 4), and viruses (Chapter 5). For all these, but particularly fungi, plant pathologists have supplemented the work of professional taxonomists by contributing more and more precise descriptions of the pathogens. They have also proposed taxonomic innovations to meet their particular needs, proposals which have sometimes necessitated new provisions in the international codes of biological nomenclature in order to ensure that as far as possible every pathogen may be designated by an internationally accepted name. In addition, plant pathologists have made major contributions to knowledge by devising new methods for the identification of the pathogens, the elucidation of life histories, and morphological, physiological, and biochemical investigations on host–pathogen relationships.

3

Fungi

Fungi are the most important pathogens affecting plants. Their role in the aetiology of plant disease is now seen to be parallel to that of bacteria in human medicine. Cereal rusts and smuts, potato blight, and Sigatoka disease of banana are diseases of world-wide incidence and importance equivalent to such bacterial diseases of man as tuberculosis and leprosy. Although there are economically important bacterial diseases of plants and serious fungus diseases of man and higher animals, medical bacteriologists at the end of the nineteenth century expressed disbelief in claims that bacteria were responsible for disease in plants (see Chapter 4) just as later medical pathologists were slow to appreciate the significance of human mycoses. Many plant pathologists still tend to believe that fungi are characteristically pathogenic for plants and only incidentally pathogenic for man and animals, with the exception of insects – the study of entomogenous fungi being a well known, if somewhat esoteric, mycological by-way. It was therefore something of a surprise to find that an examination of the index to volume 4 of the *Review of medical and veterinary mycology* showed 205 fungi to be associated with disease in man and 89 animals (all but one – a shrimp – vertebrates), that is, there was an average of 2.3 fungi per host species, while a similar analysis of the index of a comparable volume of the *Review of plant pathology* showed 1288 fungi associated with 659 host plants, again an average of approximately 2 per host.

Recently medical mycologists have realised that many human mycoses are caused by fungi which they describe as 'opportunistic', that is, normal components of the environment which only cause disease when introduced by chance under appropriate conditions into a susceptible individual. This terminology could well have been applied to many plant pathogenic fungi which, as de Bary recognised, are normally saprophytes (or saprobes) not parasites. Some plant pathogenic fungi are obligate parasites – such as the powdery mildews and the rusts (although a few of the latter have recently been cultured *in vitro*) – and are dependent on their hosts for survival but many others, even when exhibiting host specialisation, are able to live saprobically in nature.

Almost all groups of fungi include some plant pathogenic species but the greatest number of plant pathogens is to be found among the imperfect fungi, the associated perfect state (the holomorph, in contrast to the imperfect anamorph)[1] often occurring as a climax on dead tissue.

The taxonomic approach

Speciation

The early descriptions of important plant pathogenic fungi were usually very imprecise. *Puccinia graminis* (the notorious cause of black stem rust of cereals) was described by Christiaan Hendrik Persoon in his *Synopsis methodica fungorum*, 1801 (p. 228) (one of the starting-point books for the international nomenclature of fungi) simply as:

PUCCINIA GRAMINIS: conferta linearis nigrescens, sporulis subturbinatis medio constricta.

Thirty-five years later the description given by the Rev. M. J. Berkeley in his account of the fungi in J. E. Smith's *The English flora* (**7** (2):363, 1836) was very similar:

P. Graminis, Pers. (*Mildew*); spots pale diffuse, sori linear confluent amphigenous, sporidia at length black.

A. C. J. Corda, however, in 1840 (*Icones fungorum hucusque cognitorum*, **4**:11) gave a more adequate description:

PUCCINIA GRAMINIS: ... Maculis pallide-flavus, diffusis vel minutis; acervus linearibus, congestis, confluentibus, nigris; sporis clavatis obtusis, vel subfusiformibus, obtusiuscolis, medio constrictis, fuscis, hyalinis, apicula subincrassato vel obtuso, corneo; nucleo firmo, intus cavo, colorato; stipite longo, cylindrico, supra luteolo, infra fuscente, subaequali, basi truncato. Long. spor. 0,00170. p.p.p.

(which includes, it should be noted, a measurement of the spore length in parts per Paris inch) and he also illustrated the teliospores (*ibid.*, pl. 3, fig. 27).

Up to this time the aecial state of the rust was treated as a distinct species – *Aecidium berberidis* – in spite of the fact that the association of wheat rust with barberry bushes had long been noted and as early as 1805 Joseph Banks was writing 'Is it not more than possible that the parasitic fungus of the barberry and that of wheat are one and the same species ...?' Experi-

spores are colourless, comparatively small, and produced from the hyphæ in basipetal chains (Plate II. Figs. 11–14). They emit a germ-tube, which usually exhibits circumnutatory motions. The primary uredospores occur with the æcidiospores, and resemble the secondary in form, but are much more profuse. The secondary uredospores and teleutospores have localized mycelia (Plow., *Gard. Chron.*, July 25, 1885, p. 108, figs. 22, 23).

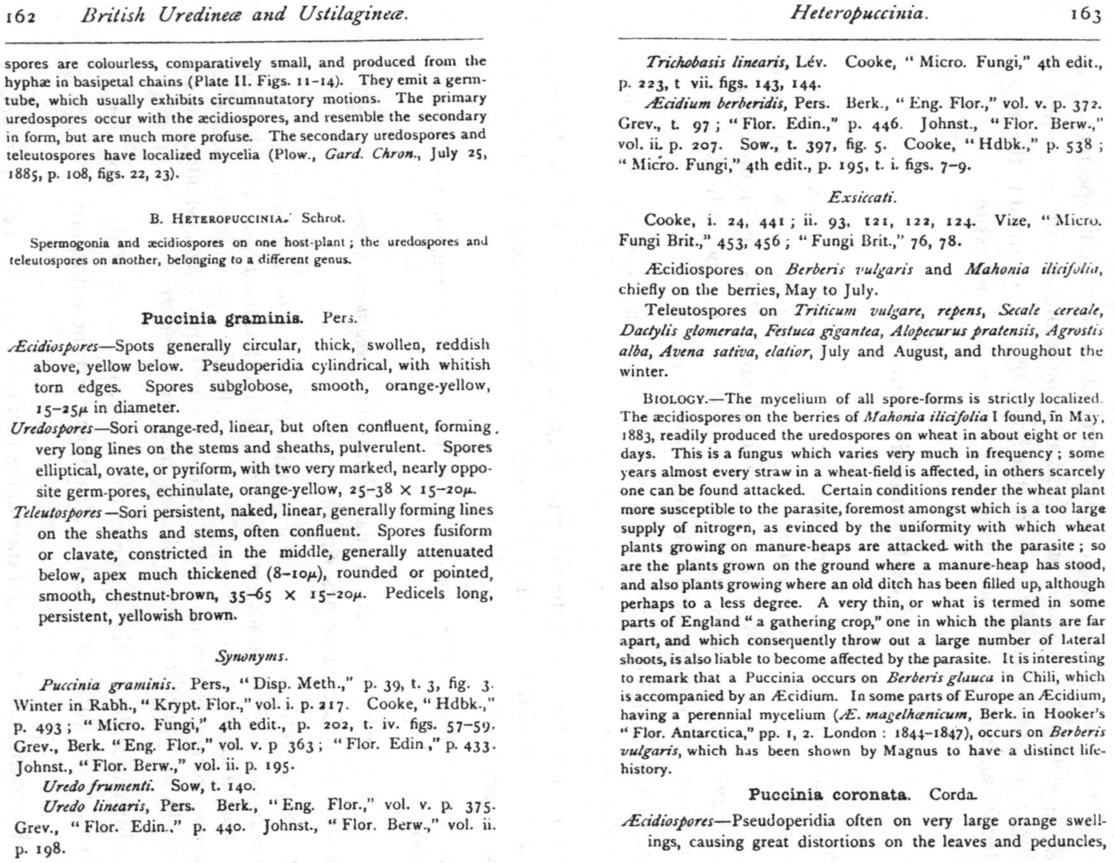

B. HETEROPUCCINIA. Schröt.

Spermogonia and æcidiospores on one host-plant; the uredospores and teleutospores on another, belonging to a different genus.

Puccinia graminis. Pers.

Æcidiospores—Spots generally circular, thick, swollen, reddish above, yellow below. Pseudoperidia cylindrical, with whitish torn edges. Spores subglobose, smooth, orange-yellow, 15–25μ in diameter.
Uredospores—Sori orange-red, linear, but often confluent, forming very long lines on the stems and sheaths, pulverulent. Spores elliptical, ovate, or pyriform, with two very marked, nearly opposite germ-pores, echinulate, orange-yellow, 25–38 × 15–20μ.
Teleutospores—Sori persistent, naked, linear, generally forming lines on the sheaths and stems, often confluent. Spores fusiform or clavate, constricted in the middle, generally attenuated below, apex much thickened (8–10μ), rounded or pointed, smooth, chestnut-brown, 35–65 × 15–20μ. Pedicels long, persistent, yellowish brown.

Synonyms.

Puccinia graminis. Pers., "Disp. Meth.," p. 39, t. 3, fig. 3. Winter in Rabh., "Krypt. Flor.," vol. i. p. 217. Cooke, "Hdbk.," p. 493 ; "Micro. Fungi," 4th edit., p. 202, t. iv. figs. 57–59. Grev., Berk. "Eng. Flor.," vol. v. p. 363 ; "Flor. Edin.," p. 433. Johnst., "Flor. Berw.," vol. ii. p. 195.
Uredo frumenti. Sow., t. 140.
Uredo linearis, Pers. Berk., "Eng. Flor.," vol. v. p. 375. Grev., "Flor. Edin." p. 440. Johnst., "Flor. Berw.," vol. ii. p. 198.

Trichobasis linearis, Lév. Cooke, "Micro. Fungi," 4th edit., p. 223, t. vii. figs. 143, 144.
Æcidium berberidis, Pers. Berk., "Eng. Flor.," vol. v. p. 372. Grev., t. 97 ; "Flor. Edin.," p. 446. Johnst., "Flor. Berw.," vol. ii. p. 207. Sow., t. 397, fig. 5. Cooke, "Hdbk.," p. 538 ; "Micro. Fungi," 4th edit., p. 195. t. i. figs. 7–9.

Exsicati.

Cooke, i. 24, 441 ; ii. 93, 121, 122, 124. Vize, "Micro. Fungi Brit.," 453, 456 ; "Fungi Brit," 76, 78.

Æcidiospores on *Berberis vulgaris* and *Mahonia ilicifolia,* chiefly on the berries, May to July.

Teleutospores on *Triticum vulgare, repens, Secale cereale, Dactylis glomerata, Festuca gigantea, Alopecurus pratensis, Agrostis alba, Avena sativa, elatior,* July and August, and throughout the winter.

BIOLOGY.—The mycelium of all spore-forms is strictly localized. The æcidiospores on the berries of *Mahonia ilicifolia* I found, in May, 1883, readily produced the uredospores on wheat in about eight or ten days. This is a fungus which varies very much in frequency ; some years almost every straw in a wheat-field is affected, in others scarcely one can be found attacked. Certain conditions render the wheat plant more susceptible to the parasite, foremost amongst which is a too large supply of nitrogen, as evinced by the uniformity with which wheat plants growing on manure-heaps are attacked with the parasite ; so are the plants grown on the ground where a manure-heap has stood, and also plants growing where an old ditch has been filled up, although perhaps to a less degree. A very thin, or what is termed in some parts of England "a gathering crop," one in which the plants are far apart, and which consequently throw out a large number of lateral shoots, is also liable to become infected by the parasite. It is interesting to remark that a Puccinia occurs on *Berberis glauca* in Chili, which is accompanied by an Æcidium. In some parts of Europe an Æcidium, having a perennial mycelium (*Æ. magelhænicum,* Berk. in Hooker's "Flor. Antarctica," pp. 1, 2. London : 1844–1847), occurs on *Berberis vulgaris,* which has been shown by Magnus to have a distinct life-history.

Puccinia coronata. Corda.

Æcidiospores—Pseudoperidia often on very large orange swellings, causing great distortions on the leaves and peduncles,

Fig. 16. Description of Puccinia graminis by Plowright (1889): 162–3.

mental proof of this conspecificity had to wait the classical study by de Bary who, during 1865–6, successfully infected barberry with sporidia from teliospores and subsequently wheat with aeciospores which developed on the barberry. He also introduced the term *heteroecism* for the phenomenon where two or more species of plants (the 'alternate hosts') are involved in the life cycle of one rust. Thereafter descriptions of *P. graminis* in standard texts, such as C. B. Plowright's *British Uredineae and Ustilagineae*, 1889, included details of aecidiospores [aeciospores], uredospores [urediniospores], and teleutospores [teliospores] (see Fig. 16).

During the first three-quarters of the nineteenth century most descriptions of plant pathogenic fungi were offered by mycologists. Then, concurrently with the advent of professionalism in plant pathology, more and more plant pathogens were described and named by phytopathologists. The first interest of pathologists in a pathogen is usually one of identification, thus the predilection they have shown to employ the host plant as a criterion for speciation; identification of the host plant being (hopefully) a short cut to the identity of the fungus. This practice is reflected in nomenclature, an effect well illustrated by a random sampling of the specific epithets. In Persoon's *Synopsis*, 1801, in round numbers, 40 per cent of the epithets are based on morphological features, 25 per cent on colour terminology, and 10 per cent on the names of higher plants. Similar examinations of the index to the first volume of the *Index of fungi* covering the years 1940–9 show the distribution of epithets to be approximately similar in these two works: between a quarter and a third are based on morphological characters, approximately 10 per cent on colour, and 30 per cent on the names of host plants. The last usage is even more pronounced among obligate parasites, such as rusts and powdery mildews, where up to 50–80 per cent of the specific epithets in some genera are derived from host names.

In addition to 'specialisation characteristics', Ciferri (1952) in an interesting analysis of the criteria for the definition of species in mycology, recognised two other categories of 'matrical characteristics' (criteria based on parasite–host relationships): (1) 'ecological characteristics' based on the effects of the fungus in or on definite organs of the host and (2) 'pathographic characteristics' based on the effect of the parasite in the host, an effect which is usually a host response. Familiar examples of the use of 'ecological characteristics' are provided by the genera *Phoma* and *Phyllosticta* which were differentiated by the occurrence of the former on the stems and the latter on the leaves of the host plants. A similar, if less clear cut, distinction was made between the allied pair of genera *Diplodia* (usually on stems, rarely on leaves) and *Ascochyta* (usually on leaves, less frequently on

stems) in which the spores, produced as in *Phoma* and *Phyllosticta* in pyc-
nidia, are two-celled instead of one-celled as in the first pair of genera.
Typical examples of specific epithets based on 'ecological' and 'pathographic' characters are *Cryptostroma corticale*, the causal agent of sooty bark
disease of sycamore and that of the peach leaf curl pathogen, *Taphrina
deformans*.

One result of the tendency to over-emphasise on the host has been that
morphological features useful for diagnosis have sometimes been neglected.
For example, the host has played a dominant role in the speciation of
Cercospora, a genus of leaf-spotting fungi of world-wide distribution and
wide host range, which was monographed by the American plant pathologist Charles Chupp (1954). *Cercospora* has a relatively complicated morphology and many of the species which have been proposed are readily
distinguished microscopically from one another. Other species are very
similar and specialisation (proved or assumed) has played a major role as a
taxonomic criterion. As a result the more than 1270 species catalogued and
described in Chupp's monograph are arranged in one series under families
of host plants from Acanthaceae (*Acanthus*) to Zingiberaceae (*Zingiber*).
Specific keys are provided when several species occur in one host family but
the chance of identifying a specimen of *Cercospora* from an unidentified or
wrongly identified host plant or from a culture in the absence of a host is
small; a very unsatisfactory state of affairs which is, however, being
remedied by the recent major revisionary studies of the genus by
F. C. Deighton.[2]

Another example of excessive multiplication of species is provided by
Ernst Gäumann's (Fig. 17) approach to *Peronospora* in the doctoral thesis on
P. parasitica (the downy mildew of brassicas) he submitted to the University
of Berne in 1918 and the monograph on the genus as a whole published five
years later (Gäumann, 1923). In the latter work 267 species are accepted.
Of these 146 were proposed by Gäumann himself and *P. parasitica* is divided
into no less than 54 species. This excessive subdivision resulted from
experimental investigations on parasitic specialisation supplemented by
histograms, derived from the measurement of the lengths and breadths of
1000 spores of each form studied, which showed the different taxa to exhibit
closely overlapping ranges of conidial dimensions, differences too small to
be of value to practising plant pathologists.

Species of both plants and animals are traditionally differentiated on
morphological criteria and since the beginning of the twentieth century
there has been increasing support for the view that fungi should be classified
at the species level 'for what they are, not for what they do or where they

Fig. 17. Ernst Albert Gäumann (1893–1963).

occur'. In other words, genera and species of fungi are best described in mycological terms, whenever possible morphology taking precedence over other considerations. By adopting this principle plant pathologists have made notable advances in taxonomic practice since E. S. Salmon at the turn of the century, in a monograph (Salmon, 1900) which is still influential, reduced to 49 the 160 species of the Erysiphaceae (powdery mildews) compiled in Saccardo's *Sylloge*. Similarly G. W. Fischer (Fischer, 1943, 1953; Duran & Fischer, 1961) did much to rationalise the taxonomy of the smuts by advocating and practising the consolidation of various well-known morphologically similar cereal and grass smuts specialised for different hosts and so generally accepted as different species, and Yerkes & Shaw (1959), in contrast to Gäumann's approach, listed over 80 synonyms for *Peronospora parasitica*. More recently still, Colin Booth, of the Commonwealth Mycological Institute, in his important monograph (Booth, 1971) accepted a mere 50 specific names of the more than 1000 proposed in the genus *Fusarium*.

Special forms and physiologic races

Concurrently with the appreciation of the advantages which accrue from a morphological approach to speciation has been recognition of the reality of intra-specific taxa for the differentiation of which morphology plays a minor role, and recognition of their major importance in epidemiological investigations and the breeding of resistant cultivars of crop plants.

It was in 1894 that Jakob Eriksson, of the Swedish Central Station for Agricultural Experiments, demonstrated experimentally the existence of a number of morphologically similar forms of *Puccinia graminis* specialised for one or more species of host plants. For example, from cereals three series of isolates were distinguished: (1) isolates from wheat able to infect barley but not oats or rye; (2) isolates from oats which did not attack wheat or rye; (3) isolates from rye to which barley but not wheat was susceptible. These categories he designated *spezialisierten Formen* (special forms; the *formae speciales* of the International Code of Botanical Nomenclature) to which Eriksson & Hennings (1896) gave trinomial names e.g. *Puccinia graminis tritici*, *P. g. avenae*, and *P. g. secalis* for the cereal forms. Later workers have sometimes accepted these taxa as varieties (*P. graminis* var. *tritici*) but they are usually treated as *formae speciales* (*P. graminis* f. sp. *tritici*) since the recognition of this category by the International Code of Botanical Nomenclature in 1930.

In 1913 Elvin Charles Stakman in the study on cereal rusts which he submitted as a doctoral thesis to the University of Minnesota and in later studies, both alone and in collaboration with colleagues (e.g. Stakman & Piemeisel, 1917) at the Agricultural Experiment Station, St Paul, Minnesota, in what became his main life's work, showed the existence within *P. graminis* f. sp. *tritici* of numerous biological or physiological races or forms. These races (for which the term 'physiologic race' was recommended by the International Botanical Congress held at Amsterdam in 1935) could be distinguished by the reactions of a selection of 12 carefully chosen, pure bred wheat cultivars (the 'differential varieties' or 'differentials') on experimental inoculation under standardised conditions. The type of infection and the varietal reaction after one to two weeks is scored on an arbitrary agreed numerical scale: 0, immune (no rust pustules); 1, very resistant (pustules very small); 2, moderately resistant (pustules small to medium); 3, moderately susceptible (pustules medium); 4, very susceptible (pustules large and often confluent); x, heterogeneous (mesothetic) (pustules variable, sometimes types 1–4 on one plant). For identifying races the five types of varietal reaction can be reduced to three: resistant (R), infection types 0, 1, 2; susceptible (S), infection types 3, 4;

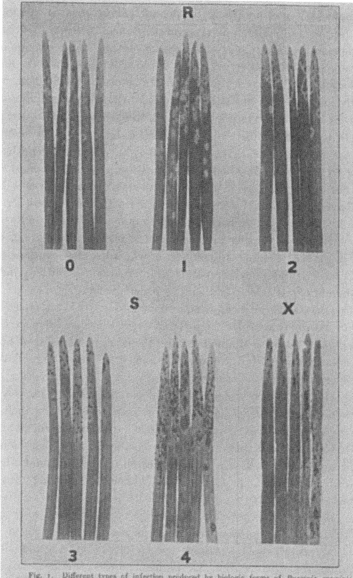

Fig. 1. Different types of infection produced by biologic forms of *Puccinia graminis* on various differential hosts of *Triticum spp.* Class R, indicating resistance, includes types 0, 1 and 2; class S, designating susceptibility, includes types 3 and 4; class X, representing the heterogeneous type of infection, has no subdivisions.

Fig. 18. Differential symptoms induced by physiologic races of Puccinia graminis. *(Stakman & Levine (1922): 4.)*

mesothetic (M), infection type x (Stakman & Levine, 1922; Stakman *et al.*, 1962) (see Fig. 18). The importance of these races to cereal pathologists investigating the epidemiology of rust necessitated the introduction of an international numerical nomenclature (outside the provisions of the Botanical Code) under the control of Stakman's laboratory at St Paul. More recently still, by employing additional differential hosts some physiologic races have been divided into 'sub-races' (distinguished by adding a capital letter to the race number, e.g. *P. graminis* f. sp. *tritici* 15B) and even subdivisions of sub-races (race 15B-1, 15B-2) have been distinguished. It should perhaps be emphasised that races, sub-races, and subdivisions of sub-races are all of equal taxonomic rank, all are physiologic races or races. They have merely been distinguished by the reaction of different series of host cultivars. It is generally recognised that the reactions of the 12 standard cultivars used as differentials have little relevance to those of some newly introduced commercial cultivars of wheat but they continue to be used because of their long acceptance as international standards and the vast amount of data associated with their deployment.

At first the stability of races was questioned. Marshall Ward (1902–3) of the University of Cambridge claimed that a race which could infect host A but not host B could be adapted to infect the latter by parasitising host C and from such findings he introduced the concept of 'bridging hosts'. Ward worked with brown rust of brome grasses (*Puccinia dispersa* [*P. recondita*]) and Salmon (1904) claimed to have obtained parallel results with powdery mildew of grasses (*Erysiphe graminis*) but later investigators, including Freeman & Johnson (1911) and Stakman *et al.* (1918) in the United States, failed to support Ward's claims which were finally disproved by a detailed 14-year study by F. T. Brooks and his research students at Cambridge on the brown rust of brome grasses when no evidence for bridging hosts was obtained (Bean *et al.*, 1954).

The categories *forma specialis* and *physiologic race* were found to be applicable to many other important plant pathogens and have proved most useful concepts. Most of Gäumann's 'species' of *Peronospora* would fall into one or other of these categories, and species of powdery mildews, the potato blight fungus (*Phytophthora infestans*), and *Plasmodiophora brassicae* (club root of brassicas) are additional examples of major pathogens for which special forms or races have been described. The taxonomy of the genus *Fusarium* has always presented much difficulty and many taxonomic innovations have been proposed. The *formae speciales* of the current taxonomic treatment of the genus (Booth, 1971) have in the past been variously designated as 'cultivars' (because exposing the pathogen to populations of cultivars of the host plant was considered to have induced variation in the pathogen by 'cultivation') or even 'clones', a usage at variance with the application of these terms to cultivated plants. Finally it may be noted that while the numbering of the physiologic races of the cereal rusts and most other pathogenic fungi has been arbitrary the designations for races of *Phytophthora infestans* are related to major genes controlling resistance to blight in the potato (W. Black, 1952; Black *et al.*, 1953).

Identification

The most important technique for the identification of plant pathogenic fungi has always been morphological examination (both macroscopic and microscopic) either of the pathogen as it occurs on the host or after isolation in pure culture. Biochemical tests have had little relevance, no true yeast having yet been proved pathogenic for plants. Neither has serology which is so useful for the identification of bacteria (see Chapter 4) and viruses (see Chapter 5). In medical mycology serological tests are frequently used to confirm a diagnosis by the reaction of antibodies produced in the host with an antigen prepared from the pathogen. In plant pathology the reverse procedure has been attempted. An antiserum prepared against the pathogen by inoculating rabbits (or other animals) with preparations of the pathogen (the antigen) should, ideally, give a specific reaction with the pathogen from which it was prepared. Unfortunately with fungi, although there have been claims for the clear-cut serological differentiation of nearly related plant pathogens such as that between *Phythium aphanidermatum* and *Phytophthora parasitica* by Morton & Dukes (1967), and serological confirmation of a number of taxonomic groupings, serological tests are not yet sufficiently reliable for routine use when making identifications.

Some side effects

The interest shown by plant pathologists in identifying and describing fungal pathogens has had several useful side effects. By compiling lists of the plant pathogens (and frequently also of saprobic forms, particularly microfungi) of their own countries plant pathologists have made notable contributions to knowledge of the geographical distribution of fungi. For example, the *Distribution maps of plant diseases*, of which more than 500 have been issued by the Commonwealth Mycological Institute since 1942, are mainly compiled and updated from regional listings by plant pathologists. Equally useful have been the many comprehensive host indexes – which complement lists of pathogens – compiled by plant pathologists (see analysis of Bibliography). Further, plant pathologists, especially during the

first half of the twentieth century when the interest in plant pathogenic
fungi was still at its height, have provided standard taxonomic monographs
on major groups of fungi – particularly orders such as the Uredinales (rusts)
and Ustilaginales (smuts) in which plant pathogenic species predominate.
Today such monographs tend to be produced by professional taxonomists.

Life histories

Knowledge of the life history of the pathogen has frequently provided
essential data for designing appropriate control measures.

Prévost (see Chapter 2) was unable to complete his observations on the
life history of wheat bunt (*Tilletia caries*). He could only deduce that seedling
infection occurred. That this deduction was correct was established by
Julius Kühn who, in his textbook *Die Krankheiten der Kulturgewächse*, 1858,
described and illustrated seedling infection. Also, while Prévost had seen
both filiform and allantoid sporidia resulting from the germination of bunt
spores it was Berkeley (1847) who first observed the fusion between filiform
sporidia which precedes production of the allantoid sporidia (by which
infection is established) and it was not until 1912 in Germany that W. Lang

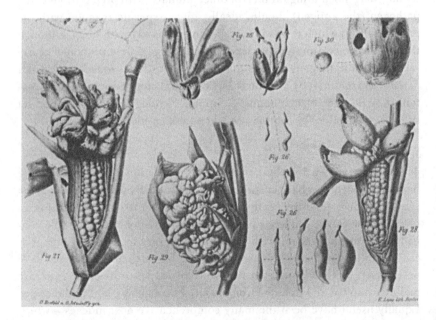

Fig. 19. Maize smut (Ustilago maydis). *(Brefeld (1872–1912), XI, Taf. 5, 1895.) Cf.*
Fig. 3.

followed the progress of the pathogen in the host plant from seedling infection to sporulation in the inflorescence. He found the mycelium to be intercellular.

Tillet (1760) failed to induce maize smut by planting seed contaminated with *Ustilago maydis* spores – the method he had found so effective with wheat bunt. This led him to the conclusion that the spores did not carry the contagion and he accounted for the disease by invoking 'too great an abundance of sap'. One hundred and twenty-five years later Oscar Brefeld, professor of botany at Munster, when investigating the same disease found that, although a small percentage (1–5 per cent) of maize seedlings sprayed with a sporidial suspension developed smut symptoms and died the survivors remained healthy at maturity. This was a surprise as he then 'held to the old view, universally current until now, that smut germs generally could penetrate into the young seedlings in order to appear later as smut beds in the full-grown plant'. Further experimentation with plants of different ages showed that meristematic tissue was susceptible to infection at all stages in the growth of the plant and inoculation of the developing inflorescence enabled him to reproduce the characteristic swollen spore-filled grains (see Fig. 19) (Brefeld, 1888).

The mode of infection of loose smut of wheat (*Ustilago tritici*), which was differentiated from bunt by both Tillet and Prévost, also proved puzzling. That the disease occurred in stands derived from some samples of seed was early established but it was not until 1896 that Frank Maddox in Tasmania announced his discovery that infection was via the flowers. He wrote:

I have never been able to cause infection and reproduce the disease with spores on the grain or in the ground, which I can so easily do with bunt spores to reproduce bunt. The only way I have been able to infect grain and reproduce smut (which seldom ever fails) is by putting the spores on the ovary of the plant at flowering time, about the same time as the pollen grains are being shed. The grain will mature without the slightest signs of being diseased . . . It is really wonderful to me how smut spores do, as the ovary is well protected by the glumes or chaff, and there is only a short period that infection seems able to take place.

This finding was confirmed, among others, by S. Nakagawa in Japan the next year (see Akai, 1977) and by Brefeld in 1903 (Brefeld (1872–1912), **13**) and the fact established that loose smut originates from mycelium of the smut within the wheat grain.

These results revealed major variations in the life histories of three related pathogens and established (as first demonstrated by Tillet, 1755) that pathogenic fungi could be seed borne. Bunt and loose smut of wheat illustrate two contrasting methods of transmission – on the seed, when the employment of a fungicide is indicated (see Chapter 7), and in the seed, when the infection may sometimes be eliminated by heat treatment (see Chapter 8).

Stem rust of wheat and many other rusts have life histories involving two or more hosts. Farmers were the first to suggest that outbreaks of wheat rust were in some way associated with barberry bushes and during the eighteenth century legislation was passed requiring the eradication of barberry adjacent to farm land (see Chapter 10). But the reality of this relationship proved controversial. Joseph Banks in 1805 believed there was a relationship but could not prove it. L. G. Windt in Germany in the following year expressed the same opinion and in 1816 a Danish schoolmaster, Niels Pedersen Scholer, succeeded in infecting rye with aeciospores from barberry. This last result was not, however, generally accepted. De Bary in 1853 treated *Puccinia graminis* on wheat and *Aecidium berberidis* on barberry as unrelated rusts but it was de Bary himself who finally settled the matter in 1865 by infecting barberry with teliospores from rye and in the following year rye with aeciospores from barberry (de Bary, 1865–6). The spermogonia which precede and accompany the aecidia on barberry had been observed by Unger (1833). Meyen (1841) suggested that the spermogonia were male and female organs but it was not until 1927 that J. H. Craigie in Canada established that the spermatia produced in the spermogonia functioned as diploidising agents which initiate the dikaryotic mycelium which bears the aecia.

Many other rusts were found to be heteroecious including *Gymnosporangium clavariiforme* (see Figs. 6, 7) and the notorious blister rust of five-needled pines (*Cronartium ribicola*), the uredinial and telial stages of which occur on wild and cultivated species of *Ribes*. These findings endorsed the correctness of barberry-eradication legislation and underlay the Ribes-eradication campaign to control pine rust (see Chapter 10).

The elucidation of the life history of the potato blight fungus was spread over many years and very confused. One strand was taxonomic. C. Montagne in France, who first described the potato blight fungus under the name *Botrytis infestans*, considered it to belong to the Mucedines (Hyphomycetes). The next year the Rev. Berkeley, to whom Montagne had sent drawings of his fungus, drew attention to its similarity to several downy mildews, and in 1847 Unger (as did Caspary in 1852) considered it to belong to the genus *Peronospora* which Corda had proposed in 1837. This opinion was accepted by de Bary (1863) but in 1876, on the basis of conidial development and the resulting conidiophore characteristics, he proposed a new genus for the potato blight fungus which has since been known as

Phytophthora infestans. The attribution of the potato blight epidemic of the 1840s to a parasitic fungus while accepted by many leading workers did not convince everybody until the observations made by de Bary (1861, 1863) on his experimental infections with zoospores derived from conidia put the question beyond doubt. How the pathogen overwintered remained unknown. From analogy with the so-called 'white rust' (white blister) of crucifers (*Albugo candida*) (in which Prévost (1807) had been the first to observe zoospores) thick-walled resting spores were sought for at the suggestion of de Bary (1863), but unsuccessfully until in July 1875 in England Worthington G. Smith, an architect by training who had turned wood engraver, botanical artist, archaeologist, and amateur naturalist, reported his discovery of such spores which, he claimed were produced in abundance when macerated potato leaves were kept submerged in water. This announcement was received with enthusiasm and later in the month Smith was honoured by the award of the Knightsian Gold Medal of the Royal Horticultural Society. His fame was short lived. The year before, the Royal Agricultural Society had commissioned de Bary, as the leading expert, to study and report on potato blight. In 1876 de Bary submitted his report in which he proposed the new genus for the potato blight fungus (as already noted) and demolished Smith's claims by showing that the resting spores were those of another fungus, a species of *Pythium*. Smith was unrepentant and nine years later was propagating the same views in his textbook (Smith, 1884). Resting spores (oospores) of *P. infestans* were eventually described from cultures on oatmeal agar by G. P. Clinton (1911) of the Connecticut Agricultural Experiment Station and Paul Murphy (1927) was the first to observe them in nature on the surface of rotten potatoes where their occurrence is rare. Germination of oospores is still unreported and it would appear that they are of minor importance in carrying the fungus over from one season to the next. This is mainly ensured by infected tubers. Badly infected tubers rot completely. Others produce infected shoots. It was Speerschneider (1857) who first claimed that tuber infection occurs by conidia developed on the foliage being washed into the soil, a finding confirmed by de Bary (1876).

A very common life history pattern is one in which the pathogenic phase is associated with the imperfect (conidial) state of the fungus, the perfect (usually ascosporic) state typically occurring as the climax on dead tissue, frequently removed in both time and space from the pathogenic phase. Apple scab is a well known example. Wallroth (1833) in Germany first reported apple scab which he attributed to the hyphomycete *Cladosporium dendriticum* (later renamed *Fusicladium dendriticum* by Fuckel) which causes more or less circular necrotic spots on the leaves and fruit (and also attacks the young wood) during the summer months while the ascosporic state (*Venturia inaequalis*) occurs on fallen infected leaves the following spring. The ascosporic state was described by M. C. Cooke in 1866 as an independent fungus in the genus *Sphaerella* and the connection between the two states established by Aderhold (1894). Similarly the anamorphic, pathogenic, state of the fungus with its chains of moniliform spores causing brown rot of fruit was first described by C. H. Persoon in 1796 from decaying fruit of pear, plum, and peach as *Torula fructigena*, a name he changed in 1801 to *Monilia fructigena*. It was not until 1905 that Aderhold and Ruhland described the perfect state as *Sclerotinia fructigena* in which the ascospores are borne in stalked, cup-like ascocarps developed in the spring on overwintered mummified fruit which was infected the previous summer.

Seed transmission

The transmission of a pathogenic fungus with, on, or in true seed (as, for example ergot sclerotia with rye seed, and the bunt and loose smut fungi of wheat, respectively) is an aspect of a life history of great significance to the practice of plant pathology. The problem has been recognised since the time of Jethro Tull and Mathieu Tillet in the eighteenth century but wheat bunt remained the only known example until 1883 when A. B. Frank, in Germany, reported the seed transmission of *Colletotrichum lindemuthianum*, the cause of bean anthracnose. Since then many instances of the seed transmission of fungi (and also of bacteria and viruses) have been established and most major crops have at least one seed-borne disease.

In passing one notorious aberration may be recalled. This is the 'mycoplasm theory' which Jacob Eriksson first offered in 1897 to explain outbreaks of cereal rust which he believed did not originate from airborne spores. He claimed that the living protoplasm of the fungus and the host could unite in a symbiotic relationship, pass from one cell to another of the host or remain viable in seed and that under appropriate conditions the fungus was reconstructed and initiated infection. Although disproved and shown to be unnecessary by Marshall Ward (1903) and others, Eriksson extended and supported this notion for another twenty years (Eriksson, 1897–1921).

There have been many attempts to free contaminated and infected seed from the pathogen by the use of fungicides or physical agents, especially heat. Today most countries have regulations to control the purity and freedom from disease of seed offered for sale and legal restrictions on the import and export of seed.

Host–parasite relationships

For the past hundred years plant pathologists have put much effort into investigations on host–parasite relationships with particular reference to pathogenic fungi. The problem has been approached from many angles. The effect of environmental factors on host–parasite interactions by affecting both the host and the pathogen and the genetic relationships between pathogen and host have been major interests. These approaches, which have contributed so much to the elucidation of the epidemiology of disease, are considered in Chapter 9. Another major area of study, among others which have greatly increased knowledge of pathogenicity, if the effect on the practice of plant pathology has been less, has been the 'physiology of parasitism' which covers not only the physiology of the infection process and subsequent invasion of the host tissue but also the biochemistry. The advent of the electron microscope stimulated a renewed interest in the histological relationships of pathogen to host (for example, on the fine structure of haustoria)[3] while recent developments have deepened understanding of the chemical effects of fungal metabolites (toxins) in inducing disease symptoms and the mechanisms of resistance due to normally occurring chemical constituents of the host (such as phenolics)[4] or to special substances (phytoalexins) produced by the host in response to infection.

An important landmark in the study of host–parasite relationships is Marshall Ward's (Fig. 20) classic report in 1888 on 'A lily disease'. First observed as a trivial defect of unknown aetiology of *Lilium candidum* characterised by spotting of the leaves, pedicels, and flower buds, the disease was more destructive during the wet summer of 1888 and this enabled Ward to determine the cause as a '*Botrytis* of the *Polyactis*-type' [*B. cinerea*]. In addition, he followed the course of infection, observed the development of appressoria by the fungus, and gave a detailed account of penetration of the host cell wall and the disintegration of the underlying tissue. He suspected that the tissue was disintegrated by a soluble ferment produced by the botrytis and he succeeded in obtaining from cultures preparations which would disintegrate tissues *in vitro* by dissolving the middle lamellae. Two years earlier de Bary had reported that the expressed sap of carrots infected by *Sclerotinia sclerotiorum* contained a thermolabile substance able to disorganise host tissue and in 1905 L. R. Jones in Wisconsin was able to demonstrate the production of a similar enzyme by the soft rot bacillus *Erwinia carotovora*. Others too, at about this time, as recorded by Jones, made similar but less detailed observations of both fungi and bacteria.

In one respect Marshall Ward was mistaken. He concluded that the

Fig. 20. Harry Marshall Ward (1854–1906).

initial penetration of the host cell by the fungal hypha was effected by enzyme action. Wiltshire (1915) favoured the same view when he invoked a cuticle-dissolving enzyme to effect the penetration of the apple scab fungus (*Venturia inaequalis*) but later strong evidence that the process is mechanical was offered by William Brown (Brown & Harvey, 1927). Brown worked (from 1912 onwards) at the Imperial College of Science & Technology in London where he maintained a life-long interest in fungal parasitism as evidenced by the notable 'Studies in the physiology of parasitism' (Brown, 1915), the results of a series of investigations (for many of which *Botrytis cinerea* was a test organism) which he and his students made during the next forty years. It was largely by the work of Brown's team that the enzyme (or enzyme mixture as it proved to be) able to disintegrate plant tissue became more and more precisely defined as is expressed by the successive terms 'cytase', 'pectinase', 'protopectinase', 'polygalacturonase', and 'polymethylgalacturonase'[5] applied to it. As a pendant to this work the elucidation of the aetiology of water spot of tomato by Ainsworth, Oyler & Read (1938) may be mentioned. This disease is characterised by minute necrotic lesions, first noted on the green fruit when each spot is surrounded by a silver-coloured circle. These spots, which can at times lead to a down grading of the fruit, were previously attributed (very reasonably) to insects or (somewhat less reasonably) to a non-parasitic agent. The true cause proved to be *Botrytis cinerea* a spore of which germinating on the surface of a young fruit (in which the epidermis is still thin-walled) during a period of high humidity penetrates the fruit but only to the extent of a small necrotic lesion while the enzyme produced diffuses outwards and separates a ring of epidermal cells from those below. With growth in size of the fruit the ring expands while the epidermal cell wall thickens so that the fruit is no longer susceptible to penetration by fungus. Another familiar example of abortive infection giving limited lesions is chocolate spot of broad beans caused by *Botrytis fabae* and *B. cinerea.*

The idea that toxic metabolites (toxins) of pathogenic fungi play a part in the symptom expression of plant disease stems back to de Bary (1886) but the toxic principle he discovered was an enzyme complex and it is not usual to treat the enzymes associated with soft rots and with the browning of vascular tissue in certain wilt diseases as toxins. Neither are growth-regulating substances such as the gibberellins produced by *Gibberella fujikuroi* and responsible for the elongated growth characteristic of bakanae disease of rice (see Akai, 1974) toxins in the usual sense of the term. In phytopathology 'toxin' is reserved for toxic chemicals ('chemopathogens') able to cause effects similar to those of disease symptoms induced by micro-organisms. One currently accepted terminology is to employ 'phytotoxin' for any product of a living organism toxic to plants and to subdivide phytotoxins into 'vivotoxins', in the sense of Dimond & Waggoner (1953) who coined the term and defined it as 'a substance produced in the infected host by the pathogen and/or the host which functions in the production of disease, but is not itself the initial inciting agent of the disease', and 'pathotoxins' for toxins proved to play an important causal role in disease.

This last category, pathotoxins, by definition the simplest, is still a small group. The first to be recognised was the toxin noted by Tanaka in Japan in 1933 from *Alternaria kikuchiana* Tanaka, the cause of black spot of Japanese pears (*Pyrus serotina*) the symptoms of which Tanaka found could be reproduced by spraying fruits of susceptible variety with a culture filtrate of the pathogen. Subsequently, the best known and most studied example of a pathotoxin, victorin from *Helminthosporium victoriae* responsible for blight of the South American oat variety Victoria, was discovered by Meehan & Murphy (1947). Later still the toxin from *Periconia circinata* was shown by Scheffer & Pringle (1961) to induce symptoms of the milo disease of grain sorghum (*Sorghum vulgare* var. *subglabrescens*).

Vivotoxins are of two main types: toxins produced by the host in response to infection, e.g. ipomearone by the sweet potato on infection by either of the two taxonomically unrelated fungi *Ceratostomella fimbriata* or *Helicobasidium mompa* (see also the alexins below) or toxins produced by the invading pathogen, as are α-picolinic acid and piriculin, metabolites of the rice blast fungus *Pyricularia oryzae* in its host; these are not specific in their action and do not induce all of the disease symptoms. Further, piriculin is more toxic to *P. oryzae* than to the rice plant but there appears to be a piriculin-binding protein which inactivates the toxic effects of piriculin on the pathogen but not on the host.

The status of most of the numerous other phytotoxins which have been investigated in greater or less detail, including alternaric acid (from *Alternaria solani*), colletotin (*Colletotrichum fuscum*), and fusaric acid, lycomarasmin, and other toxins associated with fusaria causing wilt in tomato and other plants (work on which during the 1940s and 1950s was particularly associated with Ernst Gäumann and his school at Zurich (see Gäumann, 1957)), is in varying degrees controversial. Further details of these and other phytotoxins may be obtained from the reviews by Wheeler & Luke (1963), Pringle & Scheffer (1964), and Wood (1967).

It was early suspected that one possible explanation of resistance to disease might be that certain chemical constituents (such as 'tannins') of the host plant inhibited growth of the pathogen. Wiltshire (1915) demonstrated that the conidia of the apple scab fungus (*Venturia inaequalis*) germinated better in the sap of a susceptible than a resistant variety of apple, a fact confirmed by Johnstone (1931) who found the inhibitor to be most abundant in trees deficient in nitrogen, a deficiency which increases scab resistance. Recently it has been demonstrated (by Noveroske *et al.*, 1964; Raa, 1968) that phloretin, bound as the β-glucoside phloridzin, occurs in apple leaves and on invasion by *V. inaequalis* this is hydrolysed by a host β-glucosidase to phloretin which on oxidation gives products which arrest the development of the pathogen. Another famous example is onion smudge caused by *Colletotrichum circinans*. This disease was first described from England in 1851 by the Rev. Berkeley, who named the causal fungus *Vermicularia circinans* and, like many subsequent observers, noted that the disease was most severe in white varieties of onion in which the outer scales of the bulb are unpigmented (see Frontispiece). Smudge was recorded in 1874 from the United States where a detailed study of the disease was made by J. C. Walker at the University of Wisconsin in the 1920s (Walker, 1921–4) when he demonstrated that aqueous extracts from scales of red and yellow, but not white, varieties of onion inhibited spore germination (see

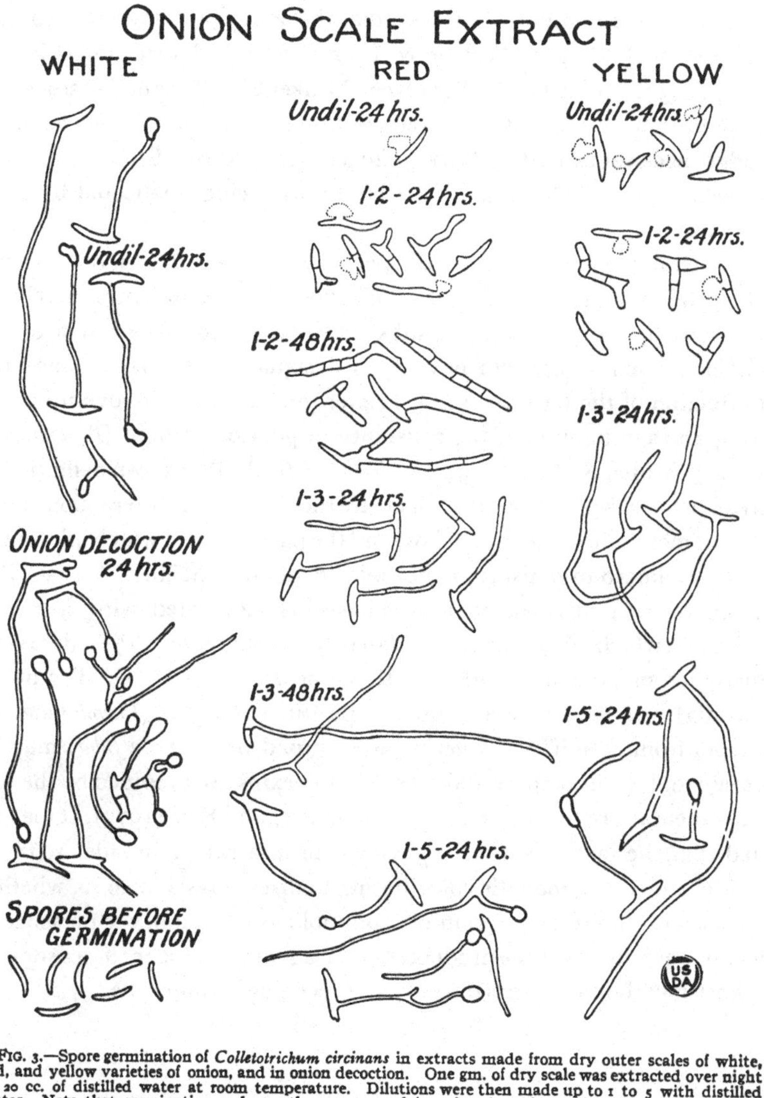

FIG. 3.—Spore germination of *Colletotrichum circinans* in extracts made from dry outer scales of white, red, and yellow varieties of onion, and in onion decoction. One gm. of dry scale was extracted over night in 20 cc. of distilled water at room temperature. Dilutions were then made up to 1 to 5 with distilled water. Note that germination and growth were normal in onion decoction and undiluted white scale extract, while in the undiluted colored scale extract typical abnormal germination occurred as shown in fig. 1. With dilution the toxicity of the extract was reduced, however, quite normal germination occurring at 1 to 5. Normal ungerminated spores are shown in the lower left-hand corner. See further explanation in the text.

Fig. 21. Inhibition of spore germination of Colletotrichum circinans *by onion scale extract.*
(*J. C. Walker (1923), fig. 3.*)

Fig. 21). Subsequently it was shown that the active principles were proto-catechuic acid (Angell, Walker & Link, 1930) and catechol (Link & Walker, 1933). In further studies Owen, Walker & Stahmann (1950) found that pungency was also directly related to resistance of onion varieties to smudge, neck rot caused by *Botrytis allii* as described by Munn (1917), and *Aspergillus niger* infection, incidence of the last being unaffected by scale pigmentation.

Another important biochemical aspect of the host–pathogen relationship is the production by the host, after infection by, or sometimes merely on contact with, the pathogen, of substances which inhibit growth of the pathogen. Such substances have been designated *phytoalexins* since the introduction of the term by K. O. Müller in relation to discoveries made during an investigation of the resistance of potato to blight (*Phytophthora infestans*) (Müller & Börger, 1941; Müller, 1956). Müller's own definition of 'phytoalexines' was 'antibiotics which are the result of the interaction of two different metabolic systems, the host and the parasite, and which inhibit the growth of micro-organisms pathogenic to plants'. Müller in 1958 also demonstrated a phytoalexin from Phaseolus inoculated with *Sclerotinia fructicola* (which was later isolated as phaseolin (7-hydroxy-3', 4'diethylcromano-coumaran) by I. Cruickshank and D. Perrin in Australia) and among many others, pisatin from pea (*Pisum sativum*), gossypol from cotton, and wyerone from broad bean (*Vicia faba*) may be mentioned. Details of these and other phytoalexins are provided by the two complementary reviews by Cruickshank (1963) and Kuć (1972). It may be noted that the last author uses phytoalexin in a rather broader sense to cover 'chemical compounds contributing to disease resistance ... whether they are found in response to injury, physiological stimuli, the presence of infectious agents or are the products of such agency' and so includes some of the host metabolites referred to in the preceding paragraphs.

4

Bacteria, including actinomycetes

Although during the first decade of the nineteenth century it had been convincingly demonstrated that fungi could cause disease in plants and by the early 1840s that they were also pathogenic for insects, higher animals, and man, the concept of infectious disease met with much resistance. It was not until the 1860s when the investigations of Pasteur and others finally refuted the doctrine of spontaneous generation and the reality of bacterial diseases of animals and man became apparent that the concept received general acceptance. No one did more to complete this change than the German bacteriologist Robert Koch who published his classic paper on anthrax in 1876 and discovered the tubercle bacillus in 1882 and the cholera vibrio the following year. He also, between 1877 and 1881, developed techniques for growing bacteria in pure culture. These results, coupled with improvements in the compound microscope, had widespread repercussions, including the recognition of bacterial diseases of plants.

It is traditional to attribute the discovery of bacterial disease in plants to the American T. J. Burrill but there are other claimants.[1] In 1850 the German chemist Emil Mitscherlich, who had interests in fermentation, recorded observations on the 'active liquid' (*wirksame Flüssigkeit*) which disintegrated potato tuber tissue in the apparent absence of fungi, an effect he attributed to 'vibrios' (motile bacteria). As Professor Estey (1975) has recently reminded historians, a stronger case can be made in support of the famous French pathologist and parasitologist Casimir-Joseph Davaine [1812–82] noted for his classical work on anthrax and septicaemia. Davaine was in general practice and had no laboratory but in 1850 he and a fellow physician and pathologist P.-F.-O. Rayer [1793–1867] were the first to report the anthrax bacillus. Subsequently Davaine published short notes on a disease of *Impatiens* and on the rotting of fruit (1886) and in 1868, in a paper on anthrax (Davaine, 1868a) wrote:

I will now say some words about a disease observed in the plant kingdom which is brought about by the invasion of bacteria. Among the succulent plants, or the vegetables with very tender and most parenchyma, many times have I seen a change that at first appeared at the root and shortly invaded the rest of the plant destroying it in a few days. This change, which reduces the tissues to a kind of rot, is caused by the development of bacteria which differ from those of anthrax in as much as they have movement. We can easily transmit this disease from one plant to another by inoculation: around the inoculation point an oil-like spot appears that grows and takes over the whole plant if the diseased part is not cut off.[2]

Later in the year Davaine gave a more detailed account (Davaine, 1868b) of his perceptive experiments with a motile bacillus (*Bacterium termo* – which may well have been the soft rot organism, *Erwinia carotovora*) isolated from rotting vegetation. He found that the day after subepidermal inoculation of succulent plants, such as *Opuntia* or *Aloe*, with a small quantity of liquid or substance containing this bacillus (Davaine did not work with pure cultures) the first sign of infection appeared, provided the wound was prevented from drying out. Further, prompted by Pasteur's example of heating wine to eradicate undesirable micro-organisms, he found that:

if one exposes a succulent plant infected by bacteria to a temperature of a little above 52 °C – for example, 55 °C which many succulent plants resist well – for long enough for the heat to penetrate the thickness of the diseased part, the bacteria cease their movement, the infection ceases to spread, the disorganised part dries, and the plant continues to grow as if it had not been attacked. (Davaine, 1868:502)[3]

and so Davaine was perhaps the first to employ heat therapy (see Chapter 8) against a plant disease.

Thomas Jonathan Burrill (Fig. 22) was a farmer's son, who as a youth during the summers learned farming and during the winters attended the nearby country school. After earning his way through high school and university he graduated in 1865 from the State Normal University of Illinois at the age of 26 when he was appointed superintendent of the Urbana public schools. His interest in collecting the local flowering plants led to his selection as botanist on Professor J. W. Powell's first expedition to the Colorado Rocky Mountains and his subsequent appointment (at first to teach algebra) at the University of Illinois where in 1870 he became professor of botany and horticulture. Burrill saw the need of studies on plant disease (see Chapter 11) and in the following years reported his investigations on a number of fungal diseases. It is, however, for his contributions to the elucidation of the bacterial nature of fire blight of the pear that he is best remembered.

Fire blight, probably indigenous in the United States, had been known since the end of the eighteenth century as a serious disease of cultivated varieties of pear, apple, and quince (K. F. Baker, 1971). William Coxe, in

Fig. 22. Thomas Jonathan Burrill (1839–1916).

his *A view of the cultivation of tree fruits*, 1817 (the first American book on fruit culture) drew attention to the depredations of fire blight (a name he introduced) which are still significant not only in America but also in England where the disease attained epidemic proportions after its introduction in 1957 (J. J. Baker, 1972:125–6). Burrill, who imported his first microscope from Germany in 1868 and later better objectives (including a 2-mm Zeiss oil-immersion apochromat), as well as objectives and oculars for locally made stands, gave his first report on fire blight, up to then thought to be caused by a fungus, to the Illinois State Horticultural Society in December 1877 (Burrill, 1878) when he accepted the prevailing view. He noted that 'a thickish, brownish, sticky matter exudes from affected limbs' and that this exudate 'is entirely made up of ... minute oscillating particles'. A year later in a second paper to the same society he tentatively identified these minute bodies as bacteria responsible for the disease and stated that 'a particle of [the] viscous fluid introduced upon the point of a knife into the bark of a healthy tree is in many cases followed by blight of the part' (Burrill, 1879). In 1882 he named this bacterium *Micrococcus amylovorus* (now *Erwinia amylovora*). Burrill did not prove that the bacterium was the pathogen. This was left to J. C. Arthur, botanist at the New York Experiment Station, Geneva, who cultured the bacterium in a sterilised infusion of corn (maize) meal and after five subculturings over a period of about four months successfully infected the green fruit of a Bartlett pear. He also filtered a strong aqueous infusion of blighted pear through a 'porous earthenware vessel such as used for small electric batteries' and obtained infection with the unfiltered infusion but not with the filtrate. These experiments were repeated with similar results (Arthur, 1885) and in 1886 Arthur submitted to Cornell University an account of his fire blight studies as a doctoral thesis.

Additional information on the epidemiology of fire blight was obtained in 1891 by Merton B. Waite (Fig. 23) (of the USDA and a former assistant to Burrill) who showed that the bacterium is carried from flower to flower by bees and other insects, initiating blossom blight, and that injury to the host is unnecessary for infection to occur. These observations constitute the first record of a plant pathogen disseminated by insects but it was not until 1936 that K. G. Parker of Cornell University showed this method of transmission to be of economic significance, in dry weather. Under wet conditions the major spread between trees and orchards is by splash, wind-driven rain, and bacterial aerosols (Glasscock, 1971).

While fire blight was attracting attention in America, in Europe the bulb growers of Haarlem asked Professor Hugo de Vries of Amsterdam University to investigate a serious disease of hyacinths. De Vries allocated the problem to his most promising student Jan Hendrick Wakker who in 1881 found the disease to be bacterial. After spending the winter of 1881–2 in de Bary's laboratory at Strasbourg Wakker isolated the pathogen of yellow disease which he named *Bacterium [Xanthomonas] hyacinthi* (Wakker, 1881). This was his major contribution to plant pathology. After a few more years' work on yellow disease and other diseases of bulbs and on diseases of tropical crops (especially sugarcane) in Java he did no more research and from 1897 until his death 30 years later taught in a secondary school at 's-Hertogenbosch. In France, in 1879, Prillieux had described a defect of wheat grains which he attributed to a bacterium (named '*Micrococcus tritici*' in his textbook (1895–7) and still of uncertain status) and in Italy Savastano (1887) reported inoculation experiments demonstrating that tuberculosis of the olive (olive knot) was caused by a bacterium named in 1908 *Bacterium savastanoi* by Erwin F. Smith, whose name dominated plant bacteriology during the first quarter of the twentieth century.

Erwin F. Smith

Erwin Frink Smith (Fig. 23) was born 21 January 1854 at Gilbert Mills, New York. His father, a tanner and shoemaker, in 1870, took up farming in

Michigan where Erwin in the intervals of helping on the farm completed his grade schooling. He also, with the aid of friends, did much to educate himself and at the age of 18 entered the Ionia high school from which he graduated in 1880. To go to the university was his objective but like Burrill and so many of his contemporaries he had to earn his way. For a time, while attending high school, he was a school teacher and a prison guard (which allowed him time for reading) and during the years between leaving high school and entering the university, after part-time employment by the US Weather Bureau, another spell at the Ionia State Reformatory (this time as keeper of the prison), and journalist responsible for the 'Scientific and Sanitary' department of the *Michigan School Moderator* (a periodical for schools), he became correspondence clerk of the Michigan Board of Health which involved him in surveys of the literature on hygiene and the epidemiology of human diseases. Smith finally entered the University of

Fig. 23. The group of botanists who planned and originated the Bureau of Plant Industry of the United States Department of Agriculture.
Back row (left to right): *W. T. Swingle, M. B. Waite, M. A. Carleton, A. F. Woods.*
Front row (left to right): *D. G. Fairchild, P. H. Dorsett, B. T. Galloway, E. F. Smith.*

Michigan at Lansing in 1885 and graduated with honours in biology in July 1886 at the age of 32.

This varied experience served him well. In addition he had acquired a good knowledge of both French and German while evidence of his deep interest in natural history dates back to the *Catalogue of the phaenogamous and vascular cryptogamous plants of Michigan*, published in 1881 in collaboration with Charles F. Wheeler (the pharmacist and amateur botanist who had introduced Smith to French). He was widely read and a versifier – after his first wife's death in 1906 he published privately a book of '104 sonnets and various other poems' as a memorial to her.

The month following the award of his BS degree Smith was offered a temporary post at Washington and he remained with the Department of Agriculture for the rest of his life. His first major task was an investigation of peach yellows on which he made two exhaustive reports (Smith, 1888–94). Subsequently he published on fungal diseases (notably the bulletins on 'Wilt disease of cotton, watermelon and cowpea' (Smith, 1899) and 'The dry rot of potato due to *Fusarium oxysporum*' (Smith & Swingle, 1904) and the elucidation of Panama disease of banana (*F. oxysporum* f. sp. *cubense* (Smith, 1910a)) but it is his more than a hundred contributions on bacterial diseases of plants by which he would wish to be remembered. At the end of his life he recalled that in his early days: 'Robert Koch was my hero, and his influence more than any other man decided me to enter pathology and bacteriology'.

The number of important diseases which were established as bacterial in nature by his own work or by that of others in his laboratory is most impressive. They include, by Smith himself, wilt of cucurbits (*Bacterium tracheiphilus* [*Erwinia tracheiphila*]) and its transmission by the cucumber beetle (*Diabrotica vittata*), 1895; brown rot of Solanaceae (*Bact.* [*Pseudomonas*] *solanacearum*) transmitted by the Colorado beetle, 1896; Stewart's disease of sweet corn (*Ps.* [*Erwinia*] *stewartii*), angular leafspot of cotton (*Ps.* [*Xanthomonas*] *malvacearum*), and bean blight (*Ps.* [*X.*] *phaseoli*), 1901; bacterial spot of plum and peach (*Ps.* [*X.*] *pruni*), 1903; crown gall (*Bact.* [*Agrobacterium*] *tumefaciens*), with C. O. Townsend, 1907; Grand Rapids disease (bacterial canker) of tomato (*Aplanobacter* [*Corynebacterium*] *michiganense*), 1910b; yellow slime disease of cocksfoot (*A.* [*C.*] *rathayi*), 1913; and angular leaf spot of cucumber (*Bact.* [*Ps.*] *lachrymans*), Smith & Bryan, 1915; and by his co-workers *Bact.* [*Ps.*] *aptata* on beet and *Bact. marginale* [*Ps. marginalis*] and *P.* [*X.*] *vitians* on lettuce (Nellie A. Brown, 1909, 1918); citrus canker (*Ps.* [*X.*] *citri*) (Clara Hasse, 1915); bean wilt (*Bact.* [*C.*] *flaccumfaciens*) (Florence Hedges, 1922); basal glume blotch of wheat (*Bact.* [*Ps.*]

Fig. 24. Some of the early women assistants at the USDA Bureau of Plant Industry.
Florence Hedges (1878–1956), Nellie A. Brown (1877–1956), Lucia McCulloch
(1873–1955), Mary K. Bryan (1887–†), Agnes J. Quirk (†), Clara H. Hasse (1880–?1924),
Flora W. Patterson (1847–1928), Vera K. Charles (1878–1954), Charlotte Elliott
(1883–1974).

atrofaciens) scab ón neck rot of gladiolus (*Bact. marginatum* [*Ps. marginata*]), and alfalfa (lucerne) wilt (*A.* [*C.*] *insidiosum*) (Lucia McCulloch, 1920, 1921, 1925). In addition Smith summarised or expanded much of this work in the three quarto volumes of *Bacteria in relation to plant diseases*, 1905–14, and his *Introduction to bacterial diseases of plants*, 1920, is the first textbook in this field. A characteristic feature of these publications is the use made of photography. Smith was a keen photographer and most of his major contributions from his bulletins on peach yellows onwards are profusely illustrated by photographs and photomicrographs. Smith was also ahead of his time in the encouragement he gave to women both as visitors to his laboratory and as assistants (Fig. 24). It was apparently at his suggestion that Effie A. Southworth was appointed in 1887 as the first woman research worker in the United States Department of Agriculture under Dr Galloway (Fig. 23). In 1901 Agnes J. Quirk became Smith's assistant and later, in addition to the women already mentioned, Mary K. Bryan and Charlotte Elliott joined his laboratory. The last, in addition to her researches, compiled the two editions of her most useful *Manual of bacterial pathogens*, 1930, 1951. The entry Smith made in his diary on 5 March 1927 summarising the current work in his laboratory (which he visited almost daily up to within ten days of his death on 6 April of that year) gives an idea of his continued drive:

Miss Fawcett has interesting studies of distilled water well along. Agnes Quirk and Miss Brown are working on crowngall oxidation phenomena. Miss Brown has a note with two plates on sweet pea fasciation due to *Bact. tumefaciens* in *Phytopathology*. Miss Elliot has written a paper on stripe disease of oats. Miss Hedges is working on diseases of beans. Miss Bryan on several things – one of which is Aplanobacter on tomato. Lucia McCulloch is working on bulb diseases. Miss Cash on miscellaneous things. Miss Fox the same. I am trying to complete the illustrations for my Ithaca paper [Smith, 1929]. Brewer [, J. F.; the departmental artist] is helping.

The early American work on bacterial diseases of plants was not universally accepted, particularly in Germany where scientific work was then in the ascendant and American work held in low esteem. Professor Albert Fischer of Leipzig, a leading European bacteriologist, in his *Vorlesungen über Bakterien*, 1897, expressed doubts as to the reliability of the American work and concluded that there were no bacterial diseases of plants. This resulted, during 1899 to 1901, in polemical exchanges between Smith and Fischer by which Fischer was silenced and angered if not convinced.[4] Smith's results did, however, stimulate others both in the United States and elsewhere. For example, in 1895, L. H. Pammel of Iowa Agricultural College described a bacteriosis of rutabaga caused by *Bacillus* [*Xanthomonas*] *campestris*, now established as a world-wide agent of disease of crucifers; Professor

L. R. Jones of the University of Vermont, who had had a five-month introduction to plant bacteriology in Smith's laboratory in 1899, described and named the soft rot bacillus (*Bacillus carotovorus* [*Erwinia carotovora*]) in 1901; and the next year C. J. J. van Hall in the Netherlands described from lilac *Pseudomonas syringae*, an important pathogen with a wide host range. Jones subsequently reported his classical investigation on the cytolytic enzyme produced by the soft rot organism (Jones, 1905, 1910).

From about 1910 one of Smith's continuing interests on which he wrote at intervals for the rest of his life was the possibility that knowledge of crown gall in plants might contribute to the solution of the problem of human cancer. This interest, which no doubt in part originated from his introduction to medical literature in the 1870s and his knowledge of medical bacteriology, tended to become somewhat obsessional.

Generic classification of phytopathogenic bacteria

During the 1920s and 1930s there was much confusion in the generic classification of plant pathogenic bacteria – which are typically Gram negative and do not form spores – and thus a corresponding confusion in their nomenclature. Two systems of bacterial classification – those of W. Migula, *System der Bakterien*, 1897–1900 and K. B. Lehmann & R. O. Neumann, *Atlas und Grundriss der Bakteriologie* ... 1896 (edn. 7, 1926–7) – were particularly popular. E. F. Smith (1905–14) had introduced a third while successive editions of *Bergey's Manual of determinative bacteriology* provided yet another alternative. The generic name *Phytomonas*, proposed for a number of plant pathogenic bacteria in the first (1923) edition of *Bergey's Manual*, had to be abandoned because, as a later homonym of a genus of Protozoa, it infringed the International Code of Botanical Nomenclature. W. J. Dowson's separation in 1939 of *Xanthomonas* from *Pseudomonas* (see Dowson, 1943) was generally endorsed as were H. J. Conn's *Agrobacterium* of 1942 (*J. Bact.* **44**:359) and the earlier *Erwinia* based on the fire blight organism by Winslow *et al.* in 1917 (*J. Bact.* **2**:560). These generic names were all accepted in the 1948 and subsequent editions of *Bergey's Manual* and have been generally followed since (see Table 2). Publication of the textbooks by Dowson in 1949 and C. Stapp in 1958 also helped to stabilise the situation.

The general trend in the approach to bacterial speciation has paralleled that for fungi. There has, in recent years, been much consolidation of species (see Stolp *et al.*, 1965; Dye *et al.*, 1975). For example, in the genus *Xanthomonas* the number of species has been reduced from over 100 to 5 with

Table 2. *Generic classification of plant pathogenic bacteria*

	Non-motile	Motile, flagella	
		Polar	Peritrichous (or bipolar)
Migula (1895–1900)	*Bacterium*	*Pseudomonas*	*Bacillus*
Lehmann & Neuman (1897–1927)	←—————————— *Bacterium* (non-sporing) ——————————→ ←—————————— *Bacillus* (sporing) ——————————→		
E. F. Smith (1905–14)	*Aplanobacter*	*Bacterium*	*Bacillus*
Bergey's Manual (1923–39)	←————————*Phytomonas* ————————→		**Erwinia** (peritrichous)
(1948–74)	**Corynebacterium** (Gram +)	**Pseudomonas** **Xanthomonas**	**Erwinia** **Agrobacterium** (bipolar)
	[*Bacillus* (Gram +; sporing); *Bacterium* (non-sporing rods of uncertain taxonomic position)]		

many names assigned to synonymy under *X. campestris*. Similarly for *Pseudomonas* in which numerous fluorescent, oxidase negative phytopathogens have been included in *Ps. syringae*.

Bacterial names, bacteria like fungi having been traditionally considered to be plants, were at first subject to regulation by the *International Code of Botanical Nomenclature*. In 1948 bacteriologists published their first international code of nomenclature and this was recognised by the Botanical Code. Revised editions of the *International Code of Nomenclature of Bacteria* appeared in 1958 and 1975. In general the Botanical and Bacteriological Codes are very similar. One difference is that while the current Bacteriological Code in an appendix on infraspecific subdivisions recognises *forma specialis* (see Chapter 3) in the mycological sense for the forms adapted to specific hosts it also recommends the use of a new term *pathovar* (of which 'pathotype' is a synonym) for infraspecific taxa characterised by a pathogenic reaction in one or more hosts. Recently, Young *et al.* (1978) have made detailed proposals for the recognition of many of the plant pathogenic species of *Xanthomonas* and *Pseudomonas* now regarded as conspecific with *X. campestris* and *Ps. syringae* to be reclassified as pathovars. The 1975 edition (the '1976 Revision') introduced one fundamental change. Up to the present the starting point for the nomenclature of bacteria has been that for plants in general, 1753 (the date of publication of the first edition of Linnaeus's *Species plantarum*), but this has resulted in much confusion and uncertainty because the first descriptions of many bacteria have been very inadequate by modern standards. There is, therefore, to be a new starting point, 1 January 1980, after which only those names which appear on an 'Approved List of Bacterial Names' (a draft of which has already (1978) been published) or are validated by publication in the *International Journal of Systematic Bacteriology* need be taken into consideration. Names not on approved lists will no longer have any standing in nomenclature but such names will be available for re-use after appropriate validation. This decision should, for the time at least, simplify bacterial nomenclature.

Identification

Morphology proved to be of limited value for the identification of phytopathogenic bacteria (including the filamentous actinomycetes, see below) which are typically rod-shaped, non-spore forming, and Gram negative. The number and arrangement of the flagella are the morphological features of greatest diagnostic value. Pigmentation is also a useful aid for identification. As with other bacteria differentiation of species is mainly determined by a range of biochemical tests but during recent years increasing use has been made of serological techniques taken over from medical bacteriology for the rapid identification of plant pathogenic bacteria.

Among early rather tentative studies on the serology of phytopathogenic bacteria were those of Ethel Doidge (1917) on *Erwinia citrimaculans*, C. O. Jensen (1918) on *Agrobacterium tumefaciens*, S. G. Paine & Margaret Lacey (1923) on *Bacillus lathyri* and *Xanthomonas phaseoli*. These were followed by a more extensive study by St John-Brooks, Nain & Mabel Rhodes (1925) who reported the results of cross-agglutination tests with a range of plant pathogens.[5]

Another useful identification technique has been 'phage-typing' by the development of bacterial viruses (phages) which are frequently specific in their action. In 1951 Katznelson and Sutton in Canada developed phage-typing for *Xanthomonas phaseoli* in bean seed (Katznelson *et al.*, 1954). This method was used for the rapid identification of the fire blight bacterium (*Erwinia amylovora*) by Billing *et al.* (1960) and by Crosse & Garrett (1961) for *Pseudomonas mors-prunorum* races. The introduction of this technique into plant pathology was reviewed by Stolp (1956).

Life histories and host–parasite relationships

The life histories of plant pathogenic bacteria are simple. Including no obligate parasites and having no spores, the survival of unfavourable periods, overwintering, and the transmission from one crop to the next are effected by dormant cells on the host plant, as in fire blight, in seed (e.g. bean blight (*Xanthomonas phaseoli*) the first seed-transmitted bacterial disease to be recognised; Beech 1892), black rot of cabbage (*X. campestris*), black arm of cotton (*X. malvacearum*) or growing saprobically in soil (*Pseudomonas solanacearum*) or other substratum. Many bacterial pathogens are spread in the field by wind and rain or by infection introduced by cultural operations.

Less attention has been paid to host–parasite relationships than for plant pathogenic fungi. L. R. Jones's study on the pectolytic enzyme of the soft rot organism (*Erwinia carotovora*) (Jones, 1905) bridged the gap between the investigations of Marshall Ward and William Brown on this enzyme complex for fungi (see Chapter 3). One topic that has received much attention is the formation of gall tissue by *Agrobacterium tumefaciens*.

Actinomycetes

Since the publication of the first edition of *Bergey's Manual* in 1923 actinomycetes have been accepted as filamentous bacteria but before then there was much uncertainty regarding their taxonomic position and they were not infrequently considered to be fungi and classified as deuteromycetes. Actinomycetes are typically soil saprobes. Many yield commercially interesting antibiotics. A few cause disease in man, animals, and plants. The most important and widespread plant disease caused by an actinomycete is common scab of potato tubers. This defect was noted by J. C. Loudon in 1825 (*Encyclopaedia of agriculture*, p. 785) who wrote:

The scab, or ulcerated surface of the tubers, has never been satisfactorily accounted for. Some attributing it to the ammonia of horse-dung, others to alkali, and some to the use of coal ashes. Change of seed and of ground, are the only resources known at present for this malady.

Lutman & Cunningham (1914) in their important monograph review other early opinions on the aetiology of scab which, in Europe was for long confused with powdery (or corky) scab caused by *Spongospora subterranea* and the term 'scab' (which was in use before 1750, fide *The Oxford English Dictionary*) has also been applied to other potato tuber defects including black scurf (*Corticium solani*) ('black scab', 'Rhizoctonia scab'), silver scurf (*Helminthosporium solani*, syn. *Spondylocladium atrovirens*) ('silver scab'), and skin spot (*Oospora pustulans*). It was the American H. L. Bolley who in 1890 first claimed potato scab to be a bacterial disease but the scab organism was described as a fungus under the name *Oospora scabies* by Roland Thaxter in

1892. G. C. Cunningham (1912) considered the scab pathogen to belong to the higher bacteria and the name change to *Actinomyces scabies* was made by Güssow two years later. Subsequently the circumscription of *Actinomyces* was restricted to anaerobic actinomycetes, such as the species causing actinomycosis in man, and the name *Streptomyces scabies* was introduced by Waksman & Henrici in the 1948 edition of *Bergey's Manual* for the aerobic potato scab pathogen. The use of formalin against potato scab was introduced by J. C. Arthur in 1897 (Arthur, 1899).

Beet is also susceptible to infection by *S. scabies* (as first shown by Bolley, in 1893) and *S. ipomoeae* is responsible for a disease of sweet potato tubers (Person & Martin, 1940).

5

Viruses and organisms confused
with viruses

By the mid 1880s the reality of bacterial diseases of man, animals, and plants had been established and, with the exception of plants, generally accepted. There was, however, a residue of infectious diseases for which no causal agent could be found and just as Pasteur and Roux failed to elucidate the aetiology of rabies so Erwin Smith suffered what he later described as the greatest disappointment of his life when his major investigation on peach yellows (Fig. 25) in the eastern United States, apart from demonstrating that the disease agent could be transmitted by bud grafting, failed to reveal the cause (Smith, 1888–94). Most of these unknown pathogens were viruses, essential clues to the nature of which were provided between

MARYLAND ORCHARD RUINED BY PEACH YELLOWS.

Fig. 25. Peach yellows. (Smith (1888) pl. 25.)

1886 and 1898 by three workers with whose names the origins of virology will always be associated.

Tobacco mosaic virus

The first of these men was the Dutch agricultural chemist Adolf Mayer, director of the Landwirtschaftlichen Versuchs-Stationen, Wageningen, who investigated the tobacco disease for which he coined the name *Mosaikkrankheit* and proved by experiment 'that the juice from diseased plants obtained by grinding was a certain infectious substance for healthy plants' (Mayer, 1886). He also demonstrated that the clear filtrate obtained by passing the juice through double paper filters was infectious and that this capacity was lost by heating at 80 °C but not at 60 °C. He concluded that bacteria were the cause.

The second was the Russian Dmitri Ivanovski who on 12 February 1892 announced to the Academy of Science, St Petersburg 'that the sap of tobacco leaves attacked by the mosaic disease retains its infectious qualities even after filtration through Chamberland filter candles'. This was the first report of any virus passing through a bacteria-proof filter.

Finally, in 1898, Martinus Willem Beijerinck (Fig. 26), botanist, bac-

Fig. 26. Martinus Willem Beijerinck (1851–1931).

teriologist and somewhat eccentric bachelor, who became professor of microbiology at the Technical High School, Delft, published a paper entitled 'Ueber ein contagium vivum fluidum als Ursache der Flecken-krankheit der Tabaksblatter' and in doing so introduced a renowned phrase. By refined experiments Beijerinck demonstrated that the infection was not caused by aerobic or anaerobic bacteria, that the infective principle spread by diffusion into an agar gel, and concluded that this filtrable pathogenic agent was a 'contagious living fluid' which, he deduced from serial inoculations, multiplied in the infected host.

These three germinal investigations laid a firm foundation although it was more than thirty years and hundreds of scientific papers later that the true nature of tobacco mosaic virus was established. During the interval there was much speculation as to the nature of plant viruses. Some, for a time, continued to attribute virus diseases to bacteria and protozoa were also invoked for this role (e.g. by Nelson, 1922). The American Albert F. Woods (1902) advocated disturbed enzyme balance as the underlying cause, an explanation which gained some popularity until disproved by his fellow countryman H. A. Allard in 1916 while James Johnson, professor of horticulture at the University of Wisconsin, in 1925, on the basis of the induction of disease in various solanaceous plants with the sap from appar-ently healthy potatoes, offered a so-called 'viroplasm' hypothesis according to which 'some part of the normal protoplasm or cytoplasm of one species, if properly introduced into the cells of another species, might find conditions compatible for growth in that species and bring about the abnormality known as virus disease'. In 1942 he made an attempt to test this hypothesis by cross inoculating 122 species of legumes (representing 50 genera) to bean (*Phaseolus vulgaris*) and other test plants. The results were on the whole negative but he did discover two new viruses which he considered to have pre-existed in the hosts. It was also at this time that a number of claims were made for virus multiplication *in vitro* but none could be substantiated.

In parallel with such opinions were increasing numbers of observations which pointed towards the correct solution. C. G. Vinson (1927) of the Boyce Thompson Institute for Plant Research, Yonkers, New York, demonstrated that tobacco mosaic virus could be precipitated from infected juice by acetone, alcohol, or ammonium salts and shortly afterwards Helen Purdy (1928, 1929) prepared antiserum which would neutralise the infec-tivity of infective sap. It was such pointers to the possible protein nature of the virus that led Wendell M. Stanley, a biochemist at the Rockefeller Institute for Medical Research, Princeton, N.J., to the investigation which resulted in his announcement in 1935 of the 'Isolation of a crystalline protein possessing the properties of tobacco-mosaic virus', an achievement for which he shared the Nobel Prize for Chemistry in 1946.

But the problem of the nature of the tobacco mosaic virus was not yet completely solved. It remained for a group of British workers to make the final contribution. In 1936 F. C. Bawden (Fig. 29), of Rothamsted Experimental Station in collaboration with the Cambridge biochemist N. W. Pirie (who later became head of the biochemistry department at Rothamsted) and J. D. Bernal and I. Fankuchen of the Crystallographic Laboratory, Cambridge, reported that Stanley's crystalline protein on further purification could be obtained in a liquid crystalline state which they considered was a nucleoprotein. The next year Bawden and Pirie detailed aspects of this work and concluded that all the available evidence favoured the view that the nucleoprotein was the virus, a conclusion Stanley accepted with some reluctance. This point marked a breakthrough for understanding the nature of plant viruses which was greatly deepened by subsequent electron microscopical and biochemical studies but twenty more years elapsed before it was demonstrated that nucleic acid preparations were infective and that the protein was inessential for virus multiplication.

Early records of virus diseases

Not all virus diseases are undesirable. The infectious variegation of Abutilon studied by Bauer in 1904 is considered advantageous by horticulturists as is the breaking of tulips which has a long history (see McKay & Warner, 1953). Broken tulips were esteemed by the Dutch in the seventeenth century and it is flower paintings of the period that provide some of the earliest evidence for the incidence of virus disease. Artists continued to show a predilection for broken tulips. Robert John Thornton, lecturer in medical botany at Guy's Hospital, London, provided a particularly splendid example (reproduced in colour as the frontispiece to Waterson & Wilkinson, 1978) in his *Temple of flora*, 1798, and a modern etching is illustrated in Fig. 27. Although it was not until the 1930s that Dorothy M. Caley's investigations at the John Innes Horticultural Institution, then in London, contributed to the elucidation of this floral abnormality (and her colleague McKenney Hughes (1930) showed the condition to be aphid-transmitted), that breaking could be transmitted by grafting broken and normal bulbs was known to Clusius in 1576.[1] Later, Joseph Blagrave in the appendix to the 1675 edition of *The epitome of the art of husbandry* (*New additions:* 10–12) gave details of this procedure.[2] Grafting has long been a

Fig. 27. Broken tulips. ('Old English Tulips 2', etching and aquatint by Rory McEwen, 1978.)

horticultural practice and there are three, apparently independent, English eighteenth-century records in which grafting resulted in virus transmission. The Reverend John Laurence in his *Clergy-man's recreation*, 1714, describes how a normal jasmine may be variegated by 'inoculating' (bud grafting) it from a striped plant. He observed 'that if the bud live one or two months, and after that happen to die, ... it will have communicated its Virtue to the sap, and the tree will become entirely strip'd. This Discovery undoubtedly proves the Circulation of the sap. Q.E.D.' In 1720, Mr Henry Cane reported that in 1692 he had transmitted the mottle of jasmine by grafting, and in the last decade of the century Erasmus Darwin in *The botanic garden* noted a case of yellow spotting of a passion tree being transmitted a fortnight after budding, even though the buds did not take.

Another interesting early record is that of nettlehead of hops which was described by Reginald Scott [1538?–99] in 1574 in the first English practical treatise on hop culture (see Fig. 28). Nettlehead was first recognised as a virus disease by Duffield at Wye College in 1925.

The peach was probably introduced into North America about 1630, according to Erwin F. Smith (1888), who put the first appearance of peach yellows at some time before 1791. By the first decade of the nineteenth century the trouble was widespread. On 11 February 1806 Judge Richard Peters in a paper to the Philadelphia Society for the Promotion of Agriculture wrote:

About fifty years ago, on the farm I now reside at Belmont [now part of Philadelphia], my father had a large peach orchard, which yielded abundantly. Until a general catastrophe befell it plentiful crops had been produced with very little attention. The trees began nearby to sicken, and finally perished ... Fifteen or sixteen years ago I lost one hundred and fifty peach trees in full bearing in the course of two summers by a disease engendered in the first summer. I attribute its origin to some morbid infection of the air ... The disorder being generally prevalent would, among animals, have been called an epidemic. From perfect verdure the leaves turn yellow in a few days, and the bodies blacken with spots. Those distant from the point of infection gradually caught the disease. I procured young trees from a distance in high health and planted them among those diseased. In a few weeks they became sickly and never recovered.

In November 1807 he was writing:

I still think that the disease so generally fatal (more so this year than any other in my memory), called the yellows, is atmospherical.

And in the same month, Dr James Tilton of Bellevue near Wilmington, Delaware, in a letter to Judge Peters wrote:

The disease and early death of our peach trees is a fertile source of observation, far from being exhausted ... Even that sickly appearance of the tree called the yellows, attended by numerous weakly shoots on the limbs generally, is attributed to insects by a late writer in our newspapers.

From the middle of the eighteenth century there were an increasing

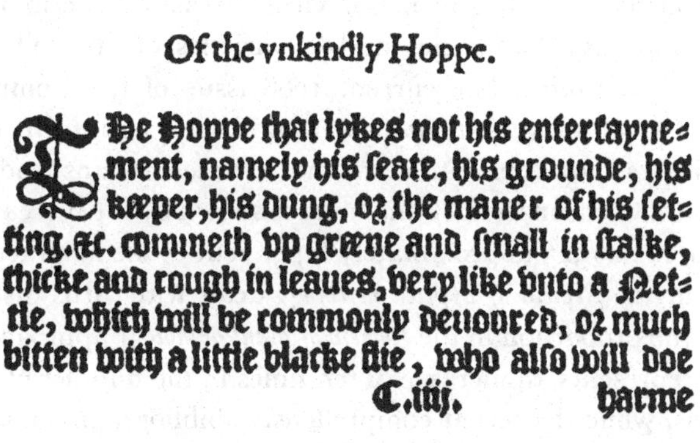

Of the vnkindly Hoppe.

He Hoppe that lykes not his entertayne-ment, namely his seate, his grounde, his keeper, his dung, oz the maner of his set-ting. &c. comneth vp greene and small in stalke, thicke and rough in leaues, very like vnto a Net-tle, which will be commonly deuoured, oz much bitten with a little blacke flie, who also will doe **C.iiij.** harme

Fig. 28. First record of nettlehead in hops. (Scott (1576):9.)

number of records both from Great Britain and the Continent of Europe of a defect of potato widely known as 'the curl', and much speculation as to its cause.[3] It was attributed to frost damage of the seed potatoes, the effects of 'blights', and to insufficiently or over-ripened tubers. The last explanation was particularly popular – in tubers allowed to remain too long in the ground 'the vegetable sap which lodges in the tuber and which constitutes the vigour of the future plant, is exhausted. Hence the young stem rises weakly, or what is usually termed curly' (Young, 1825). Dickson (1814) advocated planting potatoes intended for seed at least a fortnight later than the main crop. It was noticed in Scotland that curl was most prevalent in early districts and it was customary for farmers in such districts to purchase a great part of their seed from 'higher' (that is, more northerly) or later districts. It was not until the first decade of the twentieth century that the condition was shown to be due to a complex of potato leaf roll and other viruses (see below).

Incidence

A major world-wide trend during the twentieth century has been the increasing emphasis on virus diseases of plants due in large part to the realisation of the heavy losses they inflict, regularly or at times, on many agricultural or horticultural crops. This emphasis has been accompanied by the recognition of an increasing number of viruses proved (or assumed) to be distinct and of groups of related viruses. George Massee, the Kew mycologist, in the first edition of his *Text book of plant diseases*, published in 1910, made no reference to virus diseases and five years later in a supplement to the second edition dismissed 'Mosaic disease of potatoes, tomatoes and tobacco' in a couple of paragraphs. In America, F. D. Heald's textbook of 1926 described or listed 40 major virus diseases while in 1939 F. O. Holmes characterised 89 'species' and 39 'varieties' of virus in his *Handbook of phytopathogenic viruses*. The current 1968 issue of the Commonwealth Mycological Institute's *Plant virus names* and its 1971 supplement list well over 700 viruses – together with many more related 'strains' and suspected virus diseases. Confirmation of this trend is shown by the increasing size of the literature. In 1925 approximately 6 per cent of the abstracts on plant pathology in the *Review of applied mycology* dealt with virus diseases, fifty years later the proportion in the *Review of plant pathology* approached 20 per cent. This represents an increase of ten times in the number of individual publications, while the recent comprehensive bibliography of plant virology by Beale (1976) includes 29 000 references.

From the turn of the century until the First World War tobacco mosaic continued to attract attention in both Europe and North America. The first phase of this interest may be considered to have culminated in H. A. Allard's famous United States Department of Agriculture Bulletin No. 40 of 1914, and the contributions which followed during the next few years. These studies provided a foundation on which James Johnson and others built.

Another major development which has contributed so much to the knowledge of plant virology was the intense and sustained interest shown in virus diseases of the potato plant due to the gradual recognition and proof that viruses were responsible for the 'degeneration' of potato stocks. Potato curl was a problem of long standing. It was Hendrick Marius Quanjer at the agricultural institute at Wageningen in the Netherlands who, from 1908, pioneered the solution by an investigation of potato leaf roll (the principal component of 'curl') with special emphasis on the phloem necrosis which is a prominent symptom of infection by leaf roll and other potato viruses (see Quanjer, 1913; Pethybridge, 1939). In 1911 the American W. A. Orton visited Otto Appel in Berlin (who that year, with O. Schlumberger, had published a monograph on *Blattrollkrankheit*) and called on George H. Pethybridge (who was investigating blight and other diseases of potato in Ireland) while on a trip to Europe, during which he was the first to recognise potato mosaic in Germany and, on his return home, to establish that the same disease occurred in North America. This discovery was exploited by Orton himself (Orton, 1914) and by E. S. Schultz and D. Folsom of the Maine Experiment Station. In the meantime Pethybridge's Irish assistant Paul A. Murphy obtained a scholarship which enabled him to work at the Imperial College, London, the Biologische Reichsanstalt, Berlin, and at Cornell University. Next he obtained a post in Prince Edward Island, Canada, where, under H. T. Güssow (the Dominion Botanist) he became deeply interested in degeneration of the potato. After returning to Ireland, when Pethybridge moved to England to become chief 'mycologist' (plant pathologist) to the Ministry of Agriculture, Murphy succeeded him in 1923 as Head of the Plant Disease Division and subsequently became professor of plant pathology at University College, Dublin, where much of his outstanding work was on virus diseases of the potato. In England major research on virus diseases of potato may be traced back to the private enterprise of a very remarkable man, Redcliffe N. Salaman, who after a medical training was director of the pathological institute of the London Hospital from 1901 to 1904, a post he relinquished because of ill health. Retiring to Barley in Hertfordshire he turned to the then new subject of genetics and soon (inspired by his gardener) began to use potatoes instead of animals in his breeding experiments. His knowledge of the potato, as exemplified by *Potato varieties*, 1926, in which he summarised the results of 20 years of his own privately financed experimentation, and by the *History and social influence of the potato*, 1949, in which he brought so many aspects of his wide scholarship to bear, was unsurpassed. In 1926 Salaman became director of the new Potato Virus Station (now the Virus Research Unit of the Agricultural Research Council) at Cambridge with Kenneth M. Smith as entomologist and in 1930 the newly graduated F. C. Bawden (Fig. 29) joined the staff as Salaman's assistant. Both these young men subsequently gained international reputations as plant virologists. Kenneth Smith, who succeeded Salaman as director in 1939, has been a prolific writer of scientific papers and books on plant virology. Bawden in 1936 moved to Rothamsted Experimental Station where he later became head of the department of plant pathology and finally director of the Station and was honoured by a knighthood. Although Bawden contributed so much to more academic aspects of virology he never forgot that disease control is the final objective of phytopathological research and maintained a practical outlook.

As Bawden recalled, the facilities of the Potato Virus Research Institute were primitive, 'the "laboratory" was a wooden hut and the most sophisticated piece of apparatus, a recalcitrant "Primus" stove'. This state

Fig. 29. *Frederick Charles Bawden (1908–72) inspecting potatoes at Sutton Bonnington 1943 with P. H. Gregory (centre) and A. Beaumont (right).*

Fig. 30. Louis Otto Kunkel (1884–1960).

of affairs was to some extent alleviated by financial help from the British Empire Marketing Board which supported virus research at a number of centres including Rothamsted where a physiologist (J. Caldwell), cytologist (Frances L. Sheffield), and entomologist (Marian A. Hamilton (Mrs Watson)) were appointed to the staff in 1928. Murphy's group in Ireland was also supplemented by an entomologist (J. B. Loughnane) and a physiologist/cytologist (Phyllis E. M. Clinch) and the new insect-proofed glasshouses on the lines of those adopted by Quanjer in the Netherlands have been claimed as the first of their type in the British Isles.

A number of other economically important virus diseases, such as curly top of sugar beet (Carsner & Stahl, 1924) and cucumber mosaic (Doolittle, 1920) attracted attention in the United States before the First World War. With recovery from the effects of war interest in virus diseases in general received increasing support. In 1925, parallel with the support for studies on virus diseases of potato in Europe, L. O. Kunkel (Fig. 30) (then plant pathologist in Hawaii) joined the Boyce Thompson Institute, New York to work on virus diseases. Nine years later he became the first director of the Department of Animal and Plant Pathology of the Rockefeller Institute at Princeton which gained a world-wide reputation for plant virus research associated with the names of Kunkel, F. O. Holmes, Stanley, L. M. Black, and others.

The tally of virus diseases which have been the subject of major investigations in all parts of the world is too long to catalogue here. Representative examples are sugar-cane mosaic which achieved notoriety in the 1920s because of epidemic outbreaks in Louisiana and elsewhere which threatened the industry, diverse diseases of legumes and soft fruits (especially raspberry and strawberry) in both Europe and North America, hop mosaic and nettlehead, yellows of sugar beet (in the UK and elsewhere), and a number of diseases of cereals and grasses, while in the tropics leaf curl of cotton in the Sudan, swollen shoot of cacao in West Africa, and rosette of groundnuts in East Africa all caused much concern. During recent years diseases of apple, plum, and other tree fruits, some of them long known, have been widely recognised and more intensively studied, as have virus infections of cereals.

These investigations have yielded much detailed information which has contributed to the accuracy of diagnosis and to the elucidation of epidemiological problems. They have suggested a number of general methods for the prevention of crop losses caused by virus infection; knowledge of the host range and reaction, modes of transmission, and immunological and serological responses being particularly relevant in this connection.

Host range and reaction

It early became clear that one virus was able to induce disease in more than one host. Apparently George Clinton in 1908[4] was the first to demonstrate by cross-inoculations that tobacco mosaic virus also caused the common mosaic disease of tomato. Solanaceous plants proved to be particularly susceptible to virus infection but it gradually became evident that most plants may be attacked by one or more viruses and that while some viruses (e.g. green mottle mosaic virus of cucumber) were restricted to one or a few related plants many others had a wide host range frequently covering plants of unrelated families. Holmes in his *Handbook* (1939) listed 133 species (of 34 families) susceptible to the classical cucumber mosaic virus of Doolittle (1914), 123 species (of 27 families) for tobacco mosaic virus, and 84 species (of 19 families) for tomato spotted wilt virus which was first recognised by Samuel, Bald and Pittman in Australia in 1930 and later found to be of wide geographical distribution.

It was also soon found that closely related susceptible species exhibited very different symptoms on infection. In tobacco (*Nicotiana tabacum*) no necrotic lesions develop on the leaves inoculated with tobacco mosaic virus

Fig. 31. Five leaves from plants of Nicotiana glutinosa *used to measure the effect of diluting a tobacco mosaic virus preparation (1:1, 1:3.16, 1:10, 1:100, 1:1000). (Holmes (1929a), fig. 7.)*

which becomes systemic in the host causing mosaic symptoms. In *N. glutinosa*, on the other hand, the virus does not become systemic but necrotic local lesions form on the inoculated leaves. This last response is of particular interest because, in 1929, it was utilised by Holmes as the basis of a method for measuring virus concentration in the inoculum which he showed to be directly related to the number of necrotic lesions developed (Holmes, 1929a) (Fig. 31). Later the method was extended to many parallel examples such as the local lesions produced on French bean (*Phaseolus vulgaris*) leaves by alfalfa mosaic virus.

In some diseases, such as that induced in chrysanthemum by tomato spotted wilt virus or ringspot of tobacco, though symptoms develop soon after infection, subsequent new growth is symptom-free although the virus can still be demonstrated in the host tissues. In others, inoculation though successful results in no visible symptoms in the host; as first reported by Nishimura (1918) for *Physalis alkekengi* infected by tobacco mosaic virus and in 1925, independently, by E. S. Schultz and James Johnson for potato. These infections have been referred to as 'masked', 'latent', or even 'occult'. The more usual expression is to designate the host as a 'carrier'. One of the famous carriers of a virus was the potato cultivar King Edward VII, every individual of which though apparently healthy, when tested by grafting a scion to the cultivar Arran Victory induced a crippling disease which Salaman & Le Pelley (1930) called paracrinkle. The virus carried by King Edward (potato virus E) could not be transmitted to potato or other plants either by sap or the aphid *Myzus persicae* by Salaman and Le Pelley but later Bawden, Kassanis & Nixon (1950) succeeded in transmitting the virus to both Arran Victory potato and tomato (which is also a symptomless carrier) by rubbing sap on leaves previously dusted with diatomaceous earth or carborundum powder.

When a virus becomes systemic it frequently spreads throughout the whole plant including the apical meristems. Sometimes, however, the apical meristem is virus free and Morel (1948) and Morel & Martin (1952) obtained virus-free dahlia plants from apical meristem tissue cultures originating from infected plants. A spectacular use of this technique was that by Kassanis (1957) who obtained virus-free clones of King Edward potato from 100–250 μm blocks of apical meristem tissue from sprouts of normal (paracrinkle infected) tubers and thereby increased the yield of this popular cultivar by approximately 10 per cent (Bawden & Kassanis, 1965). Another character of King Edward is that, whether carrying the para-crinkle virus or not, it is killed by potato virus X so that stocks of King Edward are exceptional in never being infected by this virus.

Mixed infections by more than one virus proved a further complication. K. M. Smith in 1931 reported that apparently healthy potatoes of the cultivar Up-to-Date carried two viruses which he named potato viruses X and Y. The former (which can be equated with the healthy potato virus of North America and the British potato mild mosaic virus) was transmissible by needle-scratch but not by aphides, the latter by both methods. The two viruses could be separated by the use of what Smith called 'plant filters', such as *Petunia* (resistant to virus X) or *Solanum dulcamara* (virus Y resistant and double infections demonstrated by 'plant indicators') which show symptoms differing from those induced by either virus alone, e.g. the severe necrotic symptoms shown by White Burley tobacco (*Nicotiana tobacum*) after infection by X+Y viruses. Such cases, as judged by the symptom expression, are frequently taken to be new diseases; for example, the streak disease of tomato (the so-called 'mixed (or double) virus streak') caused by tobacco mosaic virus+potato virus X (Dickson, 1925) and potato crinkle caused by potato viruses A+X (Murphy & McKay, 1932).

Transmission

The oldest, most general, and sometimes only experimental method for the transmission of a virus is by grafting a bud or shoot of a diseased plant onto a healthy one.

Many viruses, like tobacco mosaic virus, are also sap transmissible. In the first investigations of tobacco mosaic the virus was inoculated into healthy plants by inserting pieces of fine capillary glass tubing containing infected sap into the healthy plant (Mayer, 1886), injecting sap with a hypodermic syringe, or pricking it in with a needle. This last relatively inefficient method continued in use long after Allard (1917) had confirmed the earlier demonstration by Clinton that tobacco mosaic virus is readily communicated by merely rubbing infected sap on the leaves of healthy plants and had shown that good transmission followed alternately crushing with fine forceps (or cutting with scissors) hairs on the leaves of diseased and healthy plants. Holmes (1929b) published a description of the now widely used method of inoculation by gently rubbing the leaf surface with a pad of butter muslin soaked in infected sap. Later this technique was made increasingly efficient by the addition of an abrasive such as carborundum powder (silicon carbide) (Rawlins & Tompkins, 1934) or celite (diatomaceous earth) which gently scarifies the surface cells of the leaf and Mackenzie, Anderson & Wernham (1966) adapted this method for accelerating mass inoculation by the use of a pressurised air brush to apply both abrasive and inoculum. Sometimes expressed plant sap inactivates a virus, e.g. chrysanthemum leaf extract inactivates tomato spotted wilt virus, a virus sensitive to oxidising agents. Such inactivation of the virus can be partially prevented by the addition of an equal quantity of 0.5 per cent solution of sodium sulphite, a procedure that greatly facilitates the detection of spotted wilt virus in chrysanthemum. In the field, sap transmission of such highly infectious viruses as tobacco mosaic is effected merely by contact between the leaves of infected and healthy plants.

Another discovery, still unexplained, which facilitates experimental infection is that by Bawden & Roberts (1948) who found that darkening tobacco and bean plants for several hours before inoculation increased their susceptibility to several viruses.

Viruses are readily transmitted by vegetative methods of propagation, that is by cuttings, tubers, bulbs, etc., but transmission by true seed is relatively rare. Probably the best established example is that of French bean mosaic virus (Reddick & Stewart, 1919). Among others are lettuce mosaic (Newhall, 1923) and cucumber mosaic (Doolittle, 1920) viruses. Tobacco mosaic virus is not transmitted by tobacco seed though there have been numerous claims, many somewhat dubious, of its transmission through tomato seed. Virus can frequently be demonstrated on the seed coat of seed from infected tomatoes but the germinated embryo from such seeds is normally virus free. This is presumably because virus travels from cell to cell via the plasmodesmata and these, as Sheffield (1936) showed, do not extend to the embryo in the tomato.

Since 1884 when Hatsuzo Hashimoto, a Japanese rice-grower of the Shiga prefecture, recorded the transmission of stunt disease of rice by the leaf hopper *Nephotettix apicalis* var. *cincticeps* (Katsura, 1936) it has become apparent that in nature the characteristic method of virus transmission is

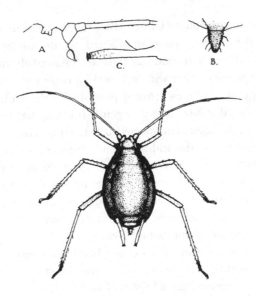

Fig. 32. Myzus persicae, *the most notorious aphid vector of plant viruses after J. P. Doncaster.*
(Smith (1931), text-fig. 4.)

by insects, especially aphides, although leaf hoppers, white flies, thrips, and
mealybugs are among the numerous groups of insects involved. Some
insects are able to transmit many different viruses. *Myzus persicae*, the peach
aphid (Fig. 32), transmits more than fifty. Others are more specific. Viruses
have been grouped according to the characteristics of their insect transmission as 'non-persistent' and 'persistent' (terms introduced by Watson &
Roberts, 1939). In the former group the insect vector becomes infective
after a short feeding period, is able to transmit the virus at once, and does
not remain viruliferous for long. Also, aphides are more easily rendered
infective when subjected to a period of starvation before being given access
to the infected tissues. Those of the second category require a longer feeding
period to become infective but they remain viruliferous for days or for life.
In the interesting case of the transmission of tomato spotted wilt virus by
Thrips tabaci, adult insects can only transmit the virus if they have fed on an
infected plant during their larval stage (Samuel *et al.*, 1930–7. II).

Other methods of transmission are by the plant parasitic dodder (*Cuscuta*) (Bennett, 1940) (which is essentially an example of graft transmission), mites (e.g. reversion in black-currants by the big bud mite, *Phytopus
ribis*), nematodes (grapevine fanleaf virus by *Xiphinema index*, first reported
by Hewitt, Raski & Goheen, 1958; also arabis mosaic and tobacco rattle
viruses, see Harrison *et al.*, 1963; Cadman, 1963), and fungi (lettuce big-

vein and tobacco necrosis viruses by zoospores of *Olpidium* (Teakle, 1960) and *Pythium*; wheat mosaic virus by *Polymyxa graminis*).

Immunological and serological relationships

Price (1932) found that tobacco (*Nicotiana tabacum*) and some other species of *Nicotiana* after infection with tobacco ringspot virus subsequently outgrew the disease symptoms and appeared healthy although still carrying the virus. Such plants were found to be immune from infection by tobacco ringspot virus. The next year Salaman (1933) reported that infection of tobacco or *Datura stramonium* with a very mild strain of potato virus X protected (immunised) the plant from infection by a highly virulent strain of the same virus but not from infection by potato virus Y or tobacco mosaic virus and in 1934 Kunkel described experiments which demonstrated a similar relationship between tobacco mosaic virus and virulent and attenuated strains of aucuba mosaic virus using *Nicotiana sylvestris* as the test plant (Fig. 33). It was from many such results that the general principle emerged

Fig. 33. Two leaves from Nicotiana sylvestris *plants five days after both were inoculated equally with aucuba mosaic virus. The leaf on the left is from a plant previously infected with an attenuated strain of tobacco mosaic; that on the right from a healthy plant. (Kunkel (1934), fig. 7.)*

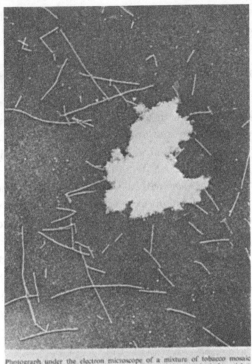

Photograph under the electron microscope of a mixture of tobacco mosaic virus (rod-shaped particles) and turnip yellow mosaic virus (spherical particles) incubated with an antiserum prepared against the latter virus. The antiserum causes a specific aggregation of the virus against which it was prepared. Shadowed with uranium (× 30,000 dia.).

Fig. 34. The agglution test. (Matthews (1957), frontispiece.)

that cross-immunity between two viruses may be taken as an indication that the two viruses are related.

The first application of serological techniques to plant viruses was by Mayme Dvorak in 1927 and shortly afterwards Helen Purdy Beale (1928, 1929, 1931) of the Boyce Thompson Institute by her studies on tobacco mosaic demonstrated that the virus-infected plant contained a specific antigen. In France, Gratia (1933) showed that plants infected by different viruses contained different specific antigens while Birkland (1934) obtained cross-reactions between plants infected with viruses believed to be related. Chester (1936) differentiated serologically strains of tobacco mosaic virus and potato virus X and offered the first serological classification of plant viruses (Chester, 1937). It was from such beginnings that serological testing (particularly the precipitin test) has become a routine and sensitive technique for the detection and identification of many plant viruses (see Fig. 34). Mathews' *Plant virus serology*, 1957, may be consulted for additional information and for recent developments, including the application of the ELISA (enzyme-linked immunosorbent assay) test to plant viruses, the review by van Regenmortel (1978).

Nomenclature and classification

Although 'nomenclature is the handmaiden of classification', for viruses nomenclature has so far taken precedence over classification and a generally acceptable remedy for this situation has yet to be found. First it was diseases that were described and named but soon it became apparent (as already noted) that one virus could affect several to many species of host plants, in other words it was responsible for more than one disease. Also, it was found that more than one virus could co-exist in one plant. These findings emphasised the need to name viruses. The initiative was taken by James Johnson in 1927 when he differentiated a series of viruses able to infect solanaceous plants by their host range, symptoms shown by a series of differential hosts, longevity *in vitro*, thermal death point for a 10-minute exposure, and resistance to certain chemicals. He then designated each virus by the common name of a principal host plant and a number. The virus of classical tobacco mosaic thus became *Tobacco virus 1* (Fig. 35). Johnson's system supplemented by additional differential criteria such as method of transmission (by sap, insects, seed, etc.), filterability, and immunological relationships, was widely adopted, sometimes, as in the first edition of Kenneth Smith's textbook (1937), with the common name of the host replaced by the Latin generic name, e.g. *Nicotiana virus 1*; a possibility suggested by Johnson himself. (One variant which has persisted ever since Kenneth Smith in 1931 distinguished potato viruses X and Y has been to designate potato viruses by capital letters in place of numbers.) The next major proposal was that of F. O. Holmes of the Rockefeller Institute who extended to viruses the traditional binomial system for the nomenclature of animals and plants. In 1939 Holmes summarised in *Phytopathology* the classification he had devised whereby plant viruses were treated in six families of Class II (Spermatophytophagi) of Division I (Phytophagi) of the Kingdom Vira. The taxonomic criteria for families and species were largely symptomological, the mosaic group of viruses being classified as the family 'Marmoraceae', tobacco mosaic virus being assigned the binomial *Marmor tabaci*. Later in the same year he illustrated his system (extended to 10 families) in his *Handbook of phytopathogenic viruses* in which 129 viruses are described and another 6-family arrangement appeared in the 6th edition of

Description of the Viruses

The following abbreviated descriptions of the viruses with which this bulletin is concerned are presented below, with the hope that they may serve as a tentative basis for classification. It is to be expected that future investigations will necessitate additional details and modifications of the present descriptions. Other valuable diagnostic features are already known to exist, such as the variations in the cytological details of infected tissues, the details of which are now being worked out by Miss Isme A. Hoggan in this laboratory.

TOBACCO MOSAIC (*Tobacco virus 1.*) Pl. I, A.
  TYPE. Allard, U. S. D. A. Bul. 40, 1914.
  HOST FAMILY. Solanaceae.
  DIFFERENTIAL HOSTS.
    On tobacco, marked mottling, malformation and stunting.
    On *N. glutinosa*, stem and leaf necrosis and stunting, no mottling.
    On tomato, mottling and stunting, no stem necrosis.
    On pokeweed, no symptoms.
  RESISTANCE TO AGING *in vitro.* Several years.
  THERMAL DEATH-POINT. 90°C. 10 minutes.
  RESISTANCE TO CHEMICALS. High (60% alcohol or 1 to 200 HNO_3[1] does not kill in one day).

CUCUMBER MOSAIC (*Cucumber virus 1.*) Pl. III, A and Pl. VII, B.
  TYPE. Doolittle, U. S. D. A. Bul. 879, 1920.
  HOST FAMILIES. Cucurbitaceae, Solanaceae, and others.
  DIFFERENTIAL HOSTS.
    On cucumber, chlorosis, mottling, stunting, malformation, necrosis.
    On *N. glutinosa*, mottling, malformation, stunting.
    On pokeweed, mottling, stunting.
    On tobacco, chlorosis, generally no malformation.
  RESISTANCE TO AGING *in vitro.* 3 days or less.
  THERMAL DEATH-POINT. 60-70°C. 10 minutes.
  RESISTANCE TO CHEMICALS. Low, 50% alcohol or 1 to 200 HNO_3, kills in one hour).

SPECKLED TOBACCO MOSAIC (*Tobacco virus 2.*)
  TYPE. Johnson, Phytopath. 16: 141, 1926.
  HOST FAMILY. Solanaceae.
  DIFFERENTIAL HOSTS.
    On tobacco, mottling or speckling.
    On petunia, mottling, stunting, malformation and necrosis.
    On henbane, chlorosis, stunting and malformation.
    No symptoms on *N. glutinosa*, pepper or pokeweed.
  RESISTANCE TO AGING *in vitro.* 3 or more months.
  THERMAL DEATH-POINT. 90°C. 10 minutes.

[1] One part nitric acid C. P. to two hundred parts water.

Fig. 35. James Johnson's method of classifying plant viruses. (Johnson (1927):9.)

Bergey's manual of determinative bacteriology, pp. 1145–224, 1948. A number of other proposals for plant virus nomenclature were made at this time and later.[5] Most of these are best forgotten. The position became very confused and the Commonwealth Mycological Institute found it increasingly difficult to treat many individual viruses under the same name in successive yearly indexes of the *Review of applied mycology*. To alleviate this the Institute decided to index viruses under names derived by adding 'virus' to the English common name of the principal disease induced. To this end virus workers in different countries were consulted on the choice of common names and on questions of the synonymy of names which were not nomenclatural ('obligate') synonyms and in 1944 a 44-page mimeographed list of *'Virus names used in the Review of applied mycology'* was published. This list was well received. Revised printed versions appeared in 1946 and 1957 and a definitive edition, 'Plant virus names. An annotated list of names and synonyms of plant virus diseases' (*Phytopath. Papers* No. 9) in 1968 to which there was a supplement in 1971.

The taxonomic (classificatory) content underlying all the systems of nomenclature so far considered is, for the most part, small and arbitrary. Usually it is limited to the attempt to consolidate different names proposed for one virus under one name and to associated strains (or suspected strains) of one virus on the basis of evidence derived from observations, more often than not, on the effects induced by the viruses or on their modes of transmission rather than of properties of the viruses themselves. Since the discovery of the nucleoprotein nature of plant viruses, biochemical investigations on the composition of viruses and observations of increasing quality on the morphology of the virus particles by the electron microscope have provided many data suitable for use as taxonomic criteria for both plant and animal viruses. Some virus particles, such as those of tobacco mosaic virus, were shown by electron microscopy to be elongated rods[6] (Fig. 34) (as long suspected).[7] Others proved to be rhombic (e.g. tobacco necrosis virus) or more or less spherical (e.g. turnip mosaic virus; Fig. 34) when they may aggregate to form true crystals. The nucleic acid can be ribonucleic acid (RNA), as in most plant viruses, or deoxyribonucleic acid (DNA) (in cauliflower mosaic virus; Shepherd, Wakeman & Romanko, 1968) and either single or double stranded. The molecular weight of the nucleic acid and the percentage of nucleic acid in the infective particle can also be determined. Such information can be used to devise more natural and informative classifications of both plant and animal viruses which hitherto had seemed to have had little in common.

It was first envisaged that the nomenclature of both bacteria and viruses should be subject to one code, the *International code of nomenclature of bacteria and viruses*, 1958, but as this was later considered to be impractical, in 1963 the Executive Committee of the International Association of Microbiological Societies set up a 'Provisional Committee for the Nomenclature of Viruses'. This committee in 1965 reported that it favoured a hierarchical system of classification with ranks similar to those employed for classifying animals, and plants, and bacteria; species being designated by Latin binomials, e.g. *Protovirus tabaci* for tobacco mosaic virus. The criteria proposed for the definition of a family were: the chemical nature of the nucleic acid; the symmetry of the nucleocapsid; the presence or absence of an envelope; and for helical viruses the diameter of the nucleocapsid and for cubical viruses the number of triangulation and the number of capsomeres.

In parallel with these proposals Gibbs, Harrison, Watson & Wildy (1966) outlined an Adansonian approach to the taxonomy of viruses whereby the different taxonomic criteria were given equal weight. They argued the case for the retention of the widespread practice among virologists to designate viruses by the addition of virus to the vernacular name of the disease incited, a name that would be given precision by an accompanying code or cryptogram of four pairs of symbols summarising the current state of knowledge about the virus. The pairs of symbols suggested were: type of nucleic acid (RNA, DNA)/strandedness of nucleic acid (single or double); molecular weight of nucleic acid in millions/percentage of nucleic acid in the infective particle; outline of particle/shape of nucleocapsid; kind of hosts infected (vertebrates, seed plants, etc.)/kind of vector. For example, their code for tobacco mosaic virus was: R/1: 2/5: E/E: S/O.[8] It was on the basis of such criteria that Gibbs & Harrison (1976) distinguished 23 groups of viruses, by such names as 'tobamovirus' (tobacco mosaic virus), 'cucumovirus' (Cucumber mosaic virus), and 'nepovirus' [*ne*matode transmitted, *po*lyhedral particles] (tobacco ringspot virus), according to whether the virus particle is helically constructed and tubular or filamentous (7 groups), isodiametric (9), variable (4), or bacilliform or bullet shaped (3) supplemented by 'viroids' (such as the causal agent of potato spindle tuber; Diener, 1972) which consist of single-stranded RNA and do not exhibit characteristic nucleoprotein particles. The merits of these proposals have yet to be assessed.

Identification

The certain proof of the identity of a virus now entails the use of the electron microscope and sophisticated biochemical and physical techniques for the elucidation of its morphology, chemical composition, and structure. This can be undertaken only by skilled staff in adequately equipped laboratories. However, for the practising plant pathologist the traditional criteria for virus identification are not yet completely outmoded. There is a high positive correlation between the reaction of well chosen differential hosts and the methods of transmission and the viruses themselves while the increasing use of specific antisera and the increasingly common access to an electron microscope frequently enables rapid and accurate diagnosis to be achieved.

Organisms confused with viruses

Virus diseases of plants have frequently been accepted as such by analogy. For many the test of filterability of the infective agent has been negative (as it is for some proved viruses) and no virus particles have been demonstrated by electron microscopy; the attribution as a virus disease being based on the failure to detect a causal agent and the transmission of the infective agent by grafting and (or) a known (or assumed) vector. It is only, however, during the last decade that evidence has been forthcoming to show that a number of hitherto widely accepted virus diseases are caused by other pathogens.

In 1967, a group of Japanese workers reported finding mycoplasma-like bodies in the phloem tissue of plants affected by mulberry dwarf, potato witches' broom, aster yellows, or *Paulownia* witches' broom and that symptoms of mulberry dwarf disease were suppressed by antibiotics of the tetracycline group to which mycoplasmas are sensitive (Doi *et al.*; Ishiie *et al.*, 1967). These results were confirmed and among additional examples discovered were such well-known diseases as clover phyllody, potato stolbur, and sandal spike while Hirumi & Maramorosch (1969) reported mycoplasma-like bodies in the salivary glands of the leafhopper vector of aster yellows. Mycoplasmas (pleuropneumonia-like organisms, PPLO) are, compared to viruses, relatively large and are composed of pleomorphic cells bounded by a membrane which contain both DNA and RNA. They can be cultured on artificial media and are resistant to penicillin but not to tetracyclines (Hopkins *et al.*, 1973). They thus differ greatly from viruses and are, indeed, themselves subject to virus infections. Of uncertain taxonomic position they were first described from man and animals in which they are common.

More recently still, the electron microscope has revealed rickettsia-like cells associated with Pierce's disease of the grapevine (Hopkins *et al.*, 1973), phoney peach (Nyland *et al.*, 1973), and clover club leaf, all insect-transmitted diseases not transmissible by sap. As long ago as 1950 Black had found that the leaf hopper (*Agalliopsis novella*) vector of clover club leaf remained infective after 21 successive generations of transovarial passage on immune plants and thereby proved conclusively that the clover club leaf pathogen multiplies in the vector. The rickettsias – like the mycoplasmas, of uncertain taxonomic status – have often been considered intermediate between bacteria and viruses but are now classified as bacteria. They are thick walled, Gram negative, and typically found associated with arthropods but some cause serious diseases of man and higher animals.

In 1972 R. E. Davis and others in the United States found a motile, helical, mycoplasma-like organism associated with corn stunt disease for which, in the following year, he and J. F. Wortley coined the name spiroplasma. Similar organisms, which may be grown *in vitro*, have been associated with little leaf and stubborn disease of citrus and in 1973 R. M. Cole and collaborators designated the latter organism *Spiroplasma citri* which they assigned to a new bacterial family, the Spiroplasmataceae. Further details and documentation of spiroplasmas, and also mycoplasmas and rickettsias, associated with plant diseases are given in a number of recent reviews.[9]

6

A note on non-parasitic disorders

Plant pathologists have always had to distinguish non-parasitic disorders from infectious diseases for which at times they may be mistaken. Paul Sorauer paid particular attention to non-parasitic disorders – in the third (1909) and subsequent editions of his *Handbuch der Pflanzenkrankheiten* they are treated in a special volume – but such emphasis is rare. Heald (1926) and Dickson (1950) in the first editions of their general textbooks devoted approximately 22 and 7 per cent, respectively, of the text to non-parasitic disorders and this proportion is down to 4 per cent in Tarr's *Principles of plant pathology*, 1972. Once an abnormality is shown not to be caused by a pathogen plant pathologists have tended to refer detailed consideration of the condition to others.

Non-parasitic disorders fall roughly into four classes – associated with climatic conditions, nutrition, toxicities, and genetical factors – under which they can be conveniently reviewed.

Climatic conditions

Since earliest times adverse effects of weather on crops have been recognised and Theophrastus (see Chapter 2) excluded such effects from his conception of plant disease. In addition to the obvious effects on plants of climatic extremes such as drought and flood, excessive heat and frost, hail, and snow, wind, and lightning the underlying environmental factors including water relationships, temperature, and light are at times able to induce abnormalities of less obvious aetiology. One of the early puzzles was the aetiology of blossom-end rot of unripe tomato fruits first described by B. T. Galloway in 1888 in the United States but a defect of world-wide incidence. Galloway attributed the black lesions on the fruit to fungal infection and in this he was followed by others. Prillieux and Delacroix in

France in 1894 favoured bacteria as the cause as did workers in America. An alternative explanation, first suggested by A. D. Selby in 1897 and endorsed by a number of later investigators, was that the condition was physiological, an explanation for which Charles Brooks (1914) (who cites the early literature) offered convincing evidence. He showed blossom-end rot to be induced by excessive watering (or a sudden check to the water supply) of rapidly growing fruit and aggravated by unbalanced manuring. Black heart of celery, which according to Dickson (1950:61) is almost co-extensive with celery culture in the United States, has features in common with blossom-end rot and is also associated with irregular water levels in the soil. Another widespread non-parasitic disorder occurring both in the orchard and during storage in which disturbed water relationships may play a part is bitter pit of apples although the cause of this disorder is still obscure. First described in Germany under the name *Stippen* in 1869 (Wortmann, 1892), it has been known by a number of other names including *Stippigkeit* (Germ.), *liège, points brun de la chair* (Fr.), *maculatura amara* (Ital.), and designated 'bitter pit' by Daniel McAlpine (Fig. 71) who made an intensive study of the condition in Australia during 1911–16.

Among disorders in the United States which have been associated with high temperature are heat canker of flax (Reddy & Brentzel, 1922), tip burn of potato (Lutman, 1919), and sunscald of beans (McMillan, 1923); high light intensity also being a factor determining the last condition. Frost not uncommonly causes cracking of branches of forest and orchard trees and also special conditions such as winter sunscald of apple which was investigated by A. J. Mix (1916). He concluded that the lesions which develop on the side of the tree facing the sun were caused by death of the bark, cambium, and sap wood induced by frost at night followed by increase in temperature of the affected area by spring sunshine during the day. Potato tubers when stored at low temperature are subject to a number of defects (see Heald (1926):165). In black heart, an internal necrosis resulting from storage in poorly ventilated bins and which can be induced experimentally by storage at high (38–45 °C) temperatures, they also provide an example of injury due to lack of oxygen (Bartholomew, 1913–15).

Nutritional disorders

Growth defects due to imbalance of the main macronutrients essential for plant growth – especially the elements nitrogen, potassium, and phosphorus – which are derived from the soil are normally corrected by the farmer or horticulturist with the assistance of the soil chemist. During the past fifty years there has been an increasing interest in disorders resulting from the lack of the so-called 'minor' or 'trace' elements which are essential for plant growth but in relatively small amounts. Lack of magnesium was shown by Garner *et al.* (1923) to be responsible for a chlorosis of tobacco known in the United States since 1912 as 'sand drown' because of its association with sandy soil leached by heavy rainfall. Subsequently, disorders due to magnesium deficiency have been recognised in many economically important crop plants including peach, apple, citrus, tomato, and other vegetables. In 1910 H. Agulhon in France found that the addition of boron to sand cultures and field plots increased the dry weight of cereals including maize, a finding which Mazé (1915–19) confirmed. Later Kathleen Warrington (1923), at Rothamsted Experimental Station, offered experimental proof that broad bean has an essential requirement for boron. It was in 1930 that Mes in Sumatra demonstrated that tobacco grown in a nutrient solution deficient in boron developed symptoms of top blight, a well-known defect of hitherto unknown aetiology. This was followed by the announcement in 1931 by Brandenburg in the Netherlands that boron deficiency induced heart and dry rot in sugar beet, an economically serious disease widespread in Europe and North America which it was found responded to application of borax (sodium tetraborate) to the soil. A period of intense interest in boron deficiency followed and important disorders in many and diverse crops were attributed to lack of this element.[1]

Other diseases of uncertain aetiology shown to be caused by nutritional deficiencies have included: grey speck of oats in Australia (Samuel & Piper, 1929) and marsh spot of peas in England (Piper, 1941) (manganese deficiencies); 'reclamation disease' in the Netherlands (Steenbjerg, 1940) and elsewhere (see Butler & Jones, (1949):312) (copper deficiency); pecan rosette (Alben, Cole & Lewis, 1932) and citrus mottle-leaf (Chapman *et al.*, 1937) (zinc deficiency); and whiptail of cauliflower in New Zealand and North America (Mitchell, 1945) (molybdenum deficiency).

The recognition of the importance of nutritional disorders was given an added stimulus by the publication in 1943 of *The diagnosis of mineral deficiencies in plants. A colour atlas and guide* by T. Wallace of Long Ashton Research Station. This influential publication was one factor encouraging a change of attitude. During the 1930s there was a tendency for a condition of unknown aetiology to be referred to virologists for study; in the 1940s a nutritional deficiency tended to be the first possibility to be explored.

Toxicities

Just as a deficiency of a basic macronutrient can induce disorders, so can an excess; as, for example, when high levels of nitrogenous fertiliser lead to lodging of cereals. So too can excess of a trace element. Boron injury to maize in the United States was reported by Conner (1918) after the use of a fertiliser containing 1.92 per cent of borax and earlier damage to crops had been sustained by the use of stable manure after treatment with borax to destroy house fly larvae. Another trace element toxicity in the States is that of crinkle leaf of cotton which was shown by Neal (1937) to be caused by an excess of manganese. The most widespread and important soil toxicity is alkali injury associated with soils in arid regions where, as a result of a rise in the water table, or irrigation with water containing salts in solution, various salts (particularly sodium chloride, sulphate, and carbonate) accumulate in the upper layers of the soil (or may even form a surface crust) thus rendering the soil infertile for most plants.

Scorching of plants in coastal areas by airborne salt during stormy weather has long been familiar. It was noted by Erasmus Darwin in his *Phytologia*, 1800, where among other disasters from external causes he drew attention to the effect on plants of 'noxious exhalations diffused in the atmosphere in the neighbourhood of some manufactories ... as the smoke from furnaces, in which lead is smelted from the ore, from potteries and lime kilns'. This type of injury is still of common occurrence and the damage may be near the source of pollution or some distance away as, for example, in the current claim that Scandinavian forests are being damaged by sulphur dioxide from industrial regions of the United Kingdom washed from the air by rain. Two other airborne toxicities that have come into prominence during recent years are fluoride damage to forests and crops originating from large phosphate, aluminium, and steel plants in the United States (particularly Florida, Washington, and Utah)[2] and the damage caused by 'smog' resulting from automobile fumes in California and other industrial parts of the world. The active principles of the smog are oxidising agents, especially ozone[3] and peroxyacyl nitrates.[4]

Finally there are the side effects of attempts to control pests and diseases. Ever since the introduction of fungicides phytotoxicity has been a major problem which has not infrequently prevented the general use of an otherwise promising fungicide. More recently there have been many records of injury caused by herbicide residues. In gardens the popularity of sodium chlorate to eliminate weeds from paths has frequently led to the chemical being washed into ponds from which water is taken for watering plants when the vein-clearing and leaf-crinkling of plants sensitive to traces of chlorate have been at times mistaken for virus infection. Even closer simulation of virus induced symptoms has been reported from the effect of hormone weedkillers such as 2.4 D when concentrations of less than one in a million can induce leaf distortion in tomato foliage which can be mistaken for cucumber mosaic virus infection (Atkinson, 1947).

Genetical disorders

The doubling of flowers, reversion of seedlings of ornamental plants to the wild type, and fasciation are familiar examples of genetical changes which are, or may be, detrimental. Many symptoms induced by viruses such as leaf variegation or deformation and the breaking of flower colour are paralleled by similar characters which are heritable. Usually it is not difficult to distinguish the two types of determinant but sometimes a genetically controlled character is virtually indistinguishable macroscopically and histologically from symptoms induced by viruses. For example, the leaf enation-bearing variant of tobacco which Honing (1923) described as a new species under the name *Nicotiana deformans* apart from being virus-free is very similar indeed to tobacco plants infected with the enation strain of tobacco mosaic virus. Again, the wiry tomato, a recessive mutant described by Lesley & Lesley (1928) resembles a tomato plant infected by cucumber mosaic virus.

In this connection it is perhaps of interest to note that Sprague *et al.* (1963) claimed that barley stripe mosaic virus has a mutagenic effect on maize. A small percentage of F_2 ears involving virus-infected P_1 males exhibited marked ratio distortions for marker genes; such distortions persisting in back-crosses.

*Control (treatment and prevention)
of plant disease*

For the last fifty years, the most popular approach to the exposition of methods for the control of plant diseases has been that advocated by Professor H. H. Whetzel in his course at Cornell University on 'Principles of plant disease control' and widely disseminated by the many students he trained. Summarising his views, Whetzel (1929) wrote:

I find that all methods as yet proposed for the control of plant diseases are based upon one of what I choose to call the four fundamental principles, exclusion, eradication, protection, and immunization ...

By exclusion is meant preventing to a profitable degree the entrance and establishment of a pathogene in an uninfested area, as in a garden, field, region, state or country. By eradication is meant the more or less complete elimination or destruction of a pathogene after it is established in a given area. By protection is meant the interposition of some effective barrier between the susceptible parts of the plant and the inoculum of the pathogene. By immunization is meant the development by natural or artificial means of an immune or highly resistant plant population in the area infested with the pathogene to be combated.

Professor Gäumann (1946, chap. 6) offered an alternative classification of control measures against infectious diseases into three main categories: 'prophylaxis against infection' (the interruption of the infection chain prior to the infection of the susceptible host); 'prophylaxis against disposition' (which attempts to reduce the 'disease proneness' of the host by breeding immune or resistant cultivars or by cultural measures); and 'therapy' (which may either be by biological measures (which usually correspond to the cultural measures of disposition prophylaxis) or physico-chemical methods).

These categorisations may usefully be employed as alternative overlays to the historical account of the control of plant diseases given in the next four chapters which deal with the control of pathogens by the use of chemicals (Chapter 7) and physical agents (Chapter 8), the epidemiological approach to disease whereby the most efficient strategy to combat disease may be devised (Chapter 9), and legislative measures designed to promote the common good (Chapter 10).

7

Chemical control

The discovery, laboratory testing, field trial, and commercial application of chemicals, particularly fungicides, for the control of plant disease – activities which involve not only plant pathologists but also chemists, agricultural engineers, and large-scale industry – make a major contribution to the practice of phytopathology. Since the beginning of this century access to the large and expanding literature on fungicides has been facilitated by a series of useful handbooks and by many review articles in which different aspects of advances in this field have been documented (see analysis of Bibliography). Many of these publications also deal, in greater or lesser detail, with historical aspects so that here it is only necessary to sketch in the salient features of the development and applications of the hundreds of fungicides in thousands of formulations which have been advocated for the treatment and prevention of plant diseases.

Fungicides

Since ancient times, the control of disease in plants and the depredations of insects has been attempted with a great variety of materials of doubtful efficacy. According to Pliny (*Nat. hist.* Bk xv, chap. 8) (citing Cato), 'amurca', the dregs of olive oil, was the therapeutant of choice for the purpose, as it was for many others. In Bk xxiii, chap. 37, Pliny lists 21 medicinal uses of amurca from a strengthener of the gums to a fomentation against gout. Other 'cures' were selected on the basis of the doctrine of humours (see Chapter 2) and it was not until the eighteenth century that chemicals now of proved toxicity to fungi and insects began to be exploited. Mercuric chloride was in use as a wood preservative by 1705 and although later in the century copper sulphate was tested as a cereal seed dressing (see

Table 4) the first fungicide to come into general use against plant disease was sulphur.

Sulphur

Sulphur was employed in Greek medicine against skin complaints and is still used in dermatological practice. Its deployment in plant pathology dates from 1802 when William Forsyth 'Gardener to His Majesty [King George IV] at Kensington and St James's' in his *Treatise on the culture and management of fruit trees* (p. 250) advised that whenever you apprehend danger from mildew:

wash or sprinkle the trees well with urine and lime-water mixed; and when the young and tender shoots are much infected, it will be necessary to wash them well with a woollen cloth dipped in the following mixture, so as to clear them of all glutinous matter, that their respiration and perspiration may not be obstructed.

Take tobacco one pound, sulphur two pounds, unslaked lime one peck, and about a pound of elder buds; pour on the above ingredients ten gallons of boiling water; cover it close and let it stand till cold; then add as much cold water as will fill a hogshead. It should then stand two or three days to settle; then take off the scum, and it is fit for use.

This last recipe is the basis of 'lime sulphur' which, in a number of variants, has since proved so popular, while elemental sulphur in a range of forms and combinations has also been widely used as a dust, spray, or fumigant (see Table 3).

Sulphur is the most important fungicide. It suffered a temporary eclipse after the discovery of Bordeaux mixture in 1882 but the re-introduction of lime sulphur after 1906 restored its popularity and although since then there has been a steady decline in the use of elemental sulphur the dithiocarbamates, the most popular organic fungicides, contain 26 per cent (thiram) to 50 per cent (nabam) bound sulphur while captan contains 10 per cent. According to McCallan (1964), although the consumption of ground sulphur as a pesticide in the United States declined from 614 million pounds (280 000 metric tons) in 1950 to 172 million (78 000) in 1961 (when the comparable figure for copper was 10 million pounds (4550)), in 1958 the total world consumption of sulphur for fungicides was 275 000 metric tons compared to 56 000 for copper and 300 for mercury.

There has been much speculation regarding the mechanism of the toxic action of sulphur which is insoluble in water. In 1930 Wilcoxon & McCallan wrote 'there is scarcely a compound of sulfur which might conceivably be formed from the element, under conditions of use, to which the toxic action has not been attributed. Among these compounds are sulfur dioxide, hydrogen sulfide, sulfuric acid, thiosulfuric acid, and pentathionic acid.' In

Table 3. *Chronology of the use of sulphur against diseases of plants*

1802	**Lime sulphur** first described for use against mildew of fruit trees by William Forsyth, England
1814	**Sulphur and lime water** against mildew of fruit trees by D. Weighton, England
1818	**Flowers of sulphur** dusting against pear scab by T. A. Knight, England
1821	**Sulphur and soap** against peach powdery mildew by J. Robertson, England (Robertson, 1824)
1834	**Self-boiled lime sulphur** (in which the heat needed is derived from the slaking quick-lime) introduced by Dr William Kendrick in the United States (Lodeman (1896):12). (Re-introduced by W. M. Scott, 1908)
1842–47	**Flowers of sulphur + lime** against peach leaf curl by T. A Knight (1842) and powdery mildew of vine by Tucker, England (Large (1940):44)
c.1840	**Sulphur dust** against mildew applied on a large scale to vine leaves when wet with dew by Count Dûchatel in France (Large (1940):51)
1852	**Eau Grison** (sulphur + slaked lime, boiled) introduced by Grison, head gardener of the vegetable houses at Versailles, Paris (Lodeman (1896):16)
	Sublimation of sulphur painted on the hot water pipes in vine houses (Bergman, 1852) or later, by the use of various types of apparatus (see Lodeman (1896):307; Martin (1928 (Edn 4, 1959):107)
1869	**Carbon disulphide** for soil injection (see soil treatment below)
1889	**Sulphur as soil treatment** against onion smut by R. Thaxter (1890)
1895	**Ceres powder** (potassium sulphide) marketed by J. L. Jensen for cereal seed treatment (*Phytopathology* 7:4, 1917)
1906	**Lime sulphur** recommended by A. B. Cordley in Oregon against apple scab (Cordley, 1908; Whetzel (1918):112)
1911–17	**Finely divided sulphur** dust introduced by Bloggett, Reddick, and Stewart of Cornell University (Whetzel (1918):113)
1916	**Ammonium polysulphide** against powdery mildews (Eyre & Salmon, 1916)
1923	**Dry-mix sulphur** (wettable sulphur) (sulphur, hydrated lime, casein) introduced by A. J. Farley
	Sulphur dioxide to prevent decay in grapes in store and transit (Jacob, 1929)
1934	**Dithiocarbamates.** See Table 5
1951	**Trichloromethylthiocarboximides.** See Table 5

1964 McCallan comprehensively reviewed and documented this work and concluded that

Sulfur is unique among fungicides. Very large amounts are taken up by spores, but it is not accumulated since practically all is given off again as hydrogen sulfide. The best evidence to date is that sulfur acts as a hydrogen acceptor and so interferes with the normal hydrogenation and dehydrogenation reactions of the cell. Also perhaps there is production of polysulfide intermediates which act as enzyme inhibitors.

The organic sulfur fungicides ... probably act by virtue of their decomposition products ... as general poisons.

Copper

The displacement of sulphur as the leading fungicide by copper after the announcement of Bordeaux mixture in 1885 was precipitated by the epidemic of downy mildew (*Plasmopara viticola*) in French vineyards which followed the introduction (about 1870) of the pathogen from North America with imported nursery stock resistant to phylloxera (*Phylloxera vastatrix*, a leaf and root infesting aphid). It was in 1882 that P. M. A. Millardet (Fig. 36), professor of botany at Bordeaux University, chanced to notice in a vineyard at Medoc the freedom from mildew of vines, adjacent to the highway, which to discourage pilfering had been rendered unattractive by treatment with a mixture of lime and copper sulphate. Following up this clue Millardet, with the help of the chemist U. Gayon, during the next three years developed the preparation which became known as *bouillie bordelaise* (Millardet, 1885). The original Bordeaux mixture was prepared by adding 15 kg of 'rich rock lime' suspended in 30 l of water to a solution of 8 kg of copper sulphate dissolved in 100 l of water and was diluted for use.[1] It was an instant success against vine mildew and soon proved itself equally

Fig. 36. Pierre-Marie-Alexis Millardet (1838–1902) and the monument erected in his memory at Bordeaux.

effective against potato blight and many other economically important diseases. Many variants of Millardet's formulation were proposed (and a number of alternatives, including Eau Celeste and Burgundy mixture, advocated; see Table 4) but for the next half century vast quantities of Bordeaux mixture were used throughout the world.

Today, it is interesting to recall the long delay in the deployment of copper as a fungicide and to note how very near observers were to an earlier breakthrough. According to Johnson (1935), Boucherie's wood preservative treatment by copper sulphate had been widely used in French vineyards for the treatment of the grape posts since 1839 and it was frequently

Table 4. *Chronology of the use of copper against diseases of plants*

1761–	**Copper sulphate** used as a seed treatment against wheat bunt by Schulthess (1761) and Tessier (1779) with inconclusive results
1807	Toxicity of copper to bunt spores proved by Prévost (1807)
1866	**Copper sulphate** successfully used as a seed treatment against bunt by J. Kühn (1873)
1882	**Bordeaux mixture** (copper sulphate + slaked lime) devised by P. M. A. Millardet in France (Millardet, 1885)
1885	**Eau Celeste** (Azurin) (copper sulphate + ammonium hydroxide) introduced by A. Audoynaud in France (Lodeman (1896):30)
1887	**Burgundy mixture** (copper sulphate + sodium carbonate) introduced by E. Masson in France (Lodeman (1896):30)
	Modified Eau Celeste (copper sulphate + sodium carbonate + ammonia) devised by G. Patrigeon
	Podechard's dust (copper sulphate, lime, flowers of sulphur, wood ashes, water; product dried and sieved) introduced
1889	**Bordeaux mixture** used in combination with an insecticide by C. M. Weed in Ohio
1891	**Copper carbonate** dissolved in ammonium carbonate solution introduced by F. D. Chester, Delaware. Also, fide Horsfall (1945), S. W. Johnson offered ammonium carbonate as the alkali in Bordeaux mixture
1900	**Bordeaux mixture in powder form** introduced by G. C. Johnson, of Kansas City, Mo.
1902	**Basic copper carbonate** used by von Tubeuf in Germany against wheat bunt; re-introduced and popularised by Darnell-Smith (1917) in Australia
1918	**Copper-lime dust** developed by Sanders & Kelsall in Nova Scotia
1921	**Cheshunt compound** (copper sulphate 2 parts, ammonium carbonate 11 parts) introduced by W. F. Bewley as a soil drench against damping-off
1925–35	Development of **fixed-copper fungicides**: basic sulphates and chlorides (e.g. Bouisol (Brit. Patent 392 556, 1931), a colloidal paste of the basic chloride ($CuCl_23Cu(OH)_2$)), the oxides (by Horsfall, 1932a), copper zeolite, hydroxide, phosphate, silicate. Also **organo-coppers** (copper oleate and resinate), see de Ong (1935)

noticed that there was less mildew on plants adjacent to treated posts. A similar clue to the toxicity of copper was also available during the period of intense interest in the potato blight epidemic of the 1840s. On 29 August 1846 under 'Home Correspondence' in the *Gardeners' chronicle* (p. 582) there was this anonymous contribution:

Copper Smoke a Preventive of Potato Disease. In the district about Neath and Swansea, 'wherever the copper smoke prevails', was the expression of an intelligent inhabitant with whom I fell into conversation, the Potatoes are sound, and the same person informed me it was also the same last year. I can verify the fact so far as the present appearance of the crop is concerned, as seen from the mail coach roof . . . the district is crowded with copper smelting furnaces.

In addition, the toxicity of copper to the spores of wheat bunt had been experimentally demonstrated by Prévost as early as 1807 (see Chapter 2) and even before this copper sulphate had, as already mentioned, been used as a seed dressing while C. F. A. Morren in Belgium in 1845, who believed potato blight to be caused by fungus, advocated a mixture of copper sulphate, lime, and common salt for the prevention of tuber rot. Unfortunately he applied this mixture to the soil and not the foliage.

Mercury
Among other fungicides which have been widely used in plant pathology are mercury compounds. Mercuric chloride, introduced during the eighteenth century as a wood preservative (as already noted) has since found various uses, especially for soil treatment; as has mercurous chloride (calomel) (see below). The most important mercury fungicides are the organo-mercury seed dressings which have been manufactured by the large chemical firms in Europe, the United Kingdom, and North America since Riehm (1913) first introduced the chlorophenol mercury compound known under the trade name 'Uspulun' for the control of bunt in wheat (see Table 5).

Organic fungicides
The first organic compound to be used as a fungicide was formaldehyde the bactericidal properties of which were apparently first recognised by Oscar Loewi in 1885 and since the end of the last century it has – as formalin, a commercially available 40 per cent aqueous solution – been much used for seed and soil treatment. Chlorophenol-mercury compounds were introduced before the First World War (see Table 5) and salicylanilide in 1926 but the era of organic fungicides as generally understood today began in 1931 when W. H. Tisdale and I. Williams of the American firm E. I. du

Table 5. *Some important organic fungicides with dates of introduction and principal uses*

Organic mercurials

1913– **Chlorophenol-mercury compound** introduced in Germany by I. G. Farbenindustrie A.-G., used as seed treatment against wheat bunt by Riehm (1913). Similar compounds for seed treatment, under a variety of trade names ('Uspulun', 'Semisan', 'Germisan'), and a series of mercurated hydrocarbons (e.g. 'Agrosan', 'Ceresan') were subsequently introduced by other large chemical firms in Germany, USA, and UK

Salicylanilide

1926 **Shirlan** (and the water-soluble sodium salt, shirlan NA) patented by the British Cotton Industry Research Association (English patent 323 579)

Dithiocarbamates (Thorn & Ludwig, 1962)

1934 W. H. Tisdale & I. Williams granted US Patent 1 972 961 for use of dithiocarbamic derivatives to control growth of fungi and microbes. H. Martin independently reported the fungicidal properties of such compounds

1937 **Thiram** (tetramethylthiuram disulphide), introduced as an insect-repellant, became a successful seed treatment chemical (Muskett & Colhoun (1940) against seedling blight of flax)

1940 **Ferbam** ('Fermate') (ferric dimethyldithiocarbamate) introduced by Palmiter & Hamilton (*Farm Res.* (NY) **9**:14, 1943) became successful for the control of orchard diseases

1943 **Nabam** ('Dithane') (disodium ethylenebisdithiocarbamate). Hester, US Patent 2 317 765, 1943. Dimond *et al.* (1943) against vegetable (esp. potato) diseases. (Zinc sulphate used as a stabiliser)

1944 **Ziram** (zinc dimethyldithiocarbamate). Heuberger & Wolfenbarger (1944) against early blight of tomato and potato

1947 **Zineb** ('Dithane Z-78', 'Parzate') (zinc ethylenebisdithiocarbamate). Barratt & Horsfall (1947)

1950 **Maneb** (manganese ethylenebisdithiocarbamate). A. L. Flenner, U.S. Patent 2 504 404, 1950. Vegetable crops

Organo-coppers

1935 See Table 3

Quinones

1942 **Chloranil** ('Spergon') (tetrachloro-1,4-benzoquinone). W. P. ter Horst, US Patent 2 349 771, 1944. Seed dressing, esp. for peas, Sharvelle *et al.* (1942)

1943 **Dichlone** ('Phygon') (2,3-dichloro-1,4-naphthoquinone). ter Horst & Felix (1943). Seed treatment

Imidazolines (Glyoxalidines)

1946 **Glyodin** (2-heptadecyl-2-imidazoline). Law & Wellman, US Patent 2 540 170, 1951. Wellman & McCallan (1946). Apple scab and cherry leaf spot control

Table 5 *cont.*
Trichloromethylthiocarboximides
1951 **Captan** (*N*-trichloromethylmercapto-4-cyclohexeximide-1,2-dicarboxi-
mide). Kittleson, US Patent, 2 553 770, 1951. Kittleson (1952). General
fungicide
1961 **Folpet** ('Phaltan') (*N*-trichloromethylthiophthalimide)

Guanidines
1959 **Dodine** ('Cyprex') (n-dodecylguanidine). Lamb, US Patent, 2 867 562,
1952. Apple scab

Arylhydrazonisoxazolones
1960 **Drazoxolon** (4-(2-chlorophenylhydrazono)-3-methyl-5-isoxalone). Seed
treatment (SALsan (grass), Mil-Col (peas and beans), 1968)

Antibiotics
See text

Systemic fungicides
See text and Table 6

Pont & Nemours Co. applied for a patent for the use of dithiocarbamic acid
derivatives to control the growth of fungi and microbes. The patent was
granted in 1934 and in the same year in England Hubert Martin of Wye
College, Kent,[2] reported the fungicidal effect of similar organic sulphur
compounds used commercially as rubber accelerators. The following year
the fiftieth anniversary of the discovery of Bordeaux mixture was celebrated
in Paris.[3] In retrospect, this may be considered to be the point at which the
eclipse of Bordeaux mixture by organic fungicides began.

The first organic fungicide to come into general use was tetramethylthiuram disulphide – later designated thiram – which after a somewhat
uncertain start gained popularity after the demonstration by A. E. Muskett
& J. Colhoun in 1940 of its usefulness against seedling blight of flax caused
by *Colletotrichum linicola*. Other fungicides, including ferbam and ziram, of
the same series followed and these were in turn supplemented by the
ethylene derivatives nabam, zineb, and maneb. (Additional information on
these and other types of organic fungicides is given in Table 5.)

This phase in the development of fungicides encouraged, and was itself
encouraged by, the elaboration of techniques for the experimental laboratory testing of fungicides and fungicidal sprays and attempts to relate the
chemical structure of fungicides to their fungicidal properties so that new
and better fungicides could be designed. It was in 1910 that Donald
Reddick and E. Wallace first described a spore germination test for evaluating fungicides by which slides or cover glasses were sprayed with the

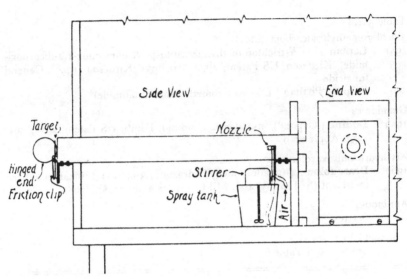

Fig. 37. Sketch of a tube-type laboratory sprayer. (Horsfall et al. (1940), fig. 3.)

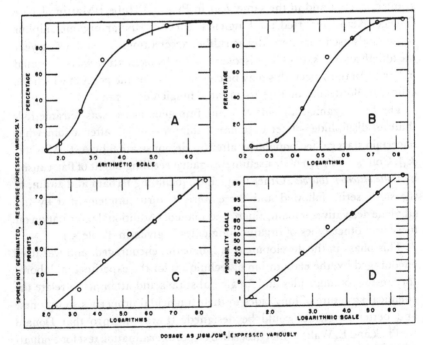

Fig. 38. Different ways of expressing the dosage–response curve. (Horsfall (1956):18.)

preparation to be tested and spores of the pathogen placed on the dry deposit in a drop of 'meteoric water'. This test was elaborated by McCallan (1930) and, because of the wide interest in the technique, in 1938 the American Phytopathological Society appointed a committee on the standardisation of fungicidal tests and the spore germination test in particular on which two reports offering recommended procedures were published in 1943 and 1947.[4] The basis of the test is the determination of the degrees of inhibition of germination in spore suspensions of equal concentration caused by serial dilutions of the fungicide. When the fungicide is not water soluble known amounts are deposited on a glass slide by dusting or by allowing a drop of a solution of the fungicide in a suitable solvent to dry and then testing the germination of spores in a suspension placed on the film of fungicide. Fungicides and sprays may also be tested in the laboratory by the use of an apparatus such as that devised by Horsfall *et al.* (1940) (Fig. 37) which standardises the technique for obtaining the deposit of fungicide. The results are expressed by plotting, in various ways (see Fig. 38), the degree of inhibition of germination against the concentration of the fungicide to give a 'dosage response curve'; a concept exploited by Dimond *et al.* (1941) and the main theme of Horsfall's 1945 textbook. The DR curve allows the toxicities of different fungicides (or of one fungicide against different pathogens) to be compared and the concentration of fungicide needed for a required degree of effectiveness to be derived. It is usual to compare the toxicities of fungicides by the concentration (effective dose, ED) which caused 50 per cent inhibition of the test organism, the ED_{50} value (which now replaces LD_{50} (= lethal dose), a terminology introduced by J. W. Trevan in 1927[5] in the medical field).

Unfortunately, the results of laboratory testing are not always confirmed by field trials and it was Hamilton (1931) who introduced a greenhouse method for evaluating fungicides against apple scab, based on a technique originally devised by G. W. Keitt and L. K. Jones at the University of Wisconsin, Madison in 1926 (see Fig. 50). This approach was further refined by Hamilton (see Holton *et al.* (1959): 253–7) and others including McCallan & Wellman (1943) who used tomato foliage.

Although Wellman (in Holton *et al.* (1959):240) estimated that large chemical firms tested perhaps 5000 different compounds a year for fungicidal (and also insecticidal and herbicidal) properties, new commercially successful fungicides have been introduced at a rate of only about one a year. One reason for this is the high cost of developing a commercial fungicide and meeting the legal requirements (see Chapter 10) regarding toxicity to man and animals and the absence of undesirable residues, a cost

Chemical control

Wellman suggested, from a breakdown of the costs of the different stages of development and manufacture, to be in excess of 1.25 million dollars. Currently the cost would be more than ten times this figure.

Antibiotics
Antibiotics, which have contributed so much to medicine, have so far played only a minor role in plant pathology. The phenomena of self-inhibition and antagonism between micro-organisms have long been known to plant pathologists. The Rev. M. C. Potter, professor of botany at Durham University, in 1908 found that the concentrated liquid in which *Pseudomonas destructans* [*Erwinia carotovora*] (which causes soft rot in turnips) had been grown would kill the organism invading turnip tissue. He also found that metabolic products of *Penicillium italicum* would eliminate the fungus from infected orange rind. Subsequently Clara A. Pratt (1934) drew attention to the staling of cultures of *Fusarium* by the development in the culture medium of metabolic products which inhibit growth and spore germination. R. Weindling (1932) demonstrated the antagonistic effect of *Trichoderma* on soil fungi and in 1936 he and Fawcett attempted the control of damping off of citrus seedlings by the addition of this fungus to the soil. In the same year Weindling and Emerson isolated the active principle which they called gliotoxin from a *Trichoderma* which is now considered to have been *Gliocladium virens*. The feverish search for new antibiotics following the recognition of the usefulness of penicillin led to the discovery of patulin from *Penicillium patulum* (Anslow, Raistrick & Smith, 1943), which is the same as expansin derived from *P. expansum* (Van Luijk, 1938). This was advocated as a cure for damping-off caused by species of *Pythium* but it was unable to compete with the fungicides already available for the purpose. Of the well established medical antibiotics, only streptomycin and actidione (cyclo-heximide) (both produced by *Streptomyces griseus*) and griseofulvin (from *Penicillium griseofulvum* and other species) have found limited use: streptomycin against bacterial infections, griseofulvin as a systemic fungicide against grey mould (*Botrytis cinerea*) of lettuce. In Japan where several agricultural antibiotics have been developed, blasticidin, discovered in 1955 has, as a wettable powder, controlled rice blast (*Pyricularia oryzae*) more successfully than organo-mercury fungicides (Marsh, 1972).

Systemic fungicides
The advantages of introducing throughout a plant a substance non-toxic to the host but toxic to a pathogen over the application of a fungicide which

remains external to the host and localised are many. During the last thirty years the search for systemic fungicides, and their commercial exploitation during the past twenty, has been a major new development. Although the chemotherapy of plant disease, as the deployment of systemic fungicides is often designated, is a modern trend, the idea of introducing substances into plants to induce new odours, tastes, or colours to the developing fruit goes back at least to the twelfth century (Ibn-al-Awam (12th cent.) **1**, chap. 15) and in the fifteenth century Leonardo da Vinci described the use of arsenic injection into fruit trees to render the fruit poisonous (as noted by Roach (1939) in his excellent historical summary). In this connection it is interesting that the versatile Richard Bradley in 1718 was writing:

I am inform'd, from a Person of Honour, that to prevent the Insults of ... *Insects*, he has known it practised, to bore a Hole with a small Gimlet sloaping downwards thro' the *Bark*, so as to reach the wood of the Tree, and pouring into it some *Quicksilver*, about half an Ounce, or more, according to the Bigness of the Tree, and stopping it up, the *Insects* then harbouring upon it will infallibly be destroy'd; ....I cannot chuse but be of his opinion, altho' I have not made the Experiment.

The recent interest in systemic fungicides has been influenced particularly by earlier work in two directions: attempts to diagnose and cure mineral deficiency disorders in fruit and other trees by the injection of appropriate salts – investigations stemming from the work of the agricultural chemist Justus von Liebig in the first half of the nineteenth century to that of W. A. Roach at the East Malling Research Station during the 1930s (Roach, 1938, 1939) – and the development of systemic insecticides (see Müller (1926) who reviews the early work in this field against both pests and diseases).

There have been many reports of systemic effects induced by conventional fungicides including George Massee's claim to have reduced the incidence of *Cercospora melonis* on cucumber leaves by watering the roots with copper sulphate solution (Massee, 1903) and in 1906 Bolley recorded the beneficial effects of the injection of copper sulphate, iron sulphate, or formaldehyde against peach leaf curl (*Taphrina deformans*). Spinks (1913) found that lithium salts prevented powdery mildew (*Erysiphe graminis*) in wheat and barley under experimental conditions and in 1920 Caroline Rumbold reported that injections of the same compounds temporarily retarded chestnut blight (*Endothia parasitica*) infections. F. T. Brooks and his students at Cambridge University inhibited silver leaf of plum by the injection of dyes and disinfectants (Brooks & Bailey, 1919) and later Roach (1939) showed injections of 8-quinolol sulphate (which Brooks and H. H. Storey had shown to be toxic to *Stereum purpureum in vitro*) to give

Table 6. *Representative systemic fungicides and their principal applications*

Benzimidazoles
1964 **Thiabendazole** (2-(4'-thiazolyl)-benzimidazole). Staron & Allard (1964).
Many pathogenic fungi (excluding phycomycetes and dematiaceous
hyphomycetes). Foliar spray, seed treatment, soil drench, post-
harvest fruit dip
1968 **Benomyl** (methyl(1-butylcarbamoyl)-2-benzimidazole-2-yl-carbamate).
Delp & Klöpping (1968). As for thiabendazole but more effective
1969 **Carbendazin** (methyl benzimidazol-2-yl carbamate) ('MBC') which can be
regarded as responsible for the activity of benomyl (Clemons & Sisler,
1969) and thiophanate-methyl, both of which are converted to it in
aqueous solution

Oxathiins (Edgington & Barron, 1967)
1966 **Carboxin** (2,3-dihydro-6-methyl-5-phenylcarbamoyl-1,4-oxanthiin).
Schmeling & Kulka (1966); also Schmeling *et al.*, US Patent 3 249
499, 1966. Barley loose smut (*Ustilago tritici*), bean rust (*Uromyces
appendiculatus*). Seed treatment, foliar spray

Pyrimidine derivatives
1963 **Dimethirimol** (5-n-butyl-2-dimethylamino-4-hydroxy-6-methyl pyrimidine). Elias *et al.* (1968), Peacock (1978). Cucumber powdery mildew
(*Sphaerotheca fuliginea*). Soil treatment
1976 **Ethirimol** (5-n-butyl-2-ethylamino-4-hydroxy-6-methylpyrimidine). Peacock (1978). Barley mildew (*Erysiphe graminis*). Seed dressing, drilled
granules

Organophosphorus compounds
1965 **Kitazin** (*O, O*-di-isopropyl-S-benzyl phosphorothiolate). Yoshinaga *et al.*
(1965). Rice blast (*Pyricularia oryzae*). Added to irrigation water or as
spray

Aromatic compounds
1971 **Thiophanate** (2,3-bis-(3-ethoxycarbonyl-2-thioureido)benzene). Thiophanate methyl (1,2-bis-(3-methyloxycarbonyl-2-thioureido)
benzene. Aelbers (1971). Cereal mildew (*Erysiphe graminis*). The
latter, like benomyl, breaks down to carbendazin (Marsh, 1972)

Triazole
1975 **Triadimefon** (1-(4-chlorophenoxy)-3,3-dimethyl-1-(1H-1,2,4-triazol-l-yl)-
2-butanone) (Kaspers *et al.*, 1975). Effective against mildews and
rusts

Acyl-alanines
1977 **Furalaxyl** (D,L-methyl *N*-(2,6-dimethylphenyl)-*N*-(2-furoyl) alaninate)
(Schwinn, Staub & Urech, 1977). The first commercial systemic
fungicide effective against oomycetes

For further details and additional fungicides see the reviews by Marsh
(1972), Erwin (1973)

effective control when injected. Finally, mention should be made of the commercially unsuccessful attempts by Horsfall, Zentmyer and Dimond to control elm disease by chemotherapy (Zentmyer *et al.*, 1946) which did, however, provide a stimulating example to others and gave Albert Dimond an interest in this field which lasted until his untimely death in 1971.

The search for systemic fungicides was further stimulated by the use of penicillin and other antibiotics in human medicine. Anderson & Nienow (1947) reported that streptomycin could be translocated in plants although they found that a concentration of 4 units per ml of sap did not inhibit infection of soybeans by *Xanthomonas phaseoli* var. *sojense*. Griseofulvin was shown to be similarly translocated (Brian *et al.*, 1951) and found application in the control of grey mould (*Botrytis cinerea*) of lettuce. Blastocidin and other antibiotics were developed commercially in Japan as systemic fungicides but the real breakthrough occurred during the 1960s with the introduction of a series of organic fungicides of which some important representatives are listed in Table 6.

These fungicides, and particularly benomyl, have been used successfully against a range of disease and the increasing and widespread attention paid to this field is well shown by the two editions of the well documented multi-author book on systemic fungicides edited by R. W. Marsh (1972). One disappointment has been the rather frequent development of resistant strains by the pathogens, a topic recently well reviewed by J. Dekker (1976) (and also in Marsh, 1972) and by S. G. Georgopoulos in Horsfall & Cowling (1977–8) 1:327–45.

Chemical inactivation of viruses

Plant viruses have been shown to be inactivated *in vitro*, and more rarely *in vivo*, by a wide range of inorganic and organic compounds of diverse types and by products from higher plants, animals, and micro-organisms. The extensive literature on the topic, aspects of which have been reviewed by R. F. Matthews (in Horsfall & Dimond (1959–60) 2, chap. 12) and T. Hirai (in Horsfall & Cowling (1977–8) 1, chap. 15), contains little that has been applied to the practice of virus disease control other than the use of soap and water, detergents, formalin or acids for cleansing the hands, tools, and apparatus contaminated by such easily sap-transmitted viruses as tobacco and cucumber mosaic viruses and potato virus X.

An interesting recent report is that by Tomlinson, Faithfull & Ward (1976), of the National Vegetable Research Station, Wellesbourne, who found that the symptoms induced by tobacco mosaic virus in tobacco (but

not local lesions on the leaves of *Nicotiana glutinosa*) and beet western yellows
virus in lettuce could be reduced or suppressed by the application of
carbendazin to the soil before infection.

Applications

Fungicides may be deployed in two general ways. They may be used either
to kill (*eradicate*) a pathogen already established on the host plant and thus
free the plant from infection and prevent the infection spreading to other
plants or to coat the exposed surfaces of healthy plants and so *prevent*
infection occurring. It seems to have been Marshall Ward (1882) from his
studies on coffee rust in Ceylon (Sri Lanka) who pioneered this last concept.
The ideal fungicide, able to serve both these functions, must therefore be
able to make contact with the pathogen and give an even protective coating
to the surface of the host plant on which it must adhere in a toxic state for as
long as possible. These properties were found to be enhanced by supple-
menting the fungicide with *spreaders* (such as, soap, other surface-active
compounds, casein, gelatine, and various oils) to ensure adequate contact
between fungicide and pathogen or host and *stickers* (e.g. flour paste, gums,
dextrines, oils,[6] and more recently, synthetic resins such as polyvinyl acet-
ate) which prolong effective protection, as in general do fine particle sprays
and dusts. In addition a fungicide must not be phytotoxic, which is a
common defect. It is this defect of Bordeaux mixture which had to be
overcome by dilution, or by formulations of increased lime content, for use
on crops sensitive to damage.

Seed treatment

Pre-sowing treatment of seed, especially of cereals, was one of the first
practices designed to mitigate losses from diseases and pests. Pliny recommended that cereal grain should be steeped in amurca (see Chapter 2) and
elsewhere writes of grain: 'It is generally supposed that if seed be first
steeped in wine, it will be less exposed to disease.' He also cites Virgil for the
opinion that it is beneficial to drench beans with nitre and amurca or steep
them for three days before they are sown in a solution of urine and water.
Such practice persisted – in 1627 Francis Bacon in his *Sylva sylvarum* (pp.
167–8) was writing: 'Though Graine, that toucheth *Oyle*, or *Fat*, receiveth
hurt, yet the *Steeping* of it, in the *Dregs* of *Oyle*, when it beginneth to Putrifie,
(which they call *Amurca*,) is thought to assure it against *Wormes*' – as did
sacrificial rites to propitiate the gods. J. G. Frazer in *The golden bough* records
the sacrifice of a young Sioux girl by the Pawnees of North America in the

1830s and the sprinkling of her blood over seed maize to ensure a good crop and notes nineteenth-century examples from India of the mixing of the ashes of sacrificed children with corn to preserve it from insects.[7]

Most of the many and diverse early recommendations for seed treatment were against unspecified conditions. The first instance of seed treatment against a particular disease may be that of Richard Remant who in 1637 reported experimentation against wheat bunt with certain steeps of whose ingredients he withheld information until satisfied as to their value. It was shortly after this that 'brining' of seed against bunt seems to have been introduced; a procedure frequently supplemented by liming which dates from classical times. The most frequently quoted reference to brining against bunt is that by Jethro Tull in *The horse-hoing husbandry*, 1733, who attributed the introduction of this method to the use of wheat salvaged from a shipwreck near Bristol which resulted in clean crops when those from home grown seed were bunted. But, as Tull suggested, this result might have been due to the cargo having originated from a district free from bunt. Brining did, however, achieve great popularity and only later in the century did it fall into disrepute.[8]

Mathieu Tillet in 1752–3, in the second of his field experiments from which he concluded that bunt originated from the bunt spores (see Chapter 2), offered a quantitative comparison from two pairs of plots of the number

Fig. 39. Treating seed grain with formalin against smut. (Arthur & Johnson (1920), fig. 3.)

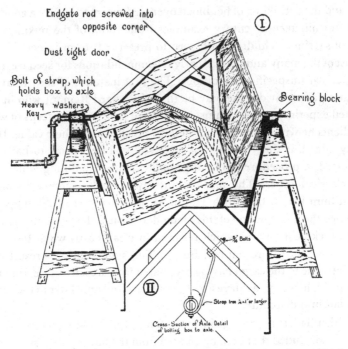

Fig. 7.—Box smut-treating machine: 1. Sketch of box dusting machine. 2. Cross-section of axle, showing how the box is bolted to the axle. Note that the strap is fitted to the axle so that when the axle is rotated, the strap pulls the box around. Direction of rotation is marked by an arrow.

Fig. 40. Design for a home-made machine for the dust treatment of cereal seed. (Güssow & Conners (1929), fig. 7.)

of bunted heads derived from seed dusted with bunt spores and similar seed after treatment with saltpetre and lime – the figures for the two being: control, 1687 and 938 bunted heads; treated seed, 10 and 84. The first fungicide to be used against bunt was copper sulphate – by Schulthess in Switzerland in 1761 and later by the Abbé Tessier in France (1779) – but the results were inconclusive. Proof of the fungicidal value of copper must be attributed to Bénédict Prévost (1807) in the course of his studies on the aetiology of bunt when he observed that bunt spores failed to germinate (or germinated poorly) after being soaked in water distilled in a copper alembic and he demonstrated that this effect was due to the presence of copper. He found that copper sulphate at 1/280 000 of the weight of water prevented spore germination and that even 1/1 000 000 retarded it perceptibly. He also noticed that filtering off the precipitate formed when copper sulphate was dissolved in ordinary water reduced the toxicity of the filtrate. Concur-

Fig. 41. 'Rotostat' seed treater. (Peacock (1978): 33.)

rently with this discovery Prévost had his attention drawn to an estate near his home in Montauban where freedom from bunt was associated with pre-sowing treatment of the seed with milk of lime in a large copper cauldron. From these observations, and a field trial in 1807 which showed that the incidence of bunt could be reduced by liming and eliminated by copper sulphate treatment of the seed, Prévost recommended the following procedure for use on a large scale:

One places in a tub as many 14 litres of water as there are hectolitres of wheat to be treated, and dissolved in it as many times nine decagrams of copper sulphate. There are two other vessels, each with a capacity of two or three hectolitres. Twelve or 14 decalitres of wheat to be treated are thrown into one, and some of the solution poured into it until it rises several decimetres above the wheat. It is stirred well, and all the floatage is carefully removed. Some wheat is placed in the second vessel, and treated in the same way. Some cross-pieces are laid upon this second vessel, and on them placed a basket of such texture that it lets the water pass freely, without letting the grain pass. When the grain in the first vessel has remained under water for half an hour, it is dipped out with a copper hand-bowl, a certain quantity of the liquid being taken at the same time. This is briskly thrown back, in order to free the wheat from any floating matter that may remain. This wheat is then thrown into the basket; and, when the basket is full and the wheat has drained sufficiently, the latter is placed in a heap. When scarcely any more wheat remains in the first vessel, more is put in, stirred, skimmed, etc. The basket is placed upon this vessel and the manipulation is from the second to the first, as it was from the first to the second, etc.[9]

To this he added the warning:

It is understood that one must avoid contaminating anew the treated wheat, either with dust that escapes from severely bunted wheat in motion, or by transporting it in bags soiled by this dust.[10]

Prévost's recommendation was not widely followed – even the Abbé Tessier did not recognise its utility – and the use of copper against bunt had to be re-introduced by Taylor (1846) in England and Julius Kühn (1886) in Germany. Later copper carbonate dressing was for a time popularised by Darnell-Smith (1917, 1919). In 1895 Th. Geuther in Germany introduced formalin as a seed treatment against smuts in general. Two years later Bolley in Dakota reported experiments on the use of formalin against wheat bunt. This became one of the most popular treatments (if unpleasant for the workers) during the first quarter of the twentieth century when the disease was a major problem in North America and elsewhere (Fig. 39). Since then the routine use of organo-mercury seed dressings (see Table 5) has reduced bunt to a disease of minor importance in the main wheat-growing regions of the world.

At first all seed treatments were undertaken on the farm and to lighten the labour of mixing seed and powdered chemicals many suggestions were made for adapting barrels or drums for the purpose or the construction of simple machines (see Fig. 40). Liquid treatments such as that with formalin were sometimes carried out by turning heaps of moistened grain with shovels (Fig. 39) or by soaking in tubs by methods reminiscent of that suggested by Prévost. Today much cereal seed is treated in bulk by seed merchants before distribution (Fig. 41).

Dusts and sprays

Most of the chemical control of plant diseases is applied as fungicidal sprays and dusts to aerial parts of growing plants and the history of the development of the necessary techniques is closely associated with the parallel problem of applying insecticides. William Forsyth in 1802 applied his lime sulphur by wiping the foliage with a cloth dipped in the fungicide and the procedure had not greatly changed by the time of the introduction of

Fig. 42. Sulphur dusting against hop mildew (Sphaerotheca humuli) *in New York State. (Blodgett (1913), fig. 104.)*

Bordeaux mixture in 1885 when Millardet's own instruction for applying his mixture read:

The workman pours part of the mixture, while stirring it, into a watering pot, which he takes in his left hand, while, with the right, by the aid of a small brush he wets the leaves, taking care constantly not to touch the grapes.[11]

Whisks of heath, straw, or similar material were used as substitutes for brushes while an alternative was to shake a broom dipped in the mixture over the plants. An early improvement was a specially designed brush with a hollow handle which could be connected by a tube to a can of fungicide carried on the operator's back.

The sulphur dusting widely employed during the first half of the nineteenth century had led to the application of the dust by means of hand bellows to the side of which a small reservoir for the dust was attached – an apparatus similar to that generally available to amateur gardeners today. According to Lodeman (1896), who gives an interesting account of early dusting and spraying machinery, the first hand-operated powder gun, in which a revolving fan forces a current of air through a tube in which the dust is mixed with the air, was invented and manufactured by Legget in the United States in 1854 where in 1895 there was available a horse-drawn machine, the 'Sirocco Dust Sprayer', in which a powerful air blast was produced by gearing the fan to the main wheels. A later example of this last type of machine is illustrated in Fig. 42.

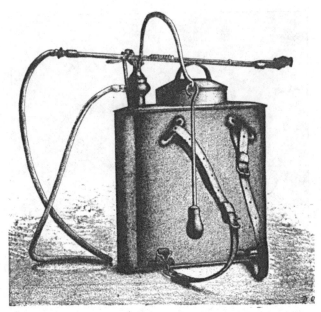

Fig. 43. The Galloway knapsack sprayer. (Galloway (1891), fig. 1.)

Fig. 44. An early powered horse-drawn spraying machine (Lodeman (1896), fig. 14.)

Fig. 45. Preparing Bordeaux mixture in the orchard. Hand operated spraying machine with long bamboo spray lances in the foreground. E. S. Salmon stands on the lower platform. (Salmon (1909), pl. 2.)

The large-scale application of liquid preparations proved more difficult to mechanise and it was the more general availability of effective dusting apparatus that led to the unsuccessful attempts by Millardet and others to develop a powder form of Bordeaux mixture. The earliest instrument for the application of sprays was the hand syringe. According to Lodeman (1896:5) in 1763 details of a lime water/tobacco preparation for the control of plant lice appeared in the newspapers of Marseilles when it was recommended that 'applications should be made by means of a small tin syringe having a rose pierced by about one thousand holes'. (Alternatively, 'many of the plant-lice may be destroyed by passing the leaves upon which they are found between two sponges wet with the tobacco water'.) From such beginnings 'garden engines' based on the force pump and able to throw larger amounts of liquid further, and the 'bucket pump' (stirrup pump) were evolved and later a range of knapsack sprayers (Fig. 43) and larger machines. These last were at first powered by the driving wheels (Fig. 44) and later by steam or internal combustion engines. Later still, the spray or dust was applied from low-flying aeroplanes or helicopters.

Bordeaux mixture in commercial practice was usually prepared by the grower, frequently in the orchard itself (Fig. 45). Larger growers centralised the operation and a limit was reached in the vast project mounted by major fruit companies to combat the Sigatoka disease (*Cerco-spora musae*) of banana in Central America which threatened to exterminate the industry. During the 1930s many square miles of plantations were sprayed every 10 to 17 days with 10-10-100 Bordeaux mixture prepared at central plants, the spray mixture being piped through the estates at an annual cost of many million dollars. Following the Second World War 80 to 90 per cent of the 87 million pounds of copper sulphate exported by the United States was used for this purpose, according to McNew (1959).

A major difficulty that had to be overcome in the development of spray-ing machinery was devising a machine that would deliver the spray evenly in small droplets of uniform size – a difficulty increased by the fact that many sprays were not solutions but particles in suspension which tended to clog the nozzle of the spray gun. Nozzle design is critical in determining droplet size. The earliest nozzles delivered a solid stream of liquid which broke up into droplets some distance from the nozzle or smaller solid jets of liquid (and hence smaller droplets) obtained by the use of a rose. A rather similar effect was obtained with a long and narrow nozzle opening. In a second class the emerging jet was made to impinge on a metal plate set at an angle, or other device, which obstructed the jet and broke it up into droplets. In the third and most successful type (the 'eddy' or 'cyclone' type; invented by the American C. V. Riley [1843–95] (the first Federal entomologist) the liquid was given a strong rotatory motion within the nozzle so that the emerging stream immediately assumed the form of a

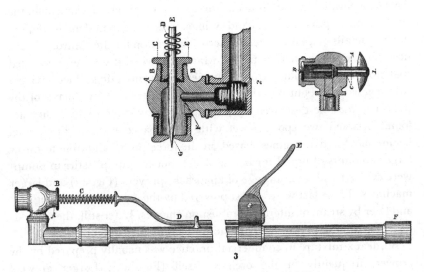

Fig. 46. Vermorel spray nozzle (Galloway (1889): 97.)

spray. One of the best known early nozzles of this design was that evolved by the brass-worker V. Vermorel of Villefranche (who collaborated with Millardet in designing spraying equipment) (see Fig. 46) and it was Vermorel who in 1886–7 introduced the 'disgorger' for cleaning the orifice of the nozzle when it became blocked.

The development of the internal combustion engine enabled much more powerful spraying machinery to be designed and in orchards and plantations the amounts of spray applied rose to levels as high as 300 gallons of spray per acre. The costs involved in providing and handling such large amounts of material encouraged the introduction of 'low volume spraying' at levels of the order of 25 gallons per acre. This economy was made possible by a better theoretical understanding of the physics of sprays and their application and deposition and the development of appropriate machinery. Hydraulic sprayers deliver droplets of a mass median diameter (MMD) of 250–400 μm which are impacted on the foliage at velocities near the terminal velocities of the droplets. Smaller droplets of a MMD of 50–100 μm applied with a supporting air blast to produce dispersion velocities which give a satisfactory dynamic catch may be produced by 'mist-blowers'. More recently, true aerosols, with dispersion of a MMD of 10–30 μm, produced by thermal, or special air-blast, aerosol generators have become available.

Rotary atomisers employing discs, gauze cages or brushes effectively applying ultra low volumes (ULV) down to less than one gallon per acre are used to improve the economics of aircraft spraying and are now being developed on ground equipment for controlled drop application (CDA) of ultra low volumes of liquid in drops of the most suitable size.

(For additional details such reviews as those by Ripper (1955) and Fulton (1965) may be consulted.)

Fumigation

Fumigation, more frequently employed against pests than diseases, has long been practised. Surapala in his Vrkssyurveda (see Chapter 2) wrote that 'Vegetables like cucumbers become free from disease if fumigated with the fumes of the bones of a cow and dog and the ordure of a cat' and he was apparently aware of phytotoxicity for he added 'A shrewd person should not administer pungent substances or apply fumigation to saplings'. The modern use of fungicidal fumigants dates from the mid-nineteenth century when sulphur was volatilised by spreading a sulphur paste on the hot water pipes of greenhouses and later by burning 'sulphur candles' or volatilising sulphur by a heated sand bath (see Table 3). Concurrently, in America,

Hays (1850) attempted the control of blight in stored potato tubers by subjecting them to the fumes of burning sulphur. Many years later sulphur dioxide was employed to prevent the spoilage (particularly by *Botrytis*) of grapes in transit to market from California (Jacob, 1929).

A more recent successful application of fumigation was the control of blue mould (*Peronospora tabacina*) in tobacco seed beds by the volatilisation of benzol (or toluol) during the night (with the beds covered by muslin) at a rate of 5 quarts per 100 square yards of bed with an evaporating pan surface of one-hundredth of the bed area. This procedure was developed in Australia by Angell *et al.* (1935) and subsequently widely employed in the United States (see Clayton *et al.*, 1942) where Clayton (1938) found para-dichlorobenzene to be more convenient and efficacious than benzol.

Soil treatment

Treatment of seed with fungicides before sowing to prevent infection by pathogens in the soil is commonly practised (see above); treatment of the soil with fungicides against specific pathogens is relatively rare. In 1891 Roland Thaxter employed elemental sulphur as a soil treatment against onion smut (*Urocystis cepulae*). Selby (1900) used formalin for the same purpose and Gifford (1911) successfully controlled damping-off of conifer seedlings by the application of 0.5–1.0 per cent formalin at 0.75 gal. per sq. foot. Another example is the use of a dilute solution of mercuric chloride (1 oz. to 10 gal., fide Chupp, 1917) as the transplanting liquid for cabbage and other brassicas to control club root (*Plasmodiophora brassicae*) (Preston, 1931; Walker, 1944). Bewley (1921) introduced Cheshunt Compound, a mixture of copper sulphate and ammonium carbonate (see Table 4) which is dissolved in water, the solution being used as a soil drench for the prevention and cure of damping-off.

A more frequent practice has been the injection of volatile chemicals into soil for the control of pests and pathogens and for effecting partial sterilisation which frequently enhances plant growth (cf. Chapter 8). Such procedure (fumigation) was first used in 1869 by Baron Arnold Paul Edward Thenard [1819–84] (son of the famous chemist L. J. Thenard) who, to control phylloxera of the vine in France, injected carbon disulphide into the soil and G. Gastine in 1876 designed a special injector or 'gun' for the purpose (see Tietz (1970) for details and illustrations). In 1901 E. Henry, also in France, used carbon disulphide to fumigate seed beds in forest nurseries and some success was obtained from 1935 with a soil injection of carbon disulphide for the control of Armillaria root rot of citrus in California (Bliss, 1951). Later the injection procedure was mechanised and a variety of other chemicals, including chloropicrin (Roark, 1934) and methyl bromide, employed.

A spectacular modern development of soil fumigation is that against verticillium wilt and other root rots of strawberry in California where specialist contractors apply a mixture of equal parts by weight of chloropicrin and methyl bromide (at 275–400 lb. per acre) by injection machines which mechanically cover the surface of the treated soil with sheets of polythene which are automatically glued together at the edges to seal in the soil. Two days later the polythene is removed (Wilhelm *et al.*, 1961; Wilhelm & Nelson in Toussoun *et al.* (1970):208–15).

Wound dressings

A minor but long-established practice is to use a wound dressing after pruning or tree surgery. Many different dressings have been advocated. Austen (1653:68) suggested clay mixed with horse dung for use after cutting out cankers on apple trees. The dressing to receive widest publicity was that devised by William Forsyth, of lime sulphur fame, who in 1791 published a pamphlet in which he disclosed the 'Directions for making a Composition for curing Diseases, Defects, and Injuries, in all Kinds of Fruit and Forest Trees, and the Method of preparing the Trees, and laying on the Composition' – a mixture of fresh cow dung, lime rubbish of old buildings, wood ashes, and sand – which he had developed for use in the Royal Gardens at Kensington. This disclosure, made under oath on 11 May 1791, resulted from an investigation of his method by a committee of both Houses of Parliament and authorisation of the payment of a reward to Forsyth, believed to have been £1500. Publicity for the method continued for the next twenty years although both its originality and efficacy were disputed. Subsequently a wide range of materials including tar, creosote, diverse chemicals and paints have been used for the treatment of wounds or pruning cuts.

Some of the success of Forsyth's method could have been due to antagonism by saprobic micro-organisms. At Long Ashton Research Station it has been found that inoculating fresh pruning cuts with micro-organisms three days to two weeks before inoculation with the canker fungus (*Nectria galligena*) reduced the number of infected shoots. Special pruning shears were developed for pruning wound inoculation (or fungicidal treatment) by which a regulated flow of liquid from a pressurised container delivered a dose of inoculum (or fungicide) at the moment of making a cut (Corke, 1978).

8

Control by physical agents

Heat

Fire

The first physical agent to be used for the control of plant disease was fire. The clearing of ground for the growing of crops by burning off the natural vegetation is an immemorial practice and fire has recently been used in Brazil on a larger scale than ever before as the first step in the conversion of millions of hectares of Amazon rain forest into cattle range. The burning of stubble and other plant residues is equally ancient. Such practices tend to elminate certain diseases and the burning of infected plant material is today employed as a routine sanitary measure on farms and market gardens while Wolf (1887) records as a long established practice among the Moors of North Africa the throwing of seeds of sorghum and sugarcane through the flames from burning straw before sowing. In spite of this long history the first comprehensive review of the use of fire for plant disease control is that by J. R. Hardison in the *Annual review of phytopathology* for 1976 where a number of beneficial and detrimental effects on plant disease of the controlled use of fire are noted. These include the routine mild ground fires which control brown spot needle blight (*Scirrhia acicola*) of long leaf pine (*Pinus palustris*) in the southern United States; similar burning, which kills the lower branches of saplings of southern pines, on which the destructive aecial cankers of *Cronartium fusiforme* develop, and also the *Quercus* species, which carry the uredinial and telial stages, controls this economically important rust. Similarly localised infections by the dwarf mistletoe (*Arceuthobium* spp.) have been eliminated from black spruce (*Picea marina*) plantings in Minnesota by cutting spruce to provide trash and burning out the infected area when the forest is wet. On the other hand in both Europe and North America outbreaks of root rot caused by *Rhizinia undulata* (*R. inflata*) have been associated with sites of bonfires where slash has been burnt.

Another spectacular success has been the control of blind seed disease (*Gloeotinia temulenta*) and ergot (*Claviceps purpurea*) in perennial rye grass (*Lolium perenne*) grown for seed in Oregon where the industry was threatened by these diseases which have however been controlled since the routine burning of the fields each year was introduced in 1948, a procedure which also increased the yield of seed.

One side effect of clean-air legislation resulting from the ecological lobby's activities has been to curtail burning in forests and arable land because of smoke pollution and this has led to the development of machinery to burn off the vegetation by oil or gas with the minimum of smoke. Hardison concluded that 'thermosanitation' is largely under-utilised in plant disease control and is a practice that might be exploited.

Baking

According to Newhall *et al.* (1934) early settlers in North America found that delicate plants such as muskmelons and watermelons could be grown successfully without damping-off on sites where piles of brushwood had been burnt, an observation basic to the modern practice of partially sterilising propagating soil by heat. Fifty years ago one widely used method, and one still practised, for the partial sterilisation of small quantities of soil by the controlled use of fire was to bake the soil in an oven consisting of a metal box with a central flue built into a brick structure so that a slow fire beneath the box heats the soil uniformly to a temperature of 100 °C for some hours. The soil must not be too moist or too dry and overheating renders it unsuitable for plant growth. Equipment is now available in which the soil is heated by electricity (first introduced for this purpose by Elvedon, 1921) either by using the damp soil as the conductor (the 'Ohio method') or by heating elements in metal tubes (the 'New York method') (Newhall & Nixon, 1935).

Steam

The most popular method for the partial sterilisation of soil has always been by steam. According to James Johnson (1946:84) the idea of applying steam to the soil as a practical control measure for a soil-infesting pest appears to have been that of W. A. Rudd in Illinois in 1893 and it soon became apparent that steaming soil released nutrient materials which enhanced plant growth – an effect of heating soil known since the report by B. Frank in Germany in 1888. It was also found that steaming had a beneficial effect on soil rendered 'sick' by intensive cropping so that from the turn of the century periodical partial steam sterilisation became a routine practice of growers of vegetables under glass in the States of Massachusetts, New York, and Ohio, by tobacco growers in the United States, and in England by glasshouse growers in the Lea Valley where tomatoes and cucumbers were grown for the London market. Sometimes, especially in America, a permanent arrangement for steaming the soil was provided by burying at a depth of about 18 inches rows of unglazed draining tiles or perforated metal tubes into which steam could be injected as required. The more popular methods developed included introducing steam under an inverted shallow metal tray laid over the patch of soil to be treated or by 'the use of movable perforated metal pipes either in the form of several pipes united as a horizontal grid or a vertical 'harrow' or as a series of individual L-shaped pipes (as in the so-called 'Hoddesdon' system, named after a town in the Lea Valley), methods which have for long been described in publications issued by Departments of Agriculture such as those by Beinhart (1918) and Bewley (1931). More recently the precision of the partial sterilisation of soil by steaming has been improved by the use of aerated steam which allows variable temperature control (Baker & Olsen, 1960; Dawson *et al.*, 1967). At lower temperatures less manganese is released and subsequent growth enhanced. Smaller quantities of soil may be steam sterilised in suitable containers made of wood, metal, or reinforced concrete and as for baking care has to be taken to prevent the soil being rendered infertile by overheating. Freshly steamed soil retards seed germination and early plant growth, due according to Johnson (1919) to the presence of ammonia, but this toxicity is temporary.

Hot air and hot water
The most important use of heat for the control of plant disease is the treatment of seed, propagating material, or even whole plants, with hot air or hot water, techniques first introduced by J. L. Jensen in Denmark.

Jens Ludwig Jensen (Fig. 47), born in 1836 in the province of Jutland, worked as a schoolmaster until 1872 when he started a company, with which he was associated until 1896, for selling scientifically tested seed. In addition, from 1868 until his death in 1904 he edited and published agricultural periodicals which from 1881 he organised, together with his work in the field and in his private laboratory, from the 'Bureau Ceres'. It was the introduction by his company of new varieties of potato into Denmark that focussed his attention on potato blight and in particular on the losses in store of infected tubers and their role in the overwintering of the pathogen. By laboratory studies he determined that *Phytophthora infestans* did not grow at temperatures below 5 °C or above 24 °C. He recommended that the

Fig. 47. Jens Ludwig Jensen (1836–1904).

storage temperature of potatoes during winter should not exceed 5 °C and also demonstrated that the pathogen in diseased tubers could be killed by subjecting them to a temperature of 40 °C for 4 hours, a treatment which did not affect subsequent growth of the potatoes as did application of the same temperature by immersing the tubers in hot water. In the light of these results, Jensen was the first to note that the introduction of potato blight into North America and Europe coincided with the replacement of sail by steam which much shortened the time of exposure of tubers to tropical temperatures. About 1885 Jensen turned his attention to cereal smuts and two years later established that although hot air was unsatisfactory smut infection could be reliably eliminated by immersing the dry wheat seed in hot water at *c.* 55 °C for 5 minutes and that the treatment was equally efficient for barley provided the seed was first soaked in cold water. These findings were given wide publicity. An English translation of Jensen's Danish paper appeared in the *Journal of the Royal Agricultural Society of England* in 1889 (and was reprinted as a pamphlet) where details of the procedure read:

The grain to be dipped is placed in a shallow cylindrical basket about twelve inches deep lined with coarse canvas, and provided with a cover made by stretching the canvas over a ring of such a diameter as will pass inside the mouth of the basket. The canvas should overlap the ring by about an inch all round. An ordinary boiler, such as is found on every farm, is filled with water and heated to boiling point.

Two vessels of sufficient size are placed near it. These may be designated 1 and 2. Supposing the boiler to contain 35 gallons of boiling water, if 12½ gallons of cold water and the same quantity of boiling water be put into each vessel, we shall have 25 gallons of water at 132 °F. in both of them. The exact temperature may be readily obtained by adding a little more hot or cold water, as the thermometer shows to be required.

A basket containing three quarters a bushel of corn, which must not be more than eight inches in depth, is now dipped into No. 1 four times; this will take rather more than a half a minute, and will reduce the temperature of the water eight or nine degrees. It is now to be rapidly dipped five or six times into No. 2, which will take about one minute, and then dipped once per minute for three minutes longer, *i.e.* five minutes altogether in the two vessels. This will reduce the temperature of the water in No. 2 from 132 ° to 129 °–130 °. If steeped barley be used the original temperature of No. 1 should be 129 °–130 °; but with unsteeped grain, for oats, barley, or rye, it does not matter if the original temperature be 133 °–136 °.

The seed-corn must now be cooled. This is best done by placing the basket on the top of a third vessel and pouring a couple of buckets of cold water upon the corn in it, taking care that the cold water falls not only upon the centre, but round the edges, so that the corn may be uniformly cooled. The basket is now emptied on the floor, and the corn spread out in a thin layer, so that it may cool completely. The water used in cooling the corn will have its temperature raised and may be employed in replenishing the boiler. The requisite temperature (132 °F.) of vessels Nos. 1 and 2 must be maintained throughout the process by adding from time to time boiling water from the boiler and transferring from them a similar amount back again to the boiler. The temperature must be regulated by a thermometer, which when used must be plunged deeply into the water.

The basket must be completely immersed each time, then lifted quite out of the water so as to allow it to drain for four or five seconds before it is dipped again.

Jensen's method was recommended by Kellerman & Swingle (1890) of the Kansas Agricultural Experiment Station in the United States where it came into general use and was later carefully re-investigated by Freeman & E. C. Johnson (1909) who advocated a 5-hour soak in cold water followed by 15 minutes at 52 °C for barley and 54 °C for wheat.

An interesting variant of the hot-water treatment was developed for use in the tropics by Luthra & Sattar (1934) in India whereby to free seed wheat from loose smut the grain was soaked in water for 4 hours (8 a.m. to noon) and then exposed to the full sun for a similar period (noon to 4 p.m.).

Hot-water treatment became a routine measure against loose smut of cereals and was widely applied against other fungus diseases including its use for the treatment of celery seed against *Septoria* blight (Krout, 1921;

Bant & Storey, 1952), the dormant runners of mint against rust (*Puccinia menthae*) (Ogilvie & Brian, 1935), and the corms of calla lilies (*Zantedeschia* spp.) for the control of *Phytophthora richardiae* root rot (Dimock, 1944), to cite a few representative examples. It has also been widely and successfully employed against eelworm infections, especially of narcissus and other bulbs.

Hot water and hot air also proved most useful for the elimination of virus infection, particularly from propagating material used to establish nuclear stocks of virus-free plants. The first example in this field seems to have been the hot-water treatment of sugarcane setts to free them from sereh disease (Wilbrink, 1923) and later the same method was used against chlorotic streak and ratoon stunt of the same host. The introduction of heat treatment against virus infections is particularly associated with the name of L. O. Kunkel of the Boyce Thompson Institute, New York, who in 1935 reported that maintaining peach trees affected by yellows at a temperature of *c.* 35 °C for two weeks or longer freed them from infection and he also found that a similar effect could be obtained by the immersion of dormant trees for 10 minutes in water at 50 °C (Kunkel, 1936). Later he used both methods to free asters and other plants from yellows (Kunkel, 1941). It is now known that the diseases so far mentioned are caused by mycoplasmas and not true viruses as was then believed and it was Kassanis, at Rothamsted Experimental Station in 1950, who can be claimed as the first to employ heat to eliminate a virus when he freed potato tubers from the leaf roll virus by storing them at 37.5 °C for 25 days. In this connection it is interesting to note that Thirumalachar (1954) reported that leaf roll of potato in the plains of India, where the seed potatoes were stored for six months during the hot season in thatched huts, only became a serious problem when cold storage facilities were installed for the seed potatoes or when seed potatoes were brought in from cooler, mountainous districts. Later, Kassanis (1954) reported the elimination of the viruses of tomato bushy stunt, carnation ring spot, cucumber mosaic, tomato aspermy, and *Abutilon* variegation at 36 °C.

A temperature high enough to inactivate a virus is often too high for the optimal growth of the crop but in France virus-free tips of carnation shoots used for propagation were obtained by raising the temperature of the glasshouses to 40 °C for a month before taking the cuttings (Ravel d'Esclapon, 1967).

During recent years this field has attracted much attention. Nyland & Goheen (1969) in a comprehensive review reported that up to 1966 approximately 1000 papers had been published on heat therapy of virus diseases. They also stated that the number of viruses and mycoplasmas successfully inactivated *in vivo* had increased from 15 up to 1950 to 120 as listed in their own review. More have been recorded since.

Other agents

Plant viruses are also inactivated by a range of physical agents including ultra violet light, X-rays, gamma rays, ultra sonic sound, and very high pressures (see the reviews by Matthews in Horsfall & Dimond (1959–60) **2**, chap. 12; Raychaudhuri & Verma in Horsfall & Cowling (1977–8) **1**, chap. 10) but heat remains the one physical agent with useful applications in the practice of disease control.

9

The epidemiological approach

Epidemiology is, strictly, the study of epidemics[1] but the term may be extended to cover the study of the dynamic and complex relationships between the host, the pathogen, and the environment which together determine the onset, progress, and outcome of disease in both individuals and populations. The epidemiological approach is basic for the rational control of disease. The epidemiology of any disease must be elucidated before the most appropriate prophylactic and therapeutic steps can be taken for its prevention and cure. The historical development of this approach is also complex. The many strands out of which the tapestry has been woven are intricately intertwined and it is only during the last two decades that important features of the general pattern have become apparent. Any attempt to analyse the evolution of the epidemiological approach tends to falsify by simplification a development in which the component elements overlie and interact with one another, both in time and space.

It might seem most logical to consider the plant–pathogen interaction before treating this involvement in relation to the environment. Historically it is more appropriate to begin with the total situation by considering aspects of the relationship of disease to the weather, the component of the environment which has for longest been claimed as a critical determinant of disease in plants.

Weather

Although deepening the epidemiological approach to disease is perhaps the most significant modern development in plant pathology, the relationship of outbreaks of disease to the environment, and particularly the weather,[2] has been a matter of comment from the earliest times. Theophrastus drew attention to the association of disease with rain and noted that crops were more commonly diseased in low situations than on high ground (see Chapter 2). In 1705 de Tournefort advocated keeping greenhouses drier during the winter months to prevent mouldiness of the plants and later in the century Stephen Hales attributed mildew of hop to excessive moisture (Hales, 1727). Prévost in 1807 satisfied himself that wheat bunt was caused by the bunt fungus but he designated the pathogen as '*la cause immédiate*' because he recognised that environmental factors played a vital role in determining outbreaks of disease.

This association of disease outbreaks with the weather provided a basis for the view (derived from the autogenetic beliefs of Unger and expressed by Hallier and Paul Sorauer, among others) that disease resulted from 'predisposition' of the host. The predispositionist's case was argued at length in 1890 by Marshall Ward, then professor of botany at the Royal Indian Engineering College, in his Croonian lecture to the Royal Society of London in which after an introductory review of the metabolic processes of the green plant he concluded that in summer after a period of cool, dull, wet weather

the parenchymatous tissues of the living plant may be in a peculiarly tender watery condition where the cell-walls are thinner and softer, the protoplasm is more permeable and less resistant, and the cell-sap contains a larger proportion of organic acids, glucose, and soluble nitrogenous materials than usual

which tend to make the plant more susceptible to fungal infection. This view recalls that of Stephen Hales who in his *Vegetable statics* of 1727 attributed the incidence of mildew of the hop to 'a rainy moist state of the air' which hindered 'in good measure the kindly perspiration of the leaves, whereby the stagnating sap corrupts and breeds moldy fen'. If Stephen Hales's notions on the relationship of plant to fungus were somewhat ambivalent Marshall Ward had no doubts regarding the reality of fungal pathogenicity and in the latter part of his lecture dealt with aspects of what has become known as 'the physiology of parasitism' which he illustrated by his study of the botrytis disease of lily (Ward, 1888) and other examples. He believed that the fungus, like the host, was influenced by environmental factors but the example he cites of his observations of botrytis on snowdrops in which he attributed infection of normal tissue to 'small invigorated mycelia developed from spores which germinated on the dead tissue of the sheaths at the base of the leaves', is now better explained on the basis of an increased 'inoculum potential' (see under Soil below).

The association of weather conditions with the onset or severity of diseases, such as potato blight with wet, and powdery mildews with hot dry, summers, had long been observed[3] but it was not until the second decade of

Fig. 48. Lewis Ralph Jones (1864–1945).

the present century that serious experimental studies were initiated to define the environmental factors which determine particular diseases. A pioneer in this field was L. R. Jones (Fig. 48) of the Agricultural Experiment Station of the University of Wisconsin at Madison who in collaboration with a series of able colleagues (including J. G. Dickson, James Johnson, G. W. Keitt, and J. C. Walker) set standards for such investigations throughout the world. L. R. Jones's first interests were in the soil but the plant pathological spectrum at Madison was wide and in the 1920s included J. C. Walker's classical studies on onion smudge (see Chapter 3) and James Johnson's in virology (see Chapter 5). Of particular interest in the present connection is the 1926 report by G. W. Keitt & Leon K. Jones of the results of their study of the epidemiology and control of apple scab (*Venturia inaequalis*) between 1919 and 1924 using a variety of different approaches.

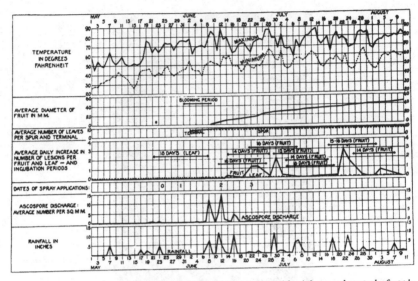

Fig. 49. A graphic summary of certain records relating to the epidemiology and control of apple scab, Sturgeon Bay, Wisconsin, 1920. (Keitt & Jones (1926), fig. 2.)

Each year the seasonal development of the host, the pathogen, and the disease were recorded in relation to the meteorological records and the application of sprays and dusts (see Fig. 49). These observations were supplemented by laboratory and greenhouse studies on the viability and longevity of spores and the relationships of their germination to moisture, temperature, and light. Infection experiments under controlled conditions were designed to study the penetration of the pathogen, the temperature, moisture and light relationships, the stage of development of the host organs, and the effect of fungicidal treatments while concurrently the development of epidemics was studied by observations on the overwintering of the fungus, the production and dissemination of ascospores, the occurrence of primary infection, the special significance of the early infection of sepals, production and dissemination of conidia, secondary infection, and the critical periods for the control of epidemics. Apple scab had been attracting much attention in America and elsewhere since its first recognition in Europe nearly a century before, particularly during the previous 25 years. Not all the findings and conclusions of Keitt and Jones's study were novel but their approach was more comprehensive than anything previously attempted and one aspect of its historical importance is in the use made of techniques, then crude, which were refined by later workers and are now standard practice.

Controlled environment

The inoculation chamber (devised by their colleagues James Johnson and J. G. Dickson) used by Keitt and L. K. Jones for infection experiments was, as seen from Fig. 50, primitive. It was, however, essentially such a chamber, but with more sophisticated control of temperature, humidity, and light, which R. H. Stoughton at Rothamsted Experimental Station combined with the 'Wisconsin soil temperature tank' (see Fig. 55) for the cabinets in which he made his investigation on the black arm disease of cotton caused by the bacterium *Xanthomonas malvacearum* (Stoughton, 1929, 1928–33). Subsequently, A. R. Wilson (1937) at Cambridge further refined one of the Rothamsted cabinets for use in his experimental study of chocolate spot of broad bean caused by *Botrytis cinerea*. Plant physiologists later developed very elaborate growth chambers, such as the 'phytotron' at the Californian Institute of Technology,[4] for experimenting with plants under controlled conditions and today as a feedback commercial systems are available to

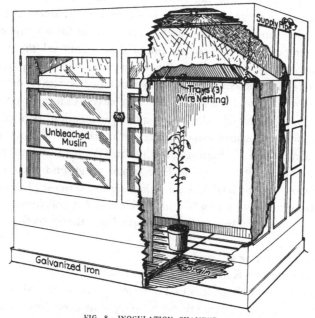

FIG. 8.—INOCULATION CHAMBER

The desired temperature and saturated humidity are maintained by spraying water of the appropriate temperature upon the inner chamber. Hot and cold water from supplies kept at suitable constant temperatures are mixed through metal valves to give the desired temperature. The cloth parts of the inner chamber are washed and sterilized at frequent intervals to preclude the possibility of injurious effects from the growth of microörganisms which might attack the cloth or give off gases which inhibit infection (see Brown, 1922).

Fig. 50. Inoculation chamber used by Keitt & Jones (1926).

growers of glasshouse crops for the automatic control of temperature, ventilation, watering, light, and nutrient supply, systems which plant pathologists have not been slow to utilise for their own ends.

Micro-climate

At the Conference of Empire Meteorologists held in London in 1929 W. B. Brierley drew attention to the importance to plant pathologists of crop meteorology – the meteorological conditions within the crop itself – as investigated by L. Roussakov in 1924 in relation to cereal rusts in the USSR and designated by him the 'micro-climate'. Brierley preferred to restrict micro-climate to the conditions immediately surrounding the pathogen and the chairman, Sir Napier Shaw, objected to micro-climate on semantic grounds but although C. E. Foister suggested the substitution of 'micro-phenology' and some later authors have used 'ecoclimate' the term micro-climate has come into general use in the sense of the meteorological conditions within or at the surface of the crop.

One of the features of the study of the micro-climate has been the development of a range of special devices to measure temperature, relative humidity, dew, and air currents within a crop. In addition to the measurement of temperature by suitably adapted conventional thermometers, thermocouples, resistor beads, and thermistors have been employed. Hair hygrometers have been supplemented by numerous devices for determining relative humidity. Dew or a film of water on leaves resulting from rain is of particular importance in relation to spore germination. Among devices available for the measurement of this factor are those based on the expansion and contraction of a lamb-gut membrane (Wallin & Polhemus, 1954), the continuous record of an indelible pencil on ground glass when the record is more pronounced when the glass surface is moist (Taylor, 1956), and the continuous weighing of a block of expanded polystyrene as in the simplified surface-wetness recorder designed by Hirst (1957).[5]

Spore trapping and spore dispersal

The method used by Keitt and L. K. Jones to trap ascospores of the apple scab fungus in the orchard was that devised by Pasteur in 1861 by which a known volume of air is filtered through a plug of nitrocellulose (gun-cotton) which is subsequently dissolved in an ether/alcohol mixture when the number of spores trapped can be estimated by microscopic examination of the residue remaining after the liquid has been allowed to evaporate. Another early technique was to trap airborne spores on untreated or sticky glass surfaces by sedimentation or impaction and later the same procedures

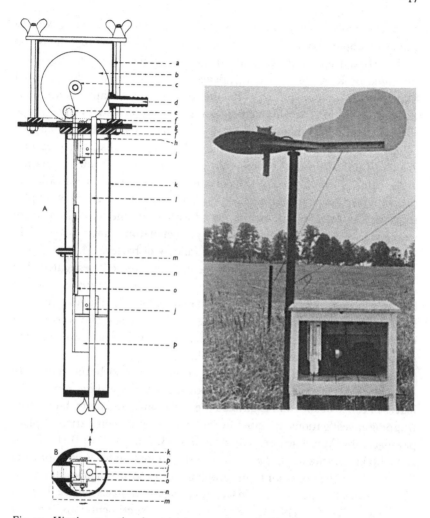

Fig. 51. Hirst's automatic volumetric spore trap in use and in diagrammatic section. (Hirst (1952), pl. 11, fig. 3; text-fig. 1.)

were used with culture plates of suitable solid media. Today a widely employed apparatus used in epidemiological studies by phytopathologists is the 'Hirst trap' (Fig. 51), an adaptation of the 'Cascade impactor' (invented by R. K. May in 1945), devised by J. M. Hirst of Rothamsted Experimental Station in 1952. By this power-driven trap a continuous 24-hour record may be obtained of spores trapped on a sticky microscope slide (in the model manufactured by Casella), a period that may be extended to seven days by replacing the slide by a plastic band on a rotating drum (as in the Burcard version of the trap). When continuous evaluation of the catch is less important the portable, hand-powered modification of the Hirst trap devised by Gregory (1954) may be employed as may simpler traps such as a strip of cellophane coated with a vaseline–paraffin mixture wrapped round a vertical metal rod and protected from rain as used by J. F. Jenkyn (1974) in his study of powdery mildew of barley.

As with climate, dispersal within the crop had to be investigated. For rain-dispersed spores the number of spores in the run-off water and splash has been estimated by devices such as those used by J. M. Waller (1972) for his investigations on coffee berry disease in which the run-off was collected in PVC gutters and drained into a reservoir and the splash drained from strategically placed vertical sheets of plastic.

Plant pathologists have long shown an interest in airborne spores. In 1882 Marshall Ward in Sri Lanka exposed microscope slides among the leaves of coffee bushes and examined the spore catch and many later spore trapping investigations are cited in the reviews on dissemination of plant pathogens by Max Gardner (1918) and J. H. Craigie (1940). Professor E. C. Stakman (Stakman *et al.*, 1923) in the United States was among the first to sample the upper air from aeroplanes by exposure of glass slides and his lead was followed by others in Canada, England, and elsewhere. In 1935 the Russian K. M. Stephanov made a valuable experimental approach to the dispersal of airborne spores in relation to the dissemination of disease in plants. Ten years later in England P. H. Gregory (1945) attempted a mathematical treatment of the topic and the following year Wolfenbarger published his independent review of the dispersal of small organisms and spores in relation to the incidence of disease. It was from Gregory's germinal paper that quantitative study of spore dispersal developed as a new branch of phytopathological investigation. In order to obtain experimental data on the deposition of airborne particles on plant and trap surfaces Gregory built at Rothamsted a small wind tunnel of 1 foot square cross-section from acrylic resin (Perspex) sheet designed to give turbulent or streamlined air flow of up to 10 m/sec (see Fig. 52). His first study was of the

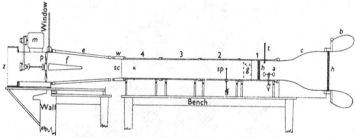

Fig. 52. Gregory's wind tunnel. 1. In diagrammatic vertical axial section. 2. Photograph of sections 3–4 with Cascade Impactor in trapping position. (Gregory (1951), text-fig. 1; pl. 9, fig. 2.)

deposition of spores on vertical cylinders the efficiency of which was compared with that of a Cascade impactor (Gregory, 1951). Subsequently the tunnel was put to a diversity of uses and after 30 years is still in use today. Many aspects of the developments in this field are reviewed in detail in the two editions of Gregory's *Microbiology of the atmosphere*, first published in 1961.

Forecasting

More precise knowledge of the relationship of disease incidence to meteorological data and host–pathogen interaction led in the early 1920s to the first attempts to forecast the likelihood of outbreaks of some of the most economically important diseases of some major crops so that growers could take appropriate precautions or in the absence of a disease warning save themselves money and effort by modifying a routine spraying programme. Keitt & L. K. Jones (1926) satisfied themselves that in Wisconsin the apple scab pathogen (*Venturia inaequalis*) overwintered to a significant extent only through the formation of perithecia on dead leaves and that ascospores, discharged in spring after intermittent wetting of the leaves by rain, were largely responsible for the initiation of the annual outbreaks of scab. They did not themselves issue forecasts but by 1926 the beginning of ascospore discharge, the time of which varies from year to year, was being notified to growers in both the United States and Canada and later this practice was extended to the Netherlands, New Zealand, and Australia.

The first routine attempts to forecast disease were begun in Italy in 1921

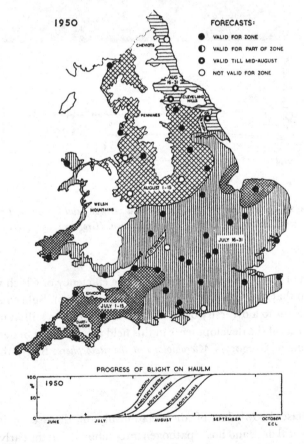

Fig. 53. *Validity of forecasts of potato blight by the Beaumont method for England and Wales in 1950 and progress curves for blight on haulm in six districts. (Large (1953), fig. 3.)*

(Voglino, 1922) and in France two years later (Chaptal, 1923) where, from reports of regional observers and meteorological data, growers were warned (in some localities by ringing the church bells) when conditions favoured the initiation of downy mildew (*Plasmopara viticola*) of the vine by germination of overwintered oospores. Subsequently schemes were developed for the forecasting of other downy mildews including those of tobacco (tobacco blue mould, *Peronospora tabacina*) (the incidence of which in North America is favoured by above normal January temperatures), lima bean (*Phytophthora phaseoli*), and cucurbits (*Pseudoperonospora cubensis*) in the United States and particularly late blight of potato (*Phytophthora infestans*) which has attracted much attention.

The first successful forecasting of potato blight was undertaken in the Netherlands where van Everdingen (1926–35) offered four conditions (the 'Dutch Rules') which enabled an outbreak of blight to be forecast: (1) a night temperature below dew point for at least 4 hours; (2) a temperature not falling below 10 °C; (3) a mean cloudiness on the following day of not less than 0.8; and (4) rainfall during the following 24 hours of at least 0.1 mm. These rules also proved satisfactory in France and in the Leningrad region of the USSR but less so in England where A. Beaumont (Fig. 29) working in South-West England offered a simplified 'Temperature–humidity rule'. He proposed two conditions (the 'Beaumont Periods'): (1) Minimum temperature not less than 50 °F, 10 °C; (2) Relative humidity (RH) not below 75 per cent for at least 2 consecutive days (Beaumont, 1947). The general validity for England and Wales of Beaumont's method is well indicated by Fig. 53. Grainger (1950) found Beaumont's rule applied to the west of Scotland after the month of June and later he claimed that a 'physiological barrier' was also involved and that no blight followed a Beaumont period when the ratio of total carbohydrate (Cp) to residual dry weight (Rs) of the plants was less than unity (Grainger, 1956). Grainger (1955) also devised the 'Auchincruive Potato Blight Forecast Recorder' – a self-calculating combination of wet and dry bulb thermometers by which Beaumont periods can be determined without special skill. L. P. Smith (1956) suggested the substitution of 90 per cent RH for at least 11 hours on each of the two days of the Beaumont period while Hirst & Stedman (1956) drew attention to the conditions within a crop where when the foliage is dense 90 per cent RH persists for longer each day than in the standard louvred screen ('Stevenson screen') 4 feet above ground level. They invariably found this relationship to be associated with rapid attacks of blight and discussed it in connection with the initiation within the crop of infection arising from the occasional above-ground infection from an infected tuber approximately a month before the outbreak of blight became

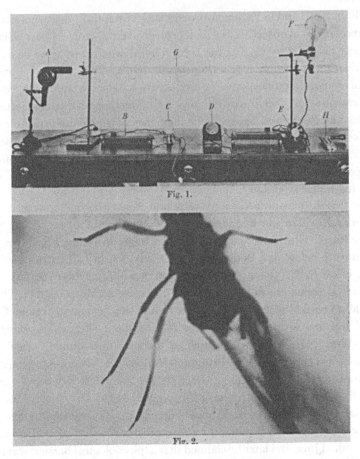

Fig. 1.

Fig. 2.

Fig. 54. 1. Wind tunnel designed by Maldwyn Davies for studying the effect of wind on aphides.
2. An aphid clinging to a smooth glass surface in the equivalent of a 70 m.p.h. gale. (Davies
(1932–5, III), pl. 1.)

general for the district (Hirst & Stedman, 1960). In 1975 R. A. Krause *et al.*
offered 'Blitecast', a computerised forecast of potato blight and Madden *et al.*
(1978) developed a similar forecast (FAST) for *Alternaria solani* on tomato.

Among other diseases for which forecasting has been attempted are rice
blast (*Pyricularia oryzae*) in Japan, wheat leaf rust (*Puccinia recondita*) in the
United States, ergot (*Claviceps purpurea*) in the USSR, and sugar beet virus
yellows in England while the topic has been regularly reviewed e.g. by
P. A. Miller & O'Brien (1952, 1957); Miller in Holton *et al.* (1959) Chap.
50; Waggoner (1960); Bourke (1970).

An interesting pendant to the topic of forecasting is the effect of weather on the dispersal of insect vectors of virus diseases as exemplified by the germinal investigations by W. Maldwyn Davies (1932–5) of the University College of North Wales, Bangor, on the factors which determine the movement and dispersal of the aphid vectors of the common potato viruses from which it was possible to evaluate the suitability of sites for the production of seed potatoes. By ingenious laboratory experiments in which he used a simple wind tunnel constructed from glass tubing and an adapted hair drier (Fig. 54) he found that winged *Myzus persicae* only fly readily when the temperature reaches 65 °F, that a relative humidity of more than 70 per cent has a very marked inhibitory effect on flight, and that flight ceased completely when the wind velocity approached only 4 miles per hour (and that the equivalent of a 70 miles an hour gale failed to remove aphides adhering to a smooth glass surface (Fig. 54)) and he was able to confirm his laboratory findings by field observations which combined continuous trapping of aphides with meteorological observations. Further he found over a period of years high aphid infestation in Flintshire where conditions are dry and the spread of pototo viruses rapid and low infestations in the humid areas of South Caernarvon and Anglesey where the spread of virus diseases is minimal. These results explained the slow degeneration of potato stocks in Scotland and enabled additional suitable areas for the production of seed potatoes to be selected.

Soil

Soil temperature and moisture were the first environmental factors affecting disease to be experimentally investigated in detail, by L. R. Jones and his colleagues and students at the University of Wisconsin during 1915 to 1925. The investigations originated from field observations on cabbage yellows (*Fusarium conglutinans*) and flax wilt (*F. lini*) which showed both diseases to be less severe at low than high soil temperatures. The experimental work was begun with plants grown in ordinary flower pots but for greater precision and convenience the famous 'Wisconsin soil temperature tanks' by which the soil containers were immersed in a constant-temperature water bath were quickly evolved (see Fig. 55 for the final version). These tanks were used to study a wide range of conditions and diseases including, in addition to cabbage yellows and flax wilt, tobacco root rot (*Thielavia basicola*), stem canker of potato (*Corticium solani*), seedling blight of cereals (*Gibberella zeae*), helminthosporium diseases of wheat, barley, and rice, take-all of wheat (*Gaeumannomyces graminis* (*Ophiobolus graminis*))[6] potato scab (*Streptomyces scabies*), cabbage club root (*Plasmodiophora brassicae*), onion smut (*Urocystis cepulae*), crown gall (*Agrobacterium tumefaciens*), and root nodules of legumes. The results are summarised and fully documented in the two classic publications by L. R. Jones (1922) and Jones, Johnson & Dickson (1926).

Normal soil contains many saprobic fungi and bacteria, as had been known since the 1880s (and in 1904 Hiltner had introduced the term rhizosphere for the zone of enhanced microbial activity he had detected in the neighbourhood of roots) but it was not until the 1920s that the reality of permanent and characteristic microbial floras of soils was generally accepted. It was in 1916–17 that the microbiologist Selman A. Waksman of Rutgers University gave affirmative answers to the two questions 'Do fungi live and produce mycelium in the soil?' and 'Is there any fungus flora of the soil?'; answers which met with approval.

Experimentation with pathogenic fungi in soil revealed inconsistencies. For example, the optimum temperature of the growth of wheat and oat seedlings is in the range 12–16 °C, that for the growth of the cereal root-rot fungi 25–30 °C. Increase in temperature of the soil by discouraging growth of the host and favouring growth of the pathogen might be expected to accentuate the severity of the disease. Sometimes experimental results were in accord with expectation; at others increase in temperature was accompanied by decrease in infection. G. B. Sanford (the first officer-in-charge of the Dominion Laboratory of Plant Pathology at Edmonton, Alberta) had suggested in 1926 that the control of potato scab (caused by the actinomycete *Streptomyces scabies*) effected by green manuring (to which attention had been drawn by W. A. Millard in England in 1921; see Millard & Taylor, 1927) was due to the inhibitory action of non-pathogenic soil micro-organisms and in 1932 A. W. Henry of the University of Alberta at Edmonton having found that with increasing temperature the infection curve for *Gaeumannomyces graminis* on wheat seedlings was downwards in unsterilised soil but upwards in sterilised soil, considered that the decrease in infection with rise in temperature in the unsterilised soil was also due to the enhanced effect of the soil microflora. This conclusion was confirmed by S. D. Garrett (1934) working on the same disease at the Waite Institute, Adelaide.

It was from such beginnings, at a time when major progress was being made in the ecology of both plants and animals, that the study of the microbial ecology of the soil arose, both as an academic exercise and in relation to root diseases of major crops. Mycological developments in the latter field are particularly associated with the name of Stephen Denis

Garrett for his own researches begun in Australia and continued at the Imperial College, London, Rothamsted Experimental Station, and the University of Cambridge, and for the series of more than a dozen review articles[7] and four books (Garrett, 1944, 1956, 1963, 1970) by which, during the past 40 years, he has documented progress in the knowledge and ideology of root-infecting fungi and kept ecological aspects of soil microbiology constantly in the public eye.

Studies on the mycology of the soil enabled a number of groups of soil fungi to be distinguished. Waksman differentiated between fungi found in almost all soils and fungi which had found their way into the soil by chance as spores or mycelium in organic debris. In 1933 O. A. Reinking & M. M. Manns from their studies of fusaria in tropical soils gave Waksman's categorisation a phytopathological twist by classifying the common saprobic fusaria found in soil as *soil inhabitants* and the pathogenic species, of which the more local distribution was associated with particular host plants, as *soil invaders*. Subsequently Garrett emphasised this approach by defining soil inhabitants as 'primitive and unspecialised parasites with a wide host range' in which 'their parasitism appears to be incidental to their saprophytic existence' and soil invaders as 'more highly specialised parasites' the presence of which in the soil 'is generally closely associated with

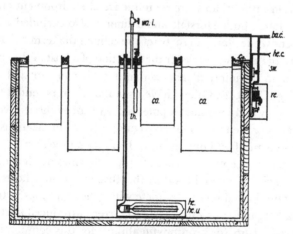

Text Fig. 3. Section of the improved Wisconsin temperature tank, showing construction and arrangement of apparatus. The soil cans (*ca.*) are suspended in the water from the rigid cover (*co.*). Water for cooling and filling the tank is run in through water inlet (*wa.i.*). The alternating 110-volt heating current is carried through heavy wires (*he.c.*) from switch (*sw.*) to relay (*re.*) and thence to the heating unit (*he.u.*) enclosed in the water-tight copper tube (*he.*) The direct current operating the relay is carried from storage batteries through line (*ba.c.*) from thermostat (*th.*) to the magnet on the relay (*re.*).

Fig. 55. Wisconsin soil tank in section. (Jones, 1922.)

that of their host plants. In the continued absence of the host plant, such fungi die out in the soil owing to their inability to compete with the soil saprophytes for an existence on non-living matter'. Later still Garrett redesignated soil invaders as *root inhabiting fungi*. He also offered an alternative classification of soil fungi according to the substrata utilised into sugar fungi, cellulose-decomposing fungi, and lignin-decomposing fungi.

Two other concepts merit notice. The first is that designated *inoculum potential* a much used and overworked term and Vanderplank has suggested that 'Any future historian of twentieth century plant pathology will have to answer the question, what was inoculum potential?' Here possibly the most prudent course is merely to indicate the origin of the term and the range of meanings subsequently imposed upon it and to refer the reader to the discussions in the Federation of British Plant Pathologists' *Guide to the use of terms in plant pathology* (1973), Horsfall & Dimond (1963), Vanderplank (1975: 84–7) and R. Baker (in Horsfall & Cowling, 1977–8, **2**, Chap. 7). Inoculum potential was introduced by J. G. Horsfall (1932b) in connection with fungicide studies in the sense of spore load with the stated implication of mass action, the greater the mass of the organism present and the more virulent the organism the more severe the disease. This view was, he continued, 'concerned only with the influence of the amount or virulence of the inoculum rather than with the influence of environment on severity of infection'. Later both Horsfall and Dimond also included effects of the environment while Garrett (1956:196) redefined the term as 'the energy of growth of a parasite available for the infection of a host at the surface of the host organ to be infected' and elaborated this in Horsfall & Dimond (1959–60, **3**, Chap. 2). Others have used inoculum potential for either or both inoculum and inoculum's potential with or without involvement of host and environment as a Humpty Dumpty word to mean just what they chose it to mean, if without Humpty Dumpty's precision.

The second concept arose from attempts to compare the ability of root-infecting fungi to survive in soil in the absence of their hosts, a survival dependent on their ability to compete with other saprobes in the colonisation of dead organic substrata. In 1950 Garrett introduced the term *competitive saprophytic ability* (or saprophytic ability) for this competition between fungi and in 1970 (p. 719) epitomised the mechanism of saprobic competition between fungi as: 'the share of a substrate [substratum] obtained by any particular fungal species will be determined partly by its intrinsic competitive saprophytic ability and partly by the balance between its inoculum potential and that of competing species'.

Root diseases

Because of the diversity and economic importance of root diseases their study has for long been popular and the amount of literature generated large. During recent years investigations on root disease have become a distinct branch of plant pathology. Root disease and other soil-borne defects were the subject of symposia at the Jubilee meetings of the Association of Applied Biologists in 1954 and the American Phytopathological Society in 1958 (Holton *et al.*, 1959:307–79) and also the topic for two important international meetings – at Berkeley, California, in 1963 (Baker & Snyder, 1965) and at the First International Congress of Plant Pathology in London five years later (Toussoun *et al.*, 1970). The reports on all these meetings are extensive and well documented.

The results of academic research together with studies on the survival of pathogenic fungi in soil have done much to explain variations in the incidence and course of many soil-borne diseases and given a better understanding of control measures. They also suggested new practices. Reduced to its simplest terms root disease is determined by the interaction between the host plant (which shows varying degrees of resistance to infection) and the pathogen (which shows varying degrees of virulence), an interaction which is influenced by the nutrient status, temperature, humidity, and microbial content of the soil. Some pathogens (the root-inhabiting fungi) are well adapted to certain host species but are of low 'saprophytic ability' while others, which may be termed 'opportunistic', are soil-inhabiting saprobes only able to infect plants under certain favourable conditions. Both these minority groups have to compete with the majority group: the non-pathogenic members of the soil microflora. Control of root diseases consists, basically, in adjusting this system in favour of the plant. Frequently this has not proved easy and much still remains to be done before unacceptable losses from many soil-borne diseases can be prevented. Here it is only possible to supplement a reminder of the few generalisations which have emerged with representative examples to illustrate their application.

In addition to such widespread practices as the use of healthy seed (or other planting material) and resistant varieties, and the observance of sanitary measures which contribute to the prevention of pathogens being carried over from one season (or crop) to the next and lower the amount of inoculum in the soil, a few generalisations are possible. For cereal and other arable crops where both the value of the individual plants and their value per acre is low crop rotation has been proved to be of basic importance, as for example, against take-all of cereals where the pathogen, *Gaeumannomyces graminis*, is restricted to Gramineae and root rot of cotton caused by *Phymatotrichum omnivorum* which has no monocotyledonous hosts. Sometimes manurial treatment or other soil amendment is indicated. Take-all of cereals is more severe when soil nutrients are inadequate (Garrett, 1942) and browning root rot (*Phythium arrhenomanes*) of cereals does little damage when the nitrogen/phosphorus balance is correct (Vanterpool, 1935), while, as already noted, club root of crucifers is reduced when the reaction of the soil is alkaline. For glasshouse, market garden, and other intensive crops one of the most effective measures against root disease is partial sterilisation of the soil either by chemicals (see Chapter 7) (which are also sometimes used for arable crops) or heat, particularly steam (see Chapter 8).

Root rots of plantation crops, especially tropical, and of forest trees have posed many intractable problems. A particularly illuminating example is that of root disease of rubber (*Hevea brasiliensis*) in Malaysia and elsewhere because since the early years of the century the evolution of measures for its control has been influenced not only by increasing knowledge of the pathogens and the dynamics of the disease but also by the economic climate and even the psychology of the planters. Rubber is subject to three major root diseases caused by *Fomes* [*Rigidoporus*] *lignosus* (white root disease), *F.* [*Phellinus*] *noxius* (brown root disease) and *Ganoderma pseudoferreum* (red root disease) which in Malaya have caused greater losses than the combined effect of all other pests and diseases of rubber. White root disease, the most serious of the three, which causes heaviest losses during the early years of a plantation, is a disease of moist conditions and therefore more prevalent in coastal than hilly districts. It was first recorded (but wrongly attributed to *F. semitostus*) from the Straits Settlements in 1904 by H. N. Ridley who suggested treating the soil of an infested area with lime and copper sulphate. Petch (1921) (Fig. 56) in Sri Lanka recommended the then current practice of incorporating lime in the soil after the removal of diseased roots (although later it was shown that the pathogen grew better in alkaline than acid media) while R. P. N. Napper in Malaya advocated the use of 2 per cent copper sulphate as a soil drench. In 1958 A. Riggenbach substituted an organic mercury fungicide for the latter. Such essentially prophylactic measures, though still practised, proved of minor significance.

Another early line of approach was the use of fungicidal wound dressings to treat roots from which diseased areas had been excised. Petch (1921) advocated tar, Napper 2 per cent copper sulphate or lime sulphur, and the Rubber Research Institute of Malaya later recommended tar acid fungicides for this purpose.

Fig. 56. Tom Petch (1870–1948).

A more successful alternative was efforts to eliminate infected material. Rubber plantations were originally established in cleared jungle but in Malaya, which has over a third of the 12 million acres of rubber planted in the Far East, in order to conserve virgin jungle, it is rubber plantations which have reached the limit of their commercial lives of 40 years or less that are now normally replanted and crop rotation is not an economic proposition. 'Absolute' clean clearing of jungle or old plantations and the careful removal and burning of stumps and all root fragments eliminates infection but this tedious and expensive procedure is impossible over large areas and there was much controversy regarding the merits of the less thorough 'ordinary' clean clearing as commercially practised, no clearing, and the selective removal of stumps. The last practice lost support as the host range of *R. lignosus* was extended but differences of opinion on the merits of clean versus no clearing (which had economic advantages) remained. Petch while recognising the hazard to rubber of root disease fungi persisting in the stumps of diseased jungle trees believed that the major source of infection of the new rubber was from the infection of the exposed surface of jungle stumps by airborne spores. He advocated control by the removal of diseased rubber trees and decaying stumps and the isolation of each infected area by a deep trench. As early as 1912 Barcroft reported his inability to infect rubber plants in soil containing the mycelium of *R. lignosus*. For successful infection the pathogen had to be growing on fragments of root. In other words, there had to be a food base (later estimated by R. A. Alston to be at least 5 cubic inches of root) to raise the inoculum potential to the required level. Another feature of the pathogen is the production of rhizomorphs which make extensive and rapid epiphytic growth on the surface of the root – commonly 5–15 feet ahead of the diseased part – and it is by such rhizomorphs when they come into contact with susceptible healthy roots that new infections are established. It was the detection of rhizomorphs that provided the basis for the successful control practice developed by Napper (1932) by which the upper tap root and laterals of every young rubber tree planted on cleared land were inspected twice or more each year and if any rhizomorphs were found they were traced back to their source which was then removed and any infected parts of the roots of the young rubber cut away. Infected areas were not isolated by trenching but digging out infected roots caused considerable disturbance to the soil. This disturbance enabled rhizomorphs to spread more quickly and more recent experimentation indicated that although Napper's method was a commercial success it did in fact increase the infection of the rubber. Napper's method is also labour intensive and since the mid 1930s the increasing cost of labour and the competition from synthetic rubber has encouraged economy in the use of labour which is one of the attractions of planting on uncleared ground.

The results of experiments set up by J. W. Weir at the Rubber Research Institute in Malaya to compare 'clean clearing' with 'no clearing' showed less infection of the young rubber two-and-a-half years after planting on uncleared than cleared blocks and less infection still where the secondary jungle had been allowed to regenerate. Napper suggested that the explanation of this result was that while the amount of inoculum in the soil may be greater, the greater quantity of roots available to the pathogen reduced the chance of infection of the rubber. The sowing of a mixed cover of creeping legumes over all the new plantation except for 6-foot wide strips along which the rubber is planted is now a standard practice. The shade from the maturing rubber results in the replacement of the cover plants by jungle grasses, etc. There is, thus, a partial rotation of the ground cover during the life of the plantation.

Under such a modern programme the health of the rubber is judged by foliage inspection. Diseased trees are rogued out, and collar inspection made of the adjacent trees in the row which are treated as recommended by Napper but rhizomorphs and roots are not followed further than the edge of the clear weeded planting strip.

In 1937 in East Africa Leach concluded that tree roots left in the ground after clearing the bush are most susceptible to attack by the honey fungus (*Armillariella mellea*) when their carbohydrate content is high and he found that bark ringing trees before felling reduced the carbohydrate level in the residual roots which were then more readily attacked by saprobic fungi than *A. mellea* and so were less potentially dangerous as sources of root rot for the commercial plantation. Napper in 1939 found that the invasion of rubber roots by saprobes rather than *Rigidoporus lignosus* was favoured either by killing the trees before felling or the stumps with sodium arsenite; this was later replaced by another arboricide, 2,4,5-T, to become notorious for its use in the Vietnam war. This too has become a standard practice when clearing before planting or thinning out a maturing plantation and as an additional precaution the cut surfaces of stumps are protected by the application of creosote.

Finally, in his comprehensive reviews of recent developments Fox (1961–70) has pointed out two other factors which affect losses and the cost of control. The first, what he termed the 'despair factor', arises when planters on collar inspection treat a tree which would have recovered if left untreated (rubber trees develop some resistance to *R. lignosus* with age), remove trees which could have been treated, or remove a source of infection which no longer has enough inoculum potential to establish disease in a young tree. The second is the 'fear factor' which leads some planters to consider all infected roots to be potentially dangerous and for the detection of which they thickly plant 'indicator' bushes.

The principles underlying control practices against white root disease of rubber are thus essentially biological. Competition from saprobic members of the soil microflora is encouraged and even if the provision of an ample supply of susceptible roots for the pathogen to attack increases its total mass, concurrent dilution reduces the inoculum concentration to less dangerous or even safe levels.

Another interesting example of biological control is provided by *Fomes annosus* [*Heterobasidion annosum*] which in Great Britain is the principal cause of root- and butt-rot diseases in the Forestry Commission's conifer plantations. Rishbeth in 1951 demonstrated conclusively that airborne basidiospores consistently infected the freshly cut pine stumps left after tree felling from which the infection spread to healthy trees. To eliminate this Rishbeth

(1959–63) at first advocated painting the freshly cut stump with creosote but for this sodium nitrite was later substituted, after trials with a range of chemicals which penetrated the stump, killed the tissues, and enabled colonisation by a range of saprobic fungi to take place. Later still Rishbeth developed a biological control whereby the stump was inoculated with the oidia of *Peniophora gigantea* which are provided for field use as dehydrated tablets (with a storage life of at least two weeks), each tablet containing approximately 10^7 viable spores.

Host–pathogen interaction

The characteristics of the three main groups of plant pathogens – fungi, bacteria, viruses – were considered in Chapters 3–5 and here it is necessary only to recall certain aspects of pathogens which are particularly relevant to epidemiology. First is the recognition that pathogens typically exhibit specialisation for one or more hosts at the generic, specific or intraspecific (especially cultivar) level; as has long been observed. In the early eighteenth century Noel Chomel was writing in his *Dictionnaire oeconomique*, 1707: 'It is observable that when *Mildews* arise or Blites fall, they generally infect one Sort of Grain ... the like befals Fruits, sometimes Apples are generally blasted, sometimes only Pears, at other times only Cherries, Walnuts, Filberts.'[8] Secondly, one pathogen may show variation in its degree of pathogenicity (that is, of virulence) to the host or to different cultivars of one botanical host species. Lastly, pathogens may develop resistance to chemicals used for their control.

Host characteristics particularly relevant to epidemiology essentially mirror those of the pathogen. Indeed, host and pathogen are at times genetically interlocked as in the gene-for-gene phenomenon (see Chapter 3) when a series of genes in the host are matched with corresponding genes in the pathogen. Most host plants are immune to most plant pathogens and typically individuals of any particular host show varying degrees of resistance (or susceptibility) to one pathogen. Further, while some host plants may be undamaged, or sustain less damage than the pathogen, by chemicals used for disease control others may be more sensitive to some chemicals than the pathogen.

As noted in Chapter 2 Theophrastus cited the belief that wild trees are not liable to disease but that cultivated kinds are and that some diseases are common to all or most trees while others are special to particular kinds. Ralph Austen (a devout 'Practiser in ye Art of Planting') in his *Treatise of fruit-trees*, 1657, noted that '*Crab-trees* ... are usually free from Canker' and recommended their use as stocks for apple grafts and the English horticulturist Thomas A. Knight in 1799 recorded the differential response of cereal varieties to rust. As a preliminary to attempts to produce new varieties of apple by breeding Knight experimented with the hybridisation of peas (and narrowly missed forestalling discoveries made by Gregor Mendel some 60 years later) and cereals. Of the last he wrote (Knight, 1799: 200–1):

The success of my endeavours to produce improved varieties of the pea, induced me to try some experiments on wheat; but these did not succeed to my expectations. I readily obtained as many varieties as I wished, by merely sowing the different kinds together; for the structure of the blossom of this plant (unlike that of the pea) freely admits the ingress of adventitious farina [pollen], and is thence very liable to sport in varieties. Some of those I obtained were excellent; others very bad; and none of them permanent. By separating the best varieties, a most abundant crop was produced; but its quality was not quite equal to the quantity, and all the discarded varieties again made their appearance. It appeared to me an extraordinary circumstance, that, in the years 1795 and 1796, when almost the whole crop in the island was blighted [rusted], the varieties thus obtained, and these only, escaped in this neighbourhood, though sown in several different soils and stations.

Varieties of cereals possessing valuable characteristics, together with other crop varieties and horticultural novelties, continued to be derived from selections made by farmers, growers or amateur specialists of outstanding individual plants, frequently after cross-breeding. By the early 1880s A. E. Blount, professor of agriculture at the Colorado Agricultural College, was producing new varieties of wheat by hybridisation and later in the decade William James Farrer (who was later appointed to a post in the Department of Agriculture of New South Wales) made similar experiments at his own expense in Canberra. In Canada this aspect of cereal improvement was put on an official footing when William Saunders became the first director of the Dominion Experimental Farms in 1888 and in the summer of 1904 his son Charles E. Saunders at the Central Experiment Farm at Ottawa, where he had been appointed cerealist for the Dominion, selected the single ear of wheat from which Marquis wheat originated. By 1918 about 20 million acres in the United States and Canada were planted with Marquis (resulting in a yield of 300 million bushels) and it would have been possible to travel 800 miles, from Nebraska to Saskatchewan, passing fields planted with this variety.[9] Marquis, one of the hard red spring wheats with good bread-making qualities, was not rust resistant but because of its earliness it was to a certain extent rust-escaping. A notable breakthrough in the breeding of rust-resistant wheat was made when Rowland H. Biffen, working at Cambridge where he subsequently became professor of agricultural botany and the first director of the Plant Breeding Institute, developed

Little Joss resistant to yellow stripe rust (*Puccinia striiformis*) by crossing the rust-susceptible Square Head's Master, a popular English wheat, with a rust-resistant Gurka wheat from Russia and demonstrated that host resistance was inherited as a Mendelian character (Biffen, 1904). Later he transferred resistance to the same rust shown by 'Club' wheat to the susceptible Michigan Bronze (and also derived susceptible 'Club' wheat from the same crossing) (Biffen, 1908). These results were treated with caution by both Biffen and others (Butler, 1905) because physiologic races of rusts had not yet been recognised and resistance sometimes varied between one locality and another.

Breeding disease-resistant cereals (which still continues with unabated vigour) was paralleled by similar work on many other crop plants after the spectacular success of the introduction, by Professor G. Foëx of the Montpellier School of Agriculture, of American phylloxera-resistant varieties of grape vine into Europe in the 1870s which saved the French vineyards from ruin. Much work on the development of resistant varieties was undertaken in the United States where the investigations by W. A. Orton (who was responsible for the inauguration within the USDA of this aspect of plant disease control as a practical measure) on cotton (Orton, 1900), cowpea (Orton, 1902), and other plants, H. L. Bolley in North Dakota on flax wilt (Bolley, 1901), and, somewhat later, L. R. Jones and his team at Madison, Wisconsin, on cabbage yellows (see Jones, 1915–23) were germinal. Today the development of resistant varieties against fungus, bacterial, and virus diseases is one of the most important aspects of plant pathological practice but because of the increasing technicalities of plant breeding and because resistance to disease is only one of the qualities plant breeders have to incorporate into crop plants most of the practitioners in this field are plant breeders by training although some plant pathologists qualify themselves as leading workers in the subject which now receives monographic treatment (e.g. Day, 1974; Nelson, 1975; Russell, 1978).

Resistance
'Resistance', as applied to the relationship of the host plant to the pathogen, has become the most overworked term of the plant pathologist's vocabulary. It carries so many meanings that its usefulness is frequently much reduced by lack of precision and grave misunderstandings can result. The Federation of British Plant Pathologists' *Guide to the use of terms in plant pathology*, 1973, recognises no less than 22 different ways in which resistance may be qualified to give precision (acquired resistance, active resistance, adult plant resistance, field resistance, etc.) and also notes a number of

synonyms for the recommended uses. One reason for this multiplicity is that the phenomenon of host resistance (which is the complement of pathogen virulence) is studied, among others, by geneticists, physiologists, and epidemiologists who describe resistance in terms appropriate to their own approach (e.g. oligogenic and polygenic for resistance controlled by a few or a series of genes; active resistance which entails a defensive reaction on the part of the host to the presence of the pathogen and passive resistance which does not).

The differentiation by Vanderplank (1963) from his epidemiological studies of 'vertical' and 'horizontal' resistance has led to much critical discussion, as for example, the series of papers by R. A. Robinson (1969–73) (and more recently a book, Robinson, 1976) and the commentary of the genetics of horizontal resistance by R. R. Nelson (1978). Vanderplank's own definitions of these terms were: 'When a variety is resistant to some races of a pathogen we shall call the resistance vertical or perpendicular. When the resistance is evenly spread against all races of the pathogen we shall call it horizontal or lateral' (Vanderplank, 1963:174). He illustrated the concept by a diagram (see Fig. 57) of the resistance of two

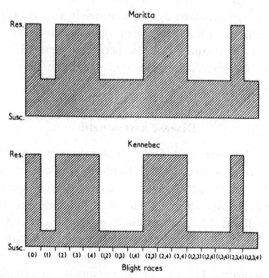

Fig. 14.1. Diagram of the resistance to blight of the foliage of two R_1 varieties: Kennebec and Maritta. The resistance is shown shaded to 16 races of blight. To races (0), (2), (3), (4), (2,3), (2,4), (3,4), and (2,3,4), the resistance of both varieties is vertical and complete. To races (1), (1,2), (1,3), (1,4), (1,2,3), (1,2,4), (1,3,4), and (1,2,3,4), resistance is horizontal, small in Kennebec and moderate in Maritta. Res. = resistant; susc. = susceptible.

Fig. 57. Vertical and horizontal resistance. (Vanderplank, 1963.)

potato cultivars to a range of races of the potato blight fungus. Vanderplank (1978: 17) stated that the definitions of horizontal and vertical resistance do not imply that horizontal and vertical resistance necessarily occur pure and that probably vertical resistance never occurs unaccompanied by horizontal resistance. This being so, Nelson (1978) took exception to Vanderplank's definition according to which neither of the two potato cultivars in Fig. 57 shows horizontal resistance because their resistance to the range of races of the pathogen is not evenly spaced. The concepts are both valid and useful and Nelson would redefine them in epidemiological terms; in terms of their effects on disease increase and disease onset. Horizontal resistance (which is typically polygenic) reduces the apparent infection rate. Vertical resistance (typically oligogenic) reduces the amount of the effective initial inoculum.

Much breeding for resistance has been, and still is, designed to incorporate vertical resistance which is usually only temporary as exposure of the pathogen population to the resistant host induces the selection of new races of the pathogen able to overcome the host's resistance. The use of multiline populations is one approach to avoid or at least delay this effect and in this connection it is interesting to recall that in 1767 Targioni-Tozzetti found it 'not so easy to render a reason, why Wheat growing seeded with Rye, or with Vetch, was not damaged by the Rust, while a Field of Wheat alone, standing between one of Rye, and one of Vetch, yielded scarcely any seed, and that the most miserable'. It is also becoming generally accepted that this breeding for vertical resistance has depleted the horizontal resistance of many crop cultivars.

Disease assessment

The first step in the assessment of the economic importance of plant disease is to determine the incidence and distribution of the diseases of any particular region. The first official survey of the occurrence of 'insect and fungus pests' on plants in England and Wales was that organised as a wartime measure by the Board of Agriculture and Fisheries in 1917 and the results of this exercise were published as a twopenny pamphlet in 1918. The survey became a permanent feature and a series of reports covering one to ten years subsequently appeared. This has been supplemented by a similar survey for Scotland covering 1924–57 (Foister, 1961). The establishment in 1950 of a Disease Measurement Section at the Ministry's Plant Pathology Laboratory with E. C. Large responsible for diseases and an entomologist for pests was a notable advance. Earlier, official plant disease surveys had been

PFLANZENSCHUTZ-NACHRICHTEN »Bayer« 20/1967, 1

Table 62: Specific annual quantity losses (in 1,000 tons)

	Actual production	Potential production	Losses due to			Total
			Insect pests	Diseases	Weeds	
Wheat	265,537	351,114	17,794	33,341	34,438	85,573
Oats	42,902	59,220	4,152	6,212	5,951	16,315
Barley	92,826	117,364	4,544	9,697	10,296	24,537
Rye	32,658	38,526	865	1,339	3,664	5,868
Rice	231,974	438,796	120,728	39,410	46,685	206,823
Millets and sorghums	76,700	122,985	11,636	12,655	21,993	46,284
Maize	218,461	339,446	44,020	32,656	44,308	120,984
Potatoes	270,784	399,972	23,758	88,895	16,535	129,188
Sugar beets	211,249	280,250	23,535	29,203	16,263	69,001
Sugar cane	483,386	1,050,220	204,873	203,087	158,874	566,834
Vegetables	201,691	279,910	23,364	31,137	23,718	78,219
Top fruits	66,567	88,001	6,147	12,825	2,462	21,434
Citrus fruits	24,395	30,937	2,505	2,851	1,186	6,542
Grapes	50,697	78,267	2,724	16,937	7,909	27,570
Coffee	3,167	5,731	741	964	860	2,565
Cocoa	1,528	2,821	368	588	337	1,293
Tea	1,125	1,653	130	252	146	528
Tobacco	4,274	6,216	645	791	506	1,942
Hops	93	119	9	10	7	26
Olives	5,180	8,137	1,493	651	813	2,957
Palm kernels	1,075	1,498	173	109	141	423
Soybeans	31,988	45,051	1,993	4,989	6,081	13,063
Groundnuts	16,688	28,037	4,920	3,104	3,325	11,349
Cottonseed	20,588	27,149	2,942	2,415	1,204	6,561
Linseed	3,398	4,219	121	339	361	821
Rapeseed	4,487	6,523	849	453	734	2,036
Sesame seed	1,656	2,231	278	74	223	575
Sunflower seed	6,346	8,242	825	247	824	1,896
Copra	3,325	5,902	862	1,128	587	2,577
Cotton	11,066	16,750	2,682	2,027	975	5,684
Other fibre crops	4,913	6,386	310	530	633	1,473
Natural rubber	2,260	3,013	151	452	150	753

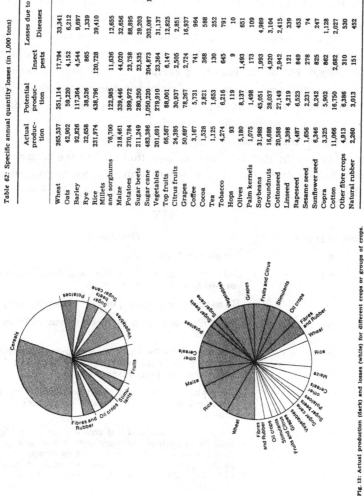

Fig. 13: Actual production (dark) and losses (white) for different crops or groups of crops

Fig. 58. Crop losses caused by pests, diseases, and weeds. (Cramer (1967): 484–5.)

instituted in the United States and it was in July 1917 that the Plant Disease Survey was given recognition as an independent enterprise within the USDA with the *Plant disease bulletin* (later the *Plant disease reporter*) as its official publication. In the following January, G. R. Lyman devoted an address to the Pittsburgh meeting of the American Phytopathological Society to a general policy statement for the survey (Lyman, 1918).

In such reports it is customary to make qualitative or comparative statements on disease incidence. For example, in the latest report for England and Wales (Baker, 1972) it is recorded that: 'In seven of the 12 years 1956–67, yellow rust was moderate or severe in winter and spring wheat trials'; 'Lettuce mosaic was in evidence annually though little was seen in 1958. In 1957 it was above average ...' – to quote at random. In order to quantify the progress of epidemiological investigations it is necessary to measure as precisely as possible the incidence and effect of disease – statistics which proved difficult to derive.

The oldest and most popular measure of the effect of disease is to estimate 'loss' of crop – the shortfall from the crop that might have been expected. This is frequently expressed in monetary terms which – as pointed out by Ordish (1952) – may be an underestimate because of failure to take into

KEY

Percentage (D.M.)	Description
0	Not seen on field
0·1	Only a few plants affected here and there; up to one or two spots in 12 yd. radius
1	Up to ten spots per plant, or general light spotting
5	About fifty spots per plant or up to one leaflet in ten attacked
25	Nearly every leaflet with lesions, plants still retaining normal form: field may smell of blight, but looks green although every plant affected
50′	Every plant affected and about one-half of leaf area destroyed by blight: field looks green flecked with brown
75	About three-quarters of leaf area destroyed by blight: field looks neither predominantly brown nor green. In some varieties the youngest leaves escape infection so that green is more conspicuous than in varieties like King Edward, which commonly shows severe shoot infection
95	Only a few leaves left green, but stems green
100	All leaves dead, stems dead or dying

Fig. 59. Key for the field assessment of potato blight. (Anon. 1947.)

account the cost of measures already taken to keep losses as low as they are or an overestimate in not allowing for the reduced cost of harvesting and marketing the crop and because a smaller crop may result in as high or higher net return than a larger disease-free crop due to scarcity escalating the market price.

Recently several careful assessments of the losses caused by plant disease have been published. The most comprehensive is that by H. H. Cramer (1967) who tackled the problem of losses caused by insect pests, diseases, and weeds on a world wide scale. He concluded that the total loss due to these three agents was approximately 35 per cent (insect pests, 13.8 per cent; diseases, 11.6 per cent; weeds, 9.5 per cent) and he summarised the results graphically both geographically and by crop (see Fig. 58). Cramer's estimate for the total loss caused by disease was very near that of 11.8 per cent made earlier by Watts Padwick (1956) for a variety of common diseases of a range of economically important crops in the then British Colonies on the basis of estimates made by local plant pathologists and published data.

One well-documented development in the field assessment of disease incidence can be traced back to the Plant Pathology Committee of the British Mycological Society which in January 1933 organised a symposium and discussion on measurement of disease intensity under the chairman-ship of Professor William Brown (Beaumont *et al.*, 1933). The topic was revived in 1941 when a subcommittee (the Disease Measurement Committee) of the Plant Pathology Committee was set up to organise a co-operative study. In the autumn of the same year the subcommittee considered the results of the first season's work and put forward a series of methods for trial in 1942 with cereal diseases, potato blight, virus diseases (or potato and sugar beet), and apple scab (Moore, 1943). The visual key for potato blight proved particularly useful and the slightly amended version offered in 1947 (Fig. 59) is still in use. So did the key for apple scab on leaves which Marsh (1947) supplemented with a standard diagram for assessing scab on the fruit, a diagram subsequently modified by Croxall *et al.* (1952) who in a direct comparison found the visual key and the standard diagram to give comparable results. Earlier Tehon & Stout (1930) in the United States had used standard diagrams for estimating apple scab on leaves (Fig. 60). Other methods were developed elsewhere. As early as 1923 McKinney devised a procedure for the estimation of helminthosporium infection of wheat seedlings which Horsfall & Heuberger (1942) adapted for assessing tomato defoliation by *Alternaria solani* by making a traverse across a field or plot and grading a known number of plants into four

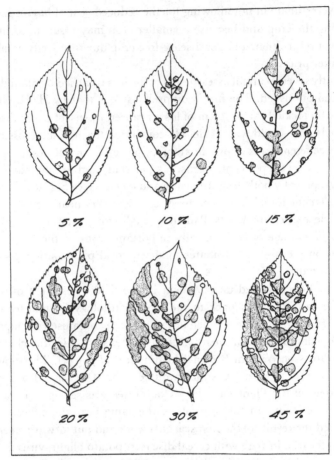

Fig. 2. Scale for estimating the intensity of the spot type of scab
attack on apple leaves

For each of the sample leaves in this scale, the total area of the spots
has been determined and expressed as a percentage of its respective leaf
area. By comparing infected leaves with these measured examples, the
average leaf area occupied by scab lesions may be computed. See the text
for details.

Fig. 60. Visual scale for estimating intensity of apple scab. (Tehon & Stout, 1930.)

recognisable categories of infection ratings when the sum of the infection
ratings divided by the number of plants times four multiplied by 100 gave
an 'index of infection'. In 1976 the UK Agricultural Development and
Advisory Service issued a loose-leaf manual of disease assessment diagrams
and FAO produced a handbook of crop loss assessment methods (1973).

Aerial photography has also been employed to study the incidence and development of plant disease, a topic reviewed by G. H. Brenchley (1968).

Epidemics

Large-scale outbreaks of plant disease have had a major influence both on the development of plant pathology and on human affairs. The epidemic of potato blight in Europe in the 1840s not only accelerated the general acceptance of the reality of the pathogenicity of fungi to plants but also as a result of the social catastrophe of the Irish famine gave *Phytophthora infestans* the distinction of having caused the fall of a British government (that of Sir Robert Peel after the repeal of the Corn Laws in 1876) and, incidentally, being responsible for the number of Irish policemen in New York. Thirty years later coffee rust in Sri Lanka in addition to giving Marshall Ward a major insight into plant disease, played its part in turning the English from coffee drinkers to tea drinkers and educating the Sri Lanka planters so that when at the turn of the century their cacao was threatened by disease they called in a plant pathologist on their own initiative.

Table 7 lists a number of diseases which at one time or another have reached epidemic proportions. Some, such as the first three, are endemic diseases which under favourable conditions have become epidemic. Bové (1970:156) cites the claim by Rudolph Kobert that 59 epidemics of ergotism occurred in Germany alone during the sixteenth to nineteenth centuries. There were also outbreaks in Russia in 1871–2, 1879–81, and 1883 and many more elsewhere. Fire blight attracted periodical attention as it spread westward during the nineteenth century from the Hudson River Highlands across the United States and by the early years of the twentieth reached California. According to McCubbin (1924) there has been a regular rise and fall in the intensity of peach yellows which has assumed epidemic proportions every 10 to 15 years. Witches' broom of cacao (first reported from Surinam in 1895 and largely responsible for the extinction of the cacao industries of Surinam and British Guiana (Guyana) in the 1920s) and South American leaf blight of *Hevea brasiliensis* (which has been one of the main factors preventing the development of an effective natural rubber industry in South America) are two tropical diseases which have played a major economic role. The most catastrophic epidemic caused by an endemic disease during recent years was that of brown spot of rice in Bengal in 1942 when during the growing season higher rainfall and higher average minimum temperature than normal, and increased cloudiness, favoured the disease so that for early cultivars at the Rice Research Stations at

Table 7. *Some representative plant disease epidemics*

10th–19th cent.	Ergot of rye (*Claviceps purpurea*)	Europe (France 994) as reflected by outbreaks of ergotism in man	Barger (1931) Fuller (1968) Bové (1970)
1780–	Fireblight of pear, etc. (*Erwinia amylovora*)	USA, 1780– UK, 1957–	Heald (1926): 307 Baker (1972): 125, 256
1791–	Peach yellows (Peach yellows 'virus')	Eastern USA	Smith (1888–94) Heald (1926): 251
1845–	Potato blight (*Phytophthora infestans*)	British Isles (esp. Ireland, 1845–50), Europe, N. and S. America, Africa, Asia, Australasia	Berkeley (1846) Woodham-Smith (1962) Cox & Large (1960)
1869–	Coffee rust (*Hemileia vastatrix*)	Sri Lanka, 1868–82 Brazil, 1969–	Ward (1882) Wellman (1970) Shieber (1972)
1872	Club root of crucifers (*Plasmodiophora brassicae*)	Russia	Woronin (1877)
1878–85	Vine mildew (*Plasmopara viticola*)	France	Heald (1926): 414
1880–1950	Panama disease of banana (*Fusarium oxysporum* f. sp. *cubense*)	C. & S. America	Wardlaw (1935–61) Stover (1962)
1890	Moko plantain disease (*Pseudomonas solanacearum*)	Trinidad	Rorer (1911) Wardlaw (1935–61)
1904–	Chestnut blight (*Endothia parasitica*)	Eastern USA	Anderson & Ranking (1914) Gravatt & Gill (1930)
1914–	South American leaf blight of Hevea rubber (*Microcyclus ulei*)	Surinam (1914), Guyana (1916), Brazil (1930–40), Costa Rica (1941), Colombia (1944), etc.	Stahel (1917) Hilton (1955) Holliday (1970)

1915–	Witches' broom of cacao (*Crinipellis perniciosus*)	Tropical S. America (Surinam, Guyana, etc.)	Baker & Holliday (1957)
1920–5	Sugarcane mosaic (Sugarcane mosaic virus)	USA (Louisiana)	Smith (1937): 429–30 Bawden (1943): 261
1920–70	Dutch elm disease (*Ceratocystis ulmi*)	Netherlands, 1921 USA (Ohio), 1931 UK, 1928, 1970	Clinton & McCormick (1936)
1924	Hop downy mildew (*Pseudoperonospora humuli*)	England C. Europe	Salmon & Ware (1925)
1933–5	Sigatoka (leafspot) of banana (*Mycosphaerella musicola*)	C. America (also Fiji, 1912; Australia, 1925)	Wardlaw (1935–61) Meredith (1970)
1936	Cacao swollen shoot (Cacao swollen shoot virus)	W. Africa, esp. Ghana	Padwick (1956): 3–6
1942	Rice brown spot *Cochliobolus miyabeanus* (anamorph *Helminthosporium oryzae*)	India (Bengal)	Padmanabhan (1973)
1949–59	Maize rust (*Puccinia polysora*)	W. Africa, 1949; then Africa generally	Cammack (1959)
1957–	Dothistroma blight of *Pinus radiata* (*Dothistroma pini*)	Tanzania, 1957 Chile, 1964–5 NZ, 1966	Gibson (1972)
1958	Tobacco blue mould (*Peronospora tabacina*)	England, 1958 then Netherlands, Germany, and Europe generally	Schmid *et al.* (1966) Baker (1972):201
1970–1	Southern corn (maize) leaf blight (*Helminthosporium turcicum*)	Southern USA	Ullstrup (1972)

Additional details about some of these epidemics are given by Klinkowski (1970).

Bankura and Chinsurah the yield per hectare was 7 to 59 per cent below that for 1941 and for medium to late cultivars at the same locations 40 to more than 90 per cent. Because of the scale of the epidemic and the exigencies of war the civil administration could not cope with the situation and it was estimated that two million people died of starvation in 1943.

Very frequently epidemics follow soon after the first introduction of a pathogen into a country or continent. The epidemic following the introduction of potato blight into Europe, and particularly Ireland, in 1845 is a notable example but there have been many others since including the introduction of vine mildew into France from North America in the 1870s, Dutch elm disease into North America from Europe in 1931 (and the return export of a virulent strain of the pathogen to England in 1970), maize rust into West Africa in 1949 (from where it spread rapidly to other parts of the continent), and the blue mould of tobacco into Europe some ten years later. The last resulted in riots by the tobacco growers of Eastern Europe when the middlemen did not pass on the benefits accruing from the increase in price of tobacco caused by the shortage of crop. Such disastrous introductions continue to occur – as for example that of coffee rust into Brazil in 1969 – and, in spite of legislation and quarantine procedures, are likely to occur in the future.

The southern corn (maize) blight epidemic of 1970–1 in the United States had a rather different origin. The increased popularity of hybrid seed corn led to the incorporation into cultivars of the 'Texas cytoplasm male sterility' (Tcms) factor which induces male sterility so that fertilisation can be effected by pollen from normal-cytoplasm plants. By 1970 approximately 85 per cent of the hybrid seed corn produced in the United States carried Tcms. It then became apparent that this type of cytoplasm also carried with it hypersusceptibility to a new race of the southern corn blight pathogen *Helminthosporium maydis* with the result that in 1970 when weather conditions favoured the disease the corn crops in the South were reduced by 50 to 100 per cent. This had both national and international repercussions. As far away as Britain farmers changed from maize to barley as a livestock feed. Less severe and more localised epidemics occurred in the following year but the outbreak was quickly brought under control by the use of normal-cytoplasm seed of which there was an adequate stock (Ullstrup, 1972). A similar but more localised outbreak of blight in oats had occurred earlier in Iowa during 1946–8 when an extensively planted crown rust (*Puccinia coronata*) resistant cultivar Victoria proved very susceptible to a helminthosporium originating from wild grasses (Scheffer & Nelson, 1967) and subsequently named *Helminthosporium victoriae* (Mehan & Murphy, 1946). Losses were so heavy that the cultivar had to be replaced.

Analysis of epidemics

There are two prominent landmarks in the recent history of our understanding of epidemiology and epidemics. The first is Ernst Gäumann's *Pflanzliche Infektionslehre* written during the Second World War in the isolation of Switzerland as an outcome of 20 years teaching of plant pathology and mycology by the author at the Swiss Federal Institute of Technology. Zurich. First published in 1946, *Pflanzliche Infektionslehre* became more generally accessible four years later in an English translation (edited by Gäumann's friend W. B. Brierley, professor of agricultural botany at the University of Reading) and proved an influential book. It is, as the author intended, 'an introduction to the principles of plant infection'; 'a contribution to the theory of plant pathology', and, in the words of Professor Brierley when reviewing the book on publication, 'it marks a decisive step towards the establishment of plant pathology as a science'.[10]

The six chapters into which the book is divided are of unequal length. More than two-thirds of the text is devoted to 'infection and infection chains' (two chapters) and the 'disease proneness of the host' (the longest chapter). 'Parasitic adaption of pathogens' and 'the disease' are dealt with more briefly and the book ends with half-a-dozen pages on 'the control of infectious diseases'. This comprehensive if somewhat idiosyncratic review will remain a classic.

Seventeen years later J. E. Vanderplank's *Plant diseases: epidemics and control*, 1963, was published. This monograph (an elaboration of the author's 'Analysis of epidemics' in Horsfall & Dimond (1959–60), **3**, Chap. 7) was also produced in relative isolation – in South Africa, far from the main centre of phytopathological endeavour in North America – by a successful potato breeder of unorthodox origins who never received any formal instruction in either plant pathology or plant breeding and who has written: 'It is my boast that in none of my books have any data of my own collecting appeared. My reliance on the experimental finding of others has been deliberately complete. But ninetenths of the discussions of these findings have been my own.'

Gäumann's critical review was basically a descriptive (qualitative) synthesis of the process of infection and the host–pathogen interaction which determines the disease. The dynamic aspects of disease, although appreciated, received less emphasis.

By assessing the incidence and severity of disease at intervals the progress of disease may be expressed as a curve which is typically sigmoid and often able to be reduced to a straight line by probit transformation as shown by E. C. Large (1945). Such curves allow comparisons to be made between one locality and another and between seasons (Fig. 61a) and also assessments of

(*a*)

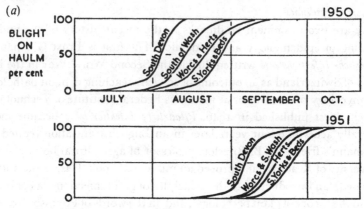

Typical blight progress curves for six regions in England for 1950 and 1951, demonstrating use of the curves for regional and seasonal comparison.

(*b*)

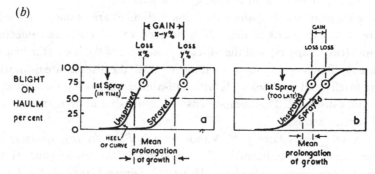

Typical pairs of progress curves for blight on sprayed and unsprayed haulm. (a) When the first spraying is given at or before the 0·1 per cent stage. (b) When it is given after that stage.

Fig. 61. *Progress curves for potato blight on haulm comparing:* (a) *six regions of England and Wales (see also Fig. 53) and two seasons;* (b) *sprayed and unsprayed crops. (Large (1952), figs. 3, 5.)*

the effects of spraying (Fig. 61b) as in the exploitation of this approach by Large (1952) and others in studies on potato blight in England from 1941. Such analyses yield much information about epidemics which is of great practical value.

   A new era in the approach to epidemiology began with the publication of Vanderplank's monograph which, in the opening words of the preface, 'describes new methods of epidemiological analysis based largely on infection rates and on the relation between the amount of inoculum and the amount of disease it produces', topics subsequently elaborated in three additional volumes (Vanderplank, 1968, 1975, 1978).

Vanderplank drew attention to two basic ways in which diseases increase with time. Diseases such as potato blight and cereal rusts where the pathogen multiplies by successive generations during the course of the epidemic he designated 'compound interest diseases' in contrast to 'simple interest diseases' such as fusarium wilt of cotton where during one season though the number of wilted plants may increase the inoculum in the soil may be considered to remain constant. He developed a series of mathematical equations relating the infection rate with the proportion of infected tissue (or number of infected plants) and time, introduced the concepts of 'vertical' and 'horizontal' resistance (see above), and drew attention to the integration of these parameters with eradicant chemicals and seed and soil disinfection which reduce the initial inoculum and protectant fungicides which reduce the infection rate. He also drew attention to the 'cryptic error' in small replicated plot experiments due to treated plots receiving too much and untreated plots too little inoculum when compared with large uniformly treated fields. In the opinion of P. H. Gregory this quantification of epidemiology gave 'for the first time a coherent and developed theory of plant epidemiology' and was 'a notable intellectual achievement'.[11]

Vanderplank's interventions have stimulated much investigation and comment both from those who have found his ideas acceptable and those from whom the response has been more critical and, although it is too early to assess the full effects, Vanderplank's contribution is certainly of permanent significance.

A recent major development in epidemiological investigation has been the use of computers for the construction of mathematical models by which epidemics may be simulated. P. E. Waggoner of the Connecticut Agricultural Experiment Station has been a leading exponent of this approach. He has devised models for early blight and tomato and potato (*Alternaria solani*) (Waggoner & Horsfall, 1969) and southern corn leaf blight (*Helminthosporium maydis*) (Waggoner, Horsfall & Lukens, 1972). Shrum (1975) has offered a similar model for stripe rust of wheat (*Puccinia striiformis*) while the topic has been reviewed by Kranz (1974) and in Horsfall & Cowling 1977–8, **2**). Computer simulation is clearly going to find a place in the forecasting and control of epidemic outbreaks of plant disease but again the current position is too fluid for any reliable assessment to be made of its final role in the practice of plant pathology.

10

Legislation and quarantine

The first legal restrictions to hinder the spread of disease were enacted against human disease. According to Garrison's *History of medicine* (Edn 4:188, 1929), it was the notorious outbreak of bubonic plague, known as the Black Death, which swept through Europe during the fourteenth century, that led the Venetian Republic to appoint three guardians of public health (1348), to exclude infected and suspected ships (1374), and to make the first quarantine of infected areas (1403). The last procedure, so called because travellers from the Levant and Egypt, where plague was endemic, were isolated in a detention hospital for 40 days (*quaranta giorni*), was first practised by Ragusa in 1377 with a detention period of one month. In 1799 an Act was passed in the United States requiring Federal officials to aid States and ports with the enforcement of local health regulations and Great Britain issued quarantine orders in 1825. Other countries issued similar legislation and, as a climax, a convention held in Paris in 1850 drew up an international quarantine code.

In passing it may be recalled that, as so frequently happens when a term proposed for use in one discipline is taken over by another, plant pathologists have broadened the original concept of 'quarantine' and currently use the term in many senses. In some countries, as Frances Sheffield (1968) noted, 'it is almost synonymous with "plant protection" and includes port inspection, field control, and even the legislation necessary to enforce these'. This is a pity for quarantine is still needed in its original and restricted sense for growing plants in isolation either at specially designed quarantine stations or at other sites under strict supervision.

National legislation

Legal restrictions against plant diseases were slow to develop and at first sporadic. It was early noted that outbreaks of black rust of wheat were associated with the presence of barberry bushes and the claim (which Klebahn (1904):205 could not substantiate) has frequently been made that in 1660 a law was passed in France at Rouen requiring the eradication of barberry. Such enactments were passed during the eighteenth century in the British colonial settlements in North America – first in Connecticut in 1726 and subsequently in Massachusetts, 1755,[1] Rhode Island, 1766, and other states. This legislation preceded the experimental proof by de Bary (1865–6) of the heteroecism of *Puccinia graminis*. Subsequently there were many, often very large scale, attempts to eradicate barberry both in North America and Europe, usually backed by legislation but not in England where barberry eradication has always been voluntary. One very successful campaign was in Denmark where, following a series of severe epidemics of rust, Professor Emil Rostrup was finally successful in stimulating the introduction in March 1903 of a law forbidding the growing of barberry in Denmark outside botanic gardens. A decade later, with most of the barberry exterminated, serious outbreaks of rust had ceased (Lind, 1915). Additional details of barberry eradication campaigns are given by Fulling (1943) in his comprehensive and well-documented review, where he also calls attention to the litigation, especially in Virginia, associated with the eradication of red cedars (the alternate host of *Gymnosporangium juniperi-virginianae*) in the neighbourhood of apple orchards in spite of compensation to the owners.

Two other successful eradication campaigns were those mounted against citrus canker (*Xanthomonas citri*) in the Gulf Coast states of the USA, where about 3 million trees were eradicated in Florida alone (Walker, 1950, Edn 2:737), and against coffee rust (*Hemileia vastatrix*) after its introduction into Papua and New Guinea (Shaw, 1968). Legislation to prevent the introduction of South American leaf blight of rubber (*Microcyclus ulei*) into South-East Asia has been evolved and in Malaysia elaborate preparations made by the Rubber Research Institute to defoliate the affected and neighbouring trees by aerial spraying should an introduction occur in that country (see Hilton, 1955; Holliday, 1970).

The first British legislation against a disease in animals or plants was an Act of 1866 granting emergency powers for the destruction of all cattle affected by rinderpest (cattle plague) which had been introduced into Britain by imported Russian cattle the previous year. The action was successful but to prevent a recurrence the Contagious Diseases (Animals) Act of 1869 was designed to regulate the import of cattle. The next legislative measure was the Destructive Insects Act, 1877 ('An Act for preventing the introduction and spreading of Insects destructive to Crops', 40 and 41 Vict. Ch. 68, 14 August 1877) designed to prevent the importation of the

[7 Edw. 7.]   *Destructive Insects and Pests Act, 1907.*   [Ch. 4.]

CHAPTER 4.

An Act to extend the Destructive Insects Act, 1877, to all Pests destructive to Crops, Trees, or Bushes. [4th July 1907.]

A.D. 1907.

BE it enacted by the King's most Excellent Majesty, by and with the advice and consent of the Lords Spiritual and Temporal, and Commons, in this present Parliament assembled, and by the authority of the same, as follows:—

Extension of 40 & 41 Vict. c. 68 to all pests.

1.—(1) The Board of Agriculture and Fisheries may, for the purpose of preventing the introduction into Great Britain of any insect, fungus, or other pest destructive to agricultural or horticultural crops or to trees or bushes, and for preventing the spreading in Great Britain of any such insect, fungus, or other pest, exercise all such powers as may be exercised by the Board in relation to the Colorado beetle under the Destructive Insects Act, 1877; and that Act shall apply accordingly as if in that Act the expression "insect" included all such insects, fungi, and other pests, and the expression "crop" included all such crops, trees, and bushes:

Provided that the Board shall not make an order directing the payment of compensation by any local authority for the removal or destruction of any crop or any trees or bushes unless the local authority consent to make the payment.

56 & 57 Vict. c. 66.

(2) Section one of the Rules Publication Act, 1893, shall not apply to any order made under the Destructive Insects Act, 1877, or this Act.

(3) This Act shall apply to Ireland as if Ireland were named therein instead of Great Britain and with the substitution of the Department of Agriculture and Technical Instruction for Ireland for the Board of Agriculture and Fisheries.

[*Price ½d.*]

1

[Ch. 4.]   *Destructive Insects and Pests Act, 1907.*   [7 Edw. 7.]

A.D. 1907.
Short title.

2. This Act may be cited as the Destructive Insects and Pests Act, 1907; and the Destructive Insects Act, 1877, and this Act may be cited together as the Destructive Insects and Pests Acts, 1877 and 1907.

Printed by EYRE and SPOTTISWOODE,
FOR
ROWLAND BAILEY, Esq., I.S.O., the King's Printer of Acts of Parliament.

And to be purchased, either directly or through any Bookseller, from
WYMAN AND SONS, LTD., FETTER LANE, E.C.; or
OLIVER AND BOYD, EDINBURGH; or
E. PONSONBY, 116, GRAFTON STREET, DUBLIN.

Fig. 62. Destructive Insects and Pests Act, 1907.

Fig. 63. Sign indicating recognition under the Agricultural Chemicals Approval Scheme.

much feared Colorado beetle (*Leptinotarsa decemlineata*), a devastating pest of potato in North America which was then causing alarm after its introduction to the continent of Europe.

In 1907 this Act was enlarged by the Destructive Insect and Pests Act ('An Act to extend the Destructive Insects Act, 1877, to all Pests destructive to Crops, Trees, or Bushes', 7 Edw. VII, Ch. 4, 4 July 1907) (Fig. 62) which was modified by a supplementary Act in 1927. Subsequently these three Acts were repealed after consolidation into the Plant Health Act of 1967.

The 1907 and 1927 Acts covered all insect pests and also diseases – in the words of the 1927 Act: 'For the purposes of the principal Act, the expression "insect" shall include bacteria and other vegetable or animal organisms, and any other agent causative of a transmissible crop disease; the expression "crop" shall include seed, plant, or any part thereof.'

At first the Privy Council, and subsequently the Board (then Ministry) of Agriculture, was empowered to implement these Acts by the issue of Orders and under the 1907 Act a small staff of inspectors was appointed to administer practical aspects of the law. The first diseases to be subject to Orders, in 1908, were black knot of plum (*Plowrightia morbosa*), white root rot (*Rosellinia necatrix*), wart disease of potato (*Synchytrium endobioticum*), and American gooseberry mildew (*Sphaerotheca mors-uvae*).[2] Wart disease, first recorded in the UK in 1896 and still a notifiable disease, attracted much attention during the 1920s but it now causes negligible losses largely due to the legal requirement to plant only resistant cultivars on infested commercial land.[3] Gooseberry mildew, known in the United States for 50 years before its introduction into Europe (Russia, 1890), was first recorded for the British Isles from Northern Ireland in 1900 and in 1906 it was found in England (at Evesham, Hereford and Worcester) where it excited interest and like wart disease was designated as a notifiable disease. Though now of common and widespread occurrence in the country it has proved less harmful than was originally feared. It is no longer notifiable although sale of diseased bushes is subject to the Sale of Diseased Plant Orders 1927–43. Subsequently many other Orders were issued including those against Silver-leaf Disease of Plum, 1923 (amending the first Order of 1919), which required fruit-growers to destroy before 15 July each year all dead woody tissue liable to harbour *Stereum purpureum*, and the Fire Blight Order of 1958 (amended in 1960 and again in 1966) following the appearance of fireblight in pear orchards in Kent in the summer of 1957 (see Baker (1972):125). The latter was the first European record of this most serious bacterial disease caused by *Erwinia amylovora* which has since become firmly established in

England where the propagation of the susceptible pear cultivar Laxton's Superb is now prohibited.

In North America (see McCubbin, 1954) the first legislative measures against plant disease were promulgated by States. Such measures included a series of laws against peach yellows. The first of these – the Michigan Yellows Law of 1875 ('An Act to prevent the spread of the contagious disease of the peach tree known as the yellows in the counties of Allegan, Van Buren, and Ottawa, and to provide measures for the eradication of the same'), which was amended in 1879 and 1881 – was followed by similar enactments by the states of California (1885), and New York (1887) and also the Canadian province of Ontario (1881, 1884).[4] As in England, the first Federal legislation – the Livestock Quarantine Act of 1905 – was against animal disease and it was not until 20 August 1912 that the Plant Quarantine Act was passed to assist the Department of Agriculture in controlling plant diseases and parasites coming into the United States as well as those already present. This Act was precipitated by the concern shown regarding the recent introduction into North America from Europe of two major diseases: wart disease of potato (*Synchytrium endobioticum*) reported from Newfoundland, in 1909 (Güssow, 1910) (which led to the passing of the Canadian Destructive Insect and Pest Act of 1910) and white pine blister rust (*Cronartium ribicola*), a serious disease of *Pinus strobus* and other five-needled pines introduced from 1905 or earlier into New York and other states and also Ontario, Canada, with German and French nursery stock. Both these diseases were described by Spaulding & Field (1912) in a Farmers' Bulletin and in the covering 'Letter of transmittal to the Secretary of Agriculture' B. T. Galloway emphasises that the 'National control of importations of plant material is necessary for adequate and prompt action in such emergencies as the present one, which are certain to occur more frequently'. The white pine blister rust was too well established for legislation to prevent further introductions to have much effect and although the disease is currently held in check this state of affairs has only been attained and maintained by large-scale eradication campaigns against *Ribes*, the alternate host, over millions of acres in the neighbourhood of five-needled pines, a topic reviewed and documented by Fulling (1943).

A further development was the United States Post Office Order of 1 July 1913 which rendered nursery stock unmailable in the International Parcel Post, unless addressed to the United States Department of Agriculture. (Nursery stock was defined to include seeds of fruit and ornamental trees or shrubs but not field, vegetable, or flower seed and later Orders had to be made against wart disease of potato tubers and other diseases.)

Other more general legislation sometimes affects plant pathology. For example, Shear (1918) drew attention to the phytopathological aspects of the Food Products Inspection Law of 10 August 1917 under which a Food Inspection Service of the Bureau of Markets of the USDA was established to provide a food products inspection service in 26 of the larger cities of the United States. At that time frequently half or more of a perishable crop was lost by careless handling. In 1920 the New York Board of Health condemned and destroyed 12 million pounds of fruit and 7 million pounds of vegetables.

International co-operation

The establishment of effective international co-operation on plant disease legislation was slow to develop. As often happened with national legislation in this field, it was a pest that provided the initial stimulus. Following its importation into Europe from America about 1859, the phylloxera gall louse (*Phylloxera vastatrix*) caused major losses in the vineyards of France and neighbouring countries and aroused much consternation and activity including the possibility of concerted international action to prevent further introductions from America and spread of the pest within Europe. This led to an international conference at Berne where on 3 November 1881 the Phylloxera Convention was signed by five countries – and in the same year the state of California introduced its first legislation against phylloxera.[5] Later the number of adhering countries increased to 12. The Convention was modified after a second conference at Berne in 1889 and remained in force until supplanted by the FAO International Plant Protection Convention of 1951 (see below).

Emil Rostrup at an international congress at The Hague in 1891 drew attention to the need to prevent the introduction of new diseases by means of infected living plants and seed. Subsequently others urged similar action but it was not until February 1914, in Rome, that the International Conference of Phytopathology, convened by the International Institute of Agriculture and attended by delegates of some 30 States and Colonies, drew up the statutes of the 'International Phytopathological Convention of Rome' for the control of inter-state circulation of horticultural plants and offered a model health certificate to accompany such exports (see Butler, 1917).

Article 2 of that Convention required each State adhering to the Convention to create a government service of phytopathology which 'will comprise as a minimum: (1) the creation of one or more establishments for study and for scientific and technical research; (2) the organisation of effective supervision of the growing plants; (3) the inspection of consignments; (4) the issue of phytopathological certificates'.

Within six months the First World War broke out. No country ratified the Convention. After the war the subject was again frequently raised at international conferences as for example at the International Phytopathological Congress held at Wageningen in the Netherlands in 1923 and the International Congress of Plant Science (4th International Botanical Congress) at Ithaca, NY, three years later, at both of which H. T. Güssow (Fig. 73) of the Canada Department of Agriculture was a principal speaker (see Güssow, 1924, 1929). He also devoted his presidential address to the American Phytopathological Society to the same topic (Güssow, 1936). This agitation resulted in a second attempt to secure international agreement – at the International Plant Protection Conference of 1929, again in Rome. The former international Institute of Agriculture submitted a draft of the 1914 Convention embodying the suggestions subsequently received from various governments for its modification when after a long discussion a new 'International Convention for the Protection of Plants' was signed by 26 of the 46 participating countries. However, only 12 countries ratified the Convention which greatly restricted its effectiveness.

The Second World War created a climate more favourable for international collaboration and a series of events during the closing years of the war and the immediate post-war period led to the achievement of the first plant protection convention to receive world-wide recognition. This general movement has been accompanied by a vast bureaucratic edifice composed of many elements commonly referred to by well known, or obscure, acronyms or abbreviations so creating a confusion which has not been lessened by some abbreviations varying with the language employed.

The first crucial step was the signing of the United Nations charter in June 1945 in San Francisco after some two years of preparative work and the United Nations (UN) – which now has permanent headquarters in New York – formally came into existence on 24 October 1945. Currently (1977) there are 144 member states. The interests of the United Nations are comprehensive and the different aspects of its work are undertaken by a series of 'Agencies' of which the three most important in relation to phytopathology are the Food and Agriculture Organisation of the United Nations (FAO) established in 1945 and based in Rome, the World Health Organisation (WHO) established in 1948 and based in Geneva, Switzerland, and the United Nations Educational, Scientific and Cultural Organisation (UNESCO) established in 1946 and based in Paris. Other relevant agencies are 'The World Bank' (International Bank for Reconstruction and

Development) from which direct subventions are made for phytopathological and plant protection projects and the World Meteorological Organisation (WMO) which has interests in aspects of the epidemiology of diseases and pests of plants.

The Fifth (1949) session of the governing Conference of FAO recommended that FAO should call a meeting to consider a new plant protection convention and as a result a meeting attended by representatives of 35 countries was held at The Hague in the spring of 1950 when a draft 'International Plant Protection Convention (IPPC)' submitted by FAO was accepted in principle. After modification in the light of recommendations of The Hague conference and a panel of specialists, and comments from member governments, a final draft was approved by the Sixth session of the FAO Conference in the late autumn of 1951. Member Nations were recommended to sign the Convention not later than 1 May 1952; by when 37 nations had signed (Ling, 1953). At the 25th anniversary of the Convention in 1976 (Brader *et al.*, 1977) the number of adhering nations had increased to 75 and the list of countries party to the agreement continues to grow.

The evolution of the IPPC from the 1914 International Phytopathological Convention of Rome is apparent from its constitution. Again, a contracting Government is required to make provision, to the best of its ability, for:

(a) an official plant protection organisation with the following main functions: (i) the inspection of growing plants ... and plant products ... particularly with the object of reporting the existence, outbreak and spread of plant pests and controlling these pests; (ii) the inspection of consignments of plants and plant products moving in international traffic ... (iii) the disinfestation or disinfection of consignments of plants and plant products ...; (iv) the issue of certificates relating to phytosanitary condition ... a model for which was appended;
(b) the distribution of information within the country regarding the pests of plants and plant products and the means of their prevention and cure;
(c) research and investigation in the field of plant protection. (Article 4)

The Convention was, however, of wider scope and covered agriculture, horticulture, and forestry while 'pests' included not only insects and other animal pests such as rodents but also pathogenic micro-organisms and weeds.

The Convention offered guidance on the measures contracting parties may take under their plant protection legislation regulating the entry of plants and plant products into their territories so as to prevent the introduction of pests and at the same time stressing that no such measures may be taken unless they are made necessary by phytosanitary considerations (Article IV).

Table 8. *Regional plant protection organisations recognised under the FAO Plant Protection Convention (IPPC) of 1951*

Europe and the Mediterranean
European & Mediterranean Plant Protection Organisation (EPPO)
  Founded: 1951 (18 April; 15 signatories)
  Membership: 35 countries (all European countries (except Albania, and Iceland)
    together with the USSR, Turkey, Israel, Iran, Tunisia, Algeria, and Morocco)
  Headquarters: Paris, France
Africa (south of the Sahara)
Inter-African Phytosanitary Council (IAPSC)
  Founded: 1954 as a result of a recommendation of the 4th Commonwealth Plant
    Pathology Conference, 1948, as IAP Commission for Africa South of the
    Sahara; 1960 integrated into the Commission for Technical Co-operation in
    Africa South of the Sahara (CCTA) which in 1965 became the Scientific,
    Technical and Research Commission (STRC) of the Organisation for African
    Unity (OAU); 1967 IAP Commission became IAP Council
  Membership: 41 countries
  Headquarters: Yaoundé, Republic of the Cameroons (since 1967; before that
    London, UK)
The Standing Committee for Plant Protection of the Southern African Regional
Commission for the Conservation and Utilisation of the Soil (SARCCUS)
  Not yet recognised by FAO
Near East
FAO Near-East Plant Protection Commission (NEPPC)
  Founded: 1963
  Membership: 16 countries
  Headquarters: Cairo, Arab Republic of Egypt
South-East Asia, Australasia

Contracting parties are further required to co-operate with FAO in establishing a world reporting service on plant pests (Article VII) and to co-operate with one another in establishing regional plant protection organisations (Article VIII).

For the implementation of the IPPC there was already one regional organisation – the European and Mediterranean Plant Protection Organisation (EPPO). This was recognised by the IPPC as its first regional organisation since when others covering the greater part of the world have been established (see Table 8).

During the aftermath of the Second World War the need to conserve food supplies, and in particular the prevention of losses during storage and the depredations of the Colorado beetle on potato crops in the field, led to international co-operation which from a beginning in 1947 resulted in the establishment of the EPPO in 1951 by agreement between 15 European countries, a year before the IPPC became effective. Subsequently the

Table 8 *cont.*

FAO Plant Protection Committee for the South-East Asia and Pacific Region (PPC/SEAP)
  Founded: 1956
  Membership: 21 countries (including India, Australia, New Zealand but not (in 1978) Japan)
  Headquarters: Bangkok, Thailand (Firman, 1978)
Central America and the Caribbean
Organismo Internacional Regional de Sanidad Agropecuaria (ORISA)
  Founded: 1955
  Membership: 7 countries (Belize has associate status)
  Headquarters: San Salvador, El Salvador
FAO Caribbean Plant Protection Commission (CPPC)
  Founded: 1967
  Membership: 14 countries
  Headquarters: Port of Spain, Trinidad
South America
Organismo Bolivariano de Sanidad Agropecuaria (OBSA)
  Founded: 1965
  Membership: 3 countries
  Headquarters: Bogota, Colombia
Comité Interamericano de Proteccion Agricola (CIPA)
  Founded: 1965
  Membership: 6 countries
  Headquarters: Buenos Aires, Argentina
North America
North American Plant Protection Agreement (NAPPA)
  Founded: 1976 (30 October)
  Membership: Canada, Mexico, USA

Data from Mathys *et al.* (1976) and Brader *et al.* (1977).

Organisation was enlarged to include the Soviet Union and other countries bordering the Mediterranean so that now 35 countries are involved. Although the Paris secretariat is still small the interests of EPPO are multifarious (see Mathys *et al.*, 1976) and its many activities are channelled through a series of committees, working parties, and panels of experts by which topics are explored and assessed and contact maintained with a multiplicity of international (and mainly non-governmental) organisations (including the International Society for Plant Pathology (ISPP)) interested in some aspect of plant protection. Phytosanitary regulations naturally have a high priority as have disease surveys, the prompt reporting of newly introduced pests and pathogens, and other topics relevant to such regulations. Member countries and plant pathologists at large are kept updated by the issue of the *EPPO Bulletin* and the latest innovation is the issue of two series of illustrated *Data sheets on quarantine organisms*, one series being devoted to potentially dangerous pests and diseases not yet introduced into the area covered by EPPO, the other to pests and diseases introduced into some EPPO countries, all organisms however being subject to zero tolerance for all countries.

The situation in Europe is complicated by activities within the European Economic Community (EEC, 'The Common Market'), a component of the European Community (based in Brussels) which since 1973 has comprised nine member states. Agriculture being one of its main interests, the EEC, in attempts to attain uniformity of plant disease legislation within its member countries in order to facilitate trade, has set up committees and working parties which at times duplicate activities of the EPPO.

Quarantine stations

Not all the major diseases of major crops are co-extensive with the geographical distributions of the crops. For example, witches' broom of cacao (*Crinipellis perniciosus*), a serious disease in South America, does not occur in West Africa. South American leaf disease (*Microcyclus ulei*), the most important disease of rubber in the Western Hemisphere, is absent from Asia, and East Africa is free from tobacco blue mould (*Peronospora tabacina*), while, if many serious virus diseases have become generally dispersed, others are still of more limited distribution. The need to introduce new crops, to improve yields by importing better cultivars of traditional crops, and the demands of plant breeders frequently necessitate the movement of seed or other propagating material from a country where a serious disease occurs into a disease-free region when very stringent precautions have to be designed by the importing country. Typically the introduced material is grown in isolation in insect-proof glasshouses under strict supervision and tested for freedom from disease before being released for general use. This procedure is sometimes undertaken at special post entry quarantine stations (a specification for which has been detailed by Frances Sheffield, 1958) but because of the cost more usually in a unit attached to a research station or other institution. For example, at the Imperial College of Tropical Agriculture in Trinidad a quarantine glasshouse was constructed in 1952 to serve the British Caribbean and financed by the ten governments concerned (Wilson, 1956).

One of the first quarantine stations was that for sugarcane in Hawaii (Pemberton, 1955) where a quarantine system was introduced in 1923 and which since 1929 has been sited on the island of Molokai where no commercial cane is grown.

A very successful more general quarantine station has been the East

African Plant Quarantine Station. The first station was built at Amani, Tanganyika in 1931 and operated by the horticultural staff of the East African Agricultural Research Station. In 1954 a new purpose built station was opened on a 5-acre site in Muguga forest, surrounded by plantations of black wattle and eucalyptus, to serve the East African territories covered by the East African Agriculture and Forestry Research Organisation (EAFRO) and ten years later the station was enlarged (Sheffield, 1935). Although situated almost on the Equator the station is at an altitude of 6800 feet so that the problems normally associated with glasshouses in the tropics are greatly reduced. There are facilities for the hot water treatment of propagating material and for testing for viruses. Many thousands of consignments of sugarcane, rice, and cacao, forest trees, and ornamental plants passed through the station. Unfortunately, this station is now by-passed with increasing frequency for political motives.

An alternative, or supplementary, precaution in the movement of planting material is 'intermediate quarantine' as has been practised at the Royal Botanic Gardens, Kew, since 1927 where two special glasshouses are now devoted to growing tropical plants in isolation under the eye of a plant pathologist. Banana, rubber, sugarcane, cotton, and especially cacao have been among the many plants quarantined at Kew in transit between West Africa and the West Indies (and vice versa) and many other destinations. Whenever a plant can be vegetatively propagated it is daughter plants that are sent on to the importing country.

Certification and approval schemes

In addition to mandatory legislation designed to prevent or minimise the spread of plant disease within or between countries there is a large body of regulations, which may be mandatory or voluntary, to ensure the purity and freedom from disease of seed or other propagating material offered for sale and to control the quality and effectiveness of chemical products used for plant protection.

Seed

The testing of seed for purity began sporadically in various countries some hundred years ago; first in Denmark and Germany in 1869. As a result of informal discussions at the International Botanical Congress at Vienna in 1905 on the desirability of establishing a regional seed testing association, regional seed testing congresses were held at Hamburg in 1906 and in Munster and Wageningen in 1910. War intervened and it was on the initiative of Sir Laurence Weaver, an administrative officer of the Ministry of Agriculture and Fisheries in London, and K. Dorph-Petersen, Director of the State Seed Testing Station, Copenhagen, that a more widely based congress was held in Copenhagen in 1921 at which 16 countries were officially represented. Three years later, on 10 July 1924, a constitution for the International Seed Testing Association (ISTA) was accepted by delegates from 28 countries at a congress at Cambridge.

'The primary purpose of the Association', in the words of the current (1971) constitution, 'is to develop, adopt and publish standard procedures for sampling and testing seeds, and to promote uniform applications of these procedures for evaluation of seeds moving in international trade.' Members and laboratory members are those designated by their respective governments to participate on behalf of their governments in the activities of the Association. The Association normally organises a triennial congress and the Association's progress during its first fifty years has been conveniently summarised by Thompson (1974).

The health of seeds was a topic to which ISTA turned its attention by establishing a Plant Disease Committee under the chairmanship of Dr Lucie Doyer of Wageningen where her appointment in 1919 as a full-time seed pathologist was the first such appointment in the world. On her death in 1949 she was succeeded as chairman of the Committee by Willard F. Crosier, professor of seed pathology of Cornell University. Four years later this office was taken over by the Canadian A. J. Skolko who was followed from 1953 by Paul Neergaard who served for the next 18 years. One most useful and educational activity of the Committee was the organisation of a series of workshops for seed pathologists throughout the world (Noble, 1979). The first of these workshops was at the Ministry of Agriculture's Seed Testing Station, Cambridge in 1958 and they were for a time held almost annually.

Background support for this world-wide interest in the examination of seed for disease has been provided by the listing of seed-borne pathogens – first by Orton (1931) in West Virginia and later, in conjunction with ISTA, by Mary Noble *et al.* (1958) – and by the writing of manuals to aid the identification of seed-borne fungi (Doyer, 1938; Malone & Muskett, 1964). The whole topic has recently been authoritatively reviewed in a massive 2-volume monograph by Paul Neergaard (1977), director of the Danish Government Institute of Seed Pathology for Developing Countries in Copenhagen, in the light of his 40 years of first-hand experience.

In England official seed testing may be considered to have begun with the

Seeds Act 1920, designed 'to amend the Law with respect to the Sale and Use of Seeds for sowing and of Seed Potatoes and to provide for the testing thereof', which replaced two wartime Testing of Seeds Orders of 1917 and 1918. The Act provided for a declaration to be made by the seller as to the variety, purity, and percentage germination of all seed offered for sale for sowing and for potatoes declaration of the class of seed, variety, size, and dressing. To provide for the testing of seed official Seed Testing Stations were established at the National Institute of Agricultural Botany, Cambridge (for England and Wales), at Corstorphine, Edinburgh (for Scotland), and Queen's University, Belfast (for Northern Ireland) together with other centres licensed under the Act.

Propagating material

The vegetative propagating material which has been most subject to certification and official regulation is seed potato tubers, at first for trueness to type (cultivar) and since about 1930 for freedom from virus diseases. One starting point for such certification was the Canadian Seed Potato Certification Service initiated by H. T. Güssow in 1914, to which Paul A. Murphy contributed during 1915–20 in Prince Edward Island and Nova Scotia. The basis of the scheme was two field inspections during the growing season for trueness to type (cultivar) and incidence of disease and later these were augmented by an inspection of the tubers; approved tubers being designated by special labels (Fig. 64). The scheme was a great success. At first a tolerance of 12 per cent for disease and up to 5 per cent for mixed varieties was allowed but by 1927 the standard had been raised to a maximum of 3 per cent for disease at the second field inspection and no mixed varieties at tuber inspection. In 1920, 3956 acres passed inspection; in 1930, 34000 acres. The average yield of potatoes in Canada for 1920–9 was approximately 133 bushels per acre, that for certified seed for the same period 275 bushels (Tucker, 1927).

Similar, and more stringent, procedures were widely adopted. In the United Kingdom, for example, where a number of grades are recognised, the highest is reserved for the development of nuclear stocks from virus-tested stem-cutting stock (VTSC) when the raiser himself applies pre-scribed serological tests to detect potato viruses S and X and grows the crop in an approved geographical region on land free from potato cyst eelworm. The crop is also subject, among other requirements, to at least two inspections and leaf testing by the Ministry and the foliage has to be destroyed by an approved date. The next grade is that of Foundation Seed (FS) –

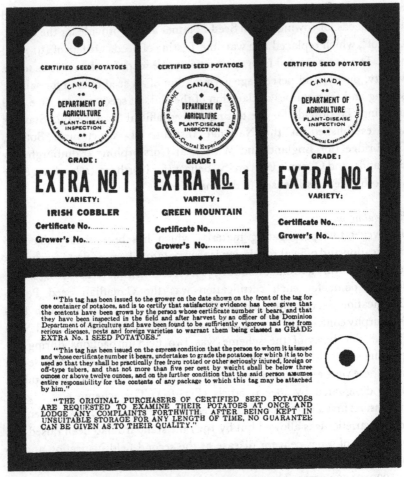

OFFICIAL CERTIFIED SEED POTATO TAGS

Yellow-coloured tags for Irish Cobbler variety; green-coloured tags for Green Mountain variety; buff manilla tags for all other varieties. Each tag bears on its reverse side the lettering shown above.

Fig. 64. Labels used for certified seed potatoes. (Tucker (1927): 5.)

normally grown from VTSC tubers one to four years after certification – which is followed by Special Stock (SS), Approved Stock (AS), and 'AA' and 'CC' Certificates.

In Great Britain the official inspection of seed potatoes was first introduced in 1918 when the Boards of Agriculture for England & Wales and for Scotland arranged for contractors in Scotland to grow certain wart-disease immune cultivars. Non-immune cultivars have also been certified in Scotland since 1922 and England and Wales since 1924.

Voluntary schemes for the certification of strawberry runners, black currant bushes, and raspberry canes were inaugurated in England and Wales in 1927 and in 1943 a scheme for the certification of hop gardens for freedom from verticillium wilt was launched. Currently there is also provision for the certification of rootstocks and trees of apple, pear, plum, and cherry, quince rootstocks, and certain ornamental plants including carnations. Details of the origins and development of all these schemes are given in the recent comprehensive historical survey by Ebbels (1979).

Plant protection products

As early as 1756 a law was passed in France banning the use of arsenic and copper compounds as seed treatments for fear of poisoning man but the general regulation of this field is a twentieth-century development.

In the United States the Insecticide and Fungicide Act of 26 April 1910, a development of the Food and Drugs Act of 1902, was intended to suppress interstate commerce in adulterated and misbranded insecticides and fungicides.

In Canada a registration scheme for fungicides was introduced by the Department of Agriculture in 1939 (with I. B. Conners responsible for registrations) and four years later a scheme for the official approval of proprietary brands of agricultural chemicals for the control of plant pests and diseases and weeds was begun in the United Kingdom (Anon., 1943). Under the latter scheme (for which a chemist was appointed to the Plant Pathology Laboratory at Harpenden in 1942) manufacturers submit evidence for the efficacy of their products against particular pests or diseases, and confidentially disclose the compositions which they agree not to change without due notification. The labelling of the products is also subject to approval and purchasers are able to recognise approved products by a distinctive mark (Fig. 63) which they carry. The scheme is currently operated on behalf of the Agricultural Departments of the UK by the Agricultural Chemicals Approval Organisation (based on the Ministry's Harpenden Laboratory) which receives full support from the British Agrochemicals Association, the British Pest Control Association, the National Association of Agricultural Contractors, the UK Agricultural Supply Trade Association, and the National Unions and Associations of farmers and growers.

The Ministry issues an annual list of approved chemicals (nearly 200 in the 1978 issue) with summaries of their uses and limitations, tabulations of the major pests and pathogens of the principal crops and the chemicals best suited for their control, together with other pertinent data.

The increasing quantities of plant protection chemicals employed and their wider, and at times uncritical, application by modern spraying and dusting machinery and by aeroplanes and helicopters has had some unfortunate side effects, especially from insecticides. For example, the now notorious large-scale campaign in the United States to eliminate the fir ant by the use of chloro-organic insecticides had a limited effect on the pest but resulted in the death of wild and domestic animals and illness in man (Ordish, 1976). Matters came to a head in 1962 with the publication of Rachel Carson's *Silent spring* which drew attention to the abuses in the use of chemicals for the control of pests, diseases, and weeds. This book, while arousing strong and often irrational emotions in both chemical manufacturers and the general public throughout the world, did draw attention to the need to regulate the use of plant protection chemicals for the common good; an attitude reinforced by the effects on streams, rivers, and potable water from the excessive use of artificial fertilisers and by the side effects of such innovations in human medicine as the drug thalidomide.

One outcome of these developments was the banning of the use of DDT in the United States. In the UK a Pesticides Safety Precautions Scheme now operates whereby manufacturers have undertaken not to market a product containing any new chemical for use in agriculture or food storage until recommendations for safe use have been agreed with the Government Department concerned, if necessary on the advice of the Advisory Committee on Pesticides and other Toxic Chemicals (which in its turn has a scientific subcommittee).

In many countries there has been long-standing legislation regulating poisons used in human and veterinary medicine and more recently legislation has been enacted which though not specifically directed against pesticides covers aspects of their use. For example, in the UK appropriate fungicides are subject to the Pharmacy and Poisons Act, 1933, which sets out provisions for the labelling, storage, and sale of scheduled poisons, the Rivers (Prevention of Pollution) Acts 1951 and 1961 (and corresponding Acts for Scotland and Northern Ireland), the Deposit of Poisonous Waste Act, 1972, and the Control of Pollution Act, 1974, while for chemicals included in the Health and Safety (Agriculture) (Poisonous substances) Regulations protective clothing has to be worn by those applying them.

A useful contribution, which greatly facilitates communication, is made by the British Standards Institution which since 1952 has published successive editions of *Recommended common names for pesticides* produced by a committee of representatives of manufacturers and users (including plant pathologists) which endeavours to ensure that the names recommended are both nationally and internationally acceptable.

Organisation and recent trends

11

Organisation for plant pathology

Farmers, growers, and gardeners have usually been reluctant to pay for advice on the diseases which affect their plants. Thus, because of the national importance of crop losses caused by disease, the tax-payer has for the past hundred years borne the greater part of the financial responsibility for providing advice on disease control, initiating the necessary research, and ensuring the availability of college and university facilities for introducing agriculturists, foresters, and horticulturists to plant protection and training phytopathologists. Organisation for plant pathology has, therefore, been dominated by the state although growers have made their contribution and, particularly during recent years, there has been increasing involvement of large industrial firms in this field. Associations of plant pathologists now also play a not unimportant part.

In the major countries the developmental pattern of the organisation for the different aspects of plant pathology has been on rather similar lines – the work of outstanding amateurs or teachers providing the foundation – and in each of the rather arbitrary, and to some extent overlapping, sections of this chapter only representative examples are offered. References are, however, where possible given to sources of more detailed accounts of the developments summarised, and additional clues will be found in the review by Chiarappa (1970) and in the analysis of the Bibliography and the list of Biographical References.

By the state[1]

Aspects of governmental involvement in plant pathology by legislation designed to enforce the implementation of control measures and to prevent the spread (both nationally and internationally) of major diseases by seed testing and schemes for the production of disease-free stocks and by regulations to ensure the quality and proper use of chemicals employed in plant protection were dealt with in Chapter 10. In addition the state makes an important contribution to the provision of educational and research facilities for phytopathology. These developments began in Europe, spread to North America and from the end of the nineteenth century plant pathology became a European and North American export to developing countries. For example, it was introduced by the British into India and elsewhere (see Butler, 1929), the Germans and French into Africa, the Dutch into the East Indies, and from the United States into Japan, Hawaii, and the Philippines. At first it was usual for workers to be seconded from their home country to investigate particular problems, as was Marshall Ward to study coffee rust in Sri Lanka (Ceylon) and Arthur E. Shipley who was sent to Bermuda in 1887 by the Colonial Office to investigate an onion disease which proved to be downy mildew caused by *Peronospora destructor* (Shipley, 1887). Later they were appointed on a career basis; for example, Butler in India, Tom Petch in Sri Lanka. Today, because of the independence of previously dependent countries there has been a tendency to revert to the earlier practice of secondment.

Europe
Germany and Austria. The birthplace of plant pathology may be considered to have been German-speaking Europe where from the middle of the nineteenth century major phytopathological work was initiated by Julius Kühn at Halle (where in 1891 he founded the Versuchsstation für Nematodenvertilgung und Pflanzenschutz), A. B. Frank (Fig. 65) in Berlin, Paul Sorauer (Fig. 66) at Proskau (and later Berlin) and Karl von Tubeuf at Munich (from 1898 as director of the Kgl. bayrische Station für Pflanzenschutz), all of whom did much to disseminate knowledge of plant disease by writing books which became the standard texts. It was not however until the turn of the century that comprehensive state phytopathological services were established in the region.

In 1889 the Deutsche Landwirtschafts-Gesellschaft set up a special committee (Sonderausschur für Pflanzenschutz) under the direction of Kühn, Frank, and Sorauer which established a network of bureaux for practical agriculturists and issued a series of annual reports. This was the forerunner of the German Official Plant Protection Service (Amtlichen Deutscher Pflanzenschutzdienst), formally instituted on 22 May 1905; for an account of the first fifty years work of which the review by Richter *et al.* (1955) may be consulted. In the same year the Kaiserlichen Gesundheitsamt (founded in 1898 with Frank as director) at Berlin-Dahlem was

Fig. 65. *Albert Bernard Frank*   Fig. 66. *Paul Sorauer*
(1839–1900).   (1838–1916).

renamed the Kaiserliche Biologische Anstalt für Land- und Forstwirtschaft under the direction of Rudolph Aderhold with Otto Appel on the staff.

In Austria, on 18 May 1901 the bacteriological section of the Landwirtschaftlich-chemischen Versuchsstation in Vienna was established as a separate organisation – the Kaiserlich-Königlich Landwirtschaftlich-bakteriologische Pflanzenschutz-Station – under the direction of Dr Karl Kornauth to study, among other topics, plant diseases. Beran (1951) has reviewed and documented the first half-century's work and development of this station which in 1945 became the Bundesanstalt für Pflanzenschutz (Federal Station for Plant Protection).

France. In 1888 E. E. Prillieux (Fig. 67), professor of botany at the Institut National Agronomique in Paris, succeeded in establishing a plant pathological laboratory annexed to his chair. In 1897, needing more space, this laboratory was transferred to the rue d'Alésia and in 1927 it was moved to Versailles to become incorporated in the Centre National de Recherches Agronomiques (now the Institut National de la Recherche Agronomique) as the Station Centrale de Pathologie Végétale. Regional stations were also established for the Sud-Ouest (at Bordeaux), Provence (Antibes), and Alsace (Colmar). On his retirement in 1901 Prillieux was succeeded by Georges Delacroix (with whom he had often collaborated) and the later directors of the Station include the well-known names of E. E. Foëx (1912–39) and Gabriel Arnaud (1939–47). Among other state financed centres for phytopathological research in France mention must be made of the Laboratoire de Cryptogamie of the Muséum National d'Histoire

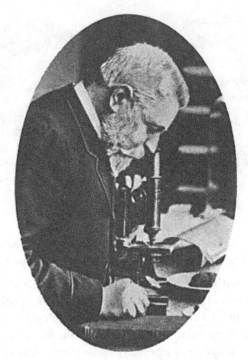

Fig. 67. Edouard E. Prillieux (1829–1915).

Naturelle in Paris where Louis Mangin throughout his tenure (1904–32) of the cryptogamic chair at the Museum (of which he was the first holder) maintained a lively interest in plant pathology as did Roger Heim (who paid special attention to diseases of tropical crops) both before and after his appointment to the chair in 1945.

Netherlands. The state Phytopathological Service (Plantenziektenkundige Dienst) of the Netherlands is based at Wageningen in the west of the country where the Agricultural University of Wageningen is also sited. These two institutions have inter-related histories associated with the name of Dr J. Ritzema Bos (Fig. 68) whose interests as a teacher and adviser in agriculture may be traced back to 1869. In 1864 the parents of Wille Commelin Scholten as a memorial to their son endowed a laboratory, which still bears his name, in Amsterdam with the triple objective of promoting extension work, research, and teaching on plant disease. The next year Ritzema Bos became the first director of this laboratory and was also appointed supernumerary professor of phytopathology in the University of Amsterdam. In 1899 Ritzema Bos was largely responsible for the establishment of the Netherlands Plant Protection Service but he remained at Amsterdam until 1906 when he was appointed director of the newly created Institute for Phytopathology (associated with the High-school for Agriculture, Horticulture & Forestry) at Wageningen where he worked for the next 16 years. In 1919 N. van Poeteren became head of the Plant Protection Service while at the High-school (by then the University College of Agriculture and today of full university status) the phytopathological work was taken over by H. M. Quanjer (Fig. 73) who had been assistant to Ritzema Bos in the Amsterdam days. The Agricultural University of Wageningen has now a world-wide reputation for its phytopathological research and teaching. The Wageningen headquarters of the advisory and research service was supplemented by other laboratories – that for Bulb Research at Lisse, built in 1920–2, being particularly well known for its phytopathological work under E. van Slogteren (Fig. 73).

Ritzema Bos's successor as director of the Wille Commelin Scholten Phytopathological Laboratory was Dr Johanna Westerdijk who made a number of phytopathological investigations and also became curator of the national collection of fungus cultures, now the Centraalbureau voor

Fig. 68. Jan Ritzema Bos (1850–1928).

Schimmelcultures. In 1920 the WCS Laboratory and the CBS were transferred to Baarn where they still share accommodation.

USSR. The most notable early Russian work on mycological plant pathology is that by Michael S. Woronin (Fig. 69) – especially his studies on club
root of crucifers (*Plasmodiophora brassicae*) (1878), smuts (in collaboration
with de Bary), and the brown rot of fruit (*Sclerotinia cinerea* and *S. fructigena*)
(see also Dunin, 1961). Woronin was an individualist of independent means
who devoted his life to the research which inspired and influenced others.
Arthur Jaczewski (Fig. 70), who did so much to lay the foundations for the
phytopathological organisation in Russia, was, before the Revolution, also
wealthy but of a very different temperament. He was born in 1863 into an

Fig. 69. Michael S. Woronin (1838–1903).

Fig. 70. Arthur Jaczewski (1863–1932).

aristocratic landowning family which, as was then the fashion, spoke French in the home and it was in Switzerland at the University of Berne that Professor Eduard Fischer introduced him to mycology. After making a number of contributions to Swiss mycology and visits to France, Italy, and Algeria Jaczewski returned in 1896 to Russia where he set up a laboratory for mycology and plant pathology (supplemented by his own private library and herbarium) at the St Petersburg Botanical Garden. In 1901 the Ministry of Agriculture established a Central Laboratory of Plant Pathology with Jaczewski as director. In 1905 he was succeeded in this post by A. A. Elekin and two years later appointed director of a new Bureau of Mycology and Phytopathology created by the Ministry. As a climax to his career the Government in 1927 designated his by then much enlarged laboratory as the 'Jaczewski Mycological and Phytopathological Institute', a component of the Academy of Agricultural Sciences in Leningrad. During the more than three decades between his return to Russia and his death in 1932 Jaczewski published many monographs, manuals, and scientific papers on diverse aspects of the fungi and plant diseases of Russia. In addition, he was one of the initiators of the Mycological Society of Russia in 1920 and helped to establish a High School of Phytopathological and Applied Zoology. On his death N. I. Vavilov, president of the Academy, cited him as being 'the founder of the Phytopathological Service and the teacher of all the phytopathologists of our country'.

British Isles. In England and Wales a Board of Agriculture was created in 1889 by the amalgamation of the Agricultural Department established in 1883 (the main function of which was to compile agricultural statistics) with other elements of government machinery. Thirty years later this became the Ministry of Agriculture and Fisheries (now Ministry of Agriculture, Fisheries & Food). The year 1889 also saw the passage of the Technical Instruction Act by which local authorities were empowered to provide technical instruction, including instruction in agriculture. Implementation of the Act was facilitated by a special grant and eventually most counties appointed full-time agricultural organisers supplemented by specialists in horticulture, dairying, poultry, etc. but not plant protection. This deficiency was made good from funds rendered available under the Development & Road Improvement Funds Act of 1909 for aiding agriculture by enabling existing or new scientific institutions to undertake research into various aspects of agricultural science and to develop advisory work. One result was the establishment of 13 Provincial Advisory Centres at each of which a team of specialists, including applied entomologists and 'mycologists', was stationed.

Although one of the first activities of the Board of Agriculture was to arrange for field trials during 1891–2 on the use of Bordeaux mixture against potato blight and in spite of the introduction of legislation against disease in plants (see Chapter 10) and the appointment of an entomologist to the Board in 1913, it was not until 1918 that the Board appointed its own full-time plant pathologist. From 1880 it had been one of the duties of M. C. Cooke, cryptogamic botanist at the Royal Botanic Gardens, Kew, and his successor George Massee, to report on specimens of plant disease submitted to the Gardens and it was at Kew that the Board's Plant Pathology Laboratory was established with A. D. Cotton as 'mycologist'. The two main functions of the new laboratory were to supervise the survey of plant diseases in England and Wales (begun as a wartime measure in 1917) and to undertake research. During the winter of 1920–1 the Kew laboratory was transferred to offices and laboratories in a private house on Milton Road, Harpenden (where it remained until moved to a purpose-built laboratory at

Hatching Green, Harpenden in 1960) and the research side was moved to Rothamsted Experimental Station, also in Harpenden, where, under W. B. Brierley, it soon became Rothamsted's plant pathology department. The system thus became two-tiered – the official and the non-official. The Ministry's Plant Pathology Laboratory at Harpenden, manned by civil servants, co-ordinated the intelligence work on plant diseases of England and Wales. The provincial advisory mycologists (financed by public money but not civil servants) supplied most of the data for the disease survey as a by-product of their advisory duties. On 1 October 1946 when the National Agricultural Advisory Service (NAAS) was initiated by the Agricultural Improvement Council (see Rae, 1955) and the Provinces replaced by eight Regions, the advisory mycologists were re-designated 'plant pathologists' and established as civil servants. In 1971 NAAS was incorporated into the Agricultural Development and Advisory Service (ADAS). Additional public money, administered by the Agricultural Research Council, is used to support a range of research stations and other special research and advisory laboratories and institutions.

A rather similar pattern developed in Scotland where phytopathological advisory work in the three regions into which the country is divided is administered from Edinburgh (where there is the equivalent of the Ministry's Harpenden laboratory) by the Department of Agriculture & Fisheries for Scotland; the advisory services for the regions being provided by agricultural colleges at Edinburgh, Auchincruive, and Aberdeen. A historical account of the state and other plant pathological services and activities for both the Republic of Ireland and Northern Ireland has recently been given by Muskett (1976).

North America

USA.[2] In the United States the now massive involvement of the Federal and State administrations in plant pathology has a rather complex history. The two main lines of development both originated during the American civil war. In the May of 1862 Congress passed an Act establishing a Department of Agriculture and in the July of that year President Abraham Lincoln signed the Land Grant Act – the so-called 'Morrill Act' after Justin S. Morrill the congressman who promoted the Bill – which made financial provision for the establishment of State colleges 'for the benefit of agriculture and the mechanic arts' by the donation to every State of Federal land on the basis of 30 000 acres for each of its Senators and Representatives in Congress. At the time the Land Grant Act was passed there were already agricultural colleges in Maryland, Michigan, and Pennsylvania and during the next decade these were supplemented by land-grant colleges in more than 25 other States. Each college had a professor of botany (and/or horticulture) but little attention was paid to plant disease. During the 1860s overproduction of agricultural products and the resulting low market prices deterred young men from taking agricultural courses and this dearth of students in the land-grant colleges gave the staff time for experimental work which they pursued with increasing zest. Representatives of agricultural colleges at a Convention in Washington in 1872 urged the establishment by States and the Federal Government of experiment stations for the promotion of agricultural knowledge. There was no immediate Federal response but agitation continued and several States founded their own experiment stations. The first was that established in 1875 at Middletown, Connecticut and moved two years later to its present site at New Haven where Rowland Thaxter was appointed as the Station's first plant pathologist ('mycologist') in 1888. It became world famous for its work on fungicides under the direction of J. G. Horsfall. Others were established in North Carolina, 1877; Cornell University, 1879; New Jersey, 1880; and the University of Tennessee, 1882. There were 17 stations by February 1887 when Congress passed the National Experiment Station Act (generally cited as the 'Hatch Act' after its author William Henry Hatch of Missouri, Chairman of the House Committee on Agriculture) under which an annual appropriation of $15 000 was granted to each state agricultural college for the purpose of establishing an agricultural experiment station among the objectives of which were 'to conduct original researches or verify experiments on the physiology of plants and animals; the diseases to which they are severally subject, with remedies for the same....' In October of the following year the Office of Experiment Stations was set up within the Department of Agriculture. At first this Office had no regulatory functions; it merely collected and diffused information regarding agricultural experiment stations of the United States at home and abroad and to facilitate this work initiated the *Experiment Station Record* of which 95 volumes were published between 1889 and the final issue in 1946. From 1894 it exerted financial control and representatives of the Office made an annual visit to each station. Subsequently the state experiment stations received increased support from Federal funds, particularly under the Adams Act of 1906, the Purnell Act, 1925, and the Jones-Bankhead Act passed ten years later, and it is the research and extension work undertaken by these experiment stations which put the United States into the van of phytopathological endeavour. The plant pathology departments of many of these stations are, like that in Connecticut, world famous. The beginnings, however, were modest. According to the statistics of station staffs there were in 1889 2 mycologists and 30 botanists in a total staff of 402. In 1905, when the total staff exceeded 800, mycologists had fallen to 4 (from 17 in 1900) but for the first time the returns included 11 plant pathologists and 15 bacteriologists in addition to 56 botanists. Today the experiment stations employ more than a thousand plant pathologists.

The Department of Agriculture was at first a Bureau under a Commissioner and it did not become an executive department with a Secretary of Cabinet rank until 1889. The first botanist to the Department (Charles Christopher Parry [1823–90] who laid the foundation of the National Herbarium now at the Smithsonian Institution) was appointed in 1869 and in 1871 Thomas Taylor, a Scottish immigrant, was instated as 'microscopist'. During the next five years Taylor reported on diseases of grapevine, pear, peach, and other plants, at the same time qualifying in medicine which he practised after official hours. Subsequently he published privately on edible mushrooms (the coloured illustrations by L. Kreiger being produced by the Government printer) and became head of a Division of Microscopy until it was abolished in 1895. His work, unfortunately, was of indifferent quality and it has become traditional for both plant pathologists and historians to pass it over in silence. Of greater significance for plant pathology was the creation in 1885 – as a result of a memorandum prepared by a committee of applied mycologists appointed by the American Association for the Advancement of Science and submitted to the Commissioner of Agriculture urging the initiation of studies on plant disease – of a Section of Mycology within the Division of Botany. This section subsequently became the Section of Vegetable Pathology (1887), Division of Vegetable Pathology (1890), and Division of Vegetable Physiology & Pathology (1895) before it was absorbed into the Bureau of Plant Industry (1901) which was in its turn later amalgamated with other bureaux as the Bureau of Plant Industry, Soils & Agricultural Engineering. The Section of Mycology originally consisted of F. Lamson-Scribner, the first Federal plant pathologist, but later in 1885 he was joined by an assistant, Beverly T. Galloway. The following year, when Lamson-Scribner left to become professor of botany (and later director) of the Tennessee Agricultural Experiment Station Galloway became head of the Section and subsequently the first Chief of the Bureau of Plant Industry.

Lamson-Scribner applied himself with vigour to plant pathological problems. He did much to encourage the use of the recently discovered Bordeaux mixture and the results of his own investigation on diseases of the grapevine were reported in the first departmental phytopathological bulletin (Lamson-Scribner, 1886) which was subsequently expanded as the first American book on plant disease (Lamson-Scribner, 1891). B. T. Galloway (Fig. 23) had great ability, both as an investigator and administrator, and this enabled him to play the major role in laying a firm foundation for Federal involvement in plant pathology and for the Bureau of Plant Industry. Galloway published more than 100 papers and reports on diverse aspects of plant diseases, devised a knapsack sprayer (see Fig. 43), and throughout his long and productive career maintained the practical outlook derived from his farming background. Among Galloway's colleagues during his term as chief Federal plant pathologist were some whose names are still remembered for their contributions to American plant pathology, such as the bacteriologists Erwin F. Smith (see Fig. 23) and T. J. Burrill (Fig. 22), W. T. Swingle (Fig. 23), H. J. Webber, N. B. Pierce, and also A. F. Woods (Fig. 23) who succeeded him. During the twentieth century the Federal involvement in plant pathology has steadily increased. In 1885 Lamson-Scribner had a share in the botanist's annual allowance of $5000, by 1904 $130000 were allocated for plant pathological projects, while in 1977 the allocation for the control of diseases of forest trees, fruit, vegetable and field crops exceeded 56 million dollars[3] with headquarters provided with research stations, laboratories, and facilities for field trials and experimentation at Beltsville on the outskirts of Washington, DC.

Another aspect of the Federal plant pathological organisation must not be forgotten – the National Fungus Collections, which is a component of Mycology Investigations, a unit of the Crops Protection Research Branch, Crops Research Division, of the Agricultural Research Service of the USDA. There were, from the first, specimens of fungi in the herbarium of the Smithsonian Institution. Other specimens were accumulated by the Section of Mycology which in 1887 – according to Lamson-Scribner's report – held 9300 labelled specimens of fungi. By 1891 the total had grown to 16397 and in 1896 Mrs Flora W. Patterson (Fig. 24) was appointed as assistant pathologist to take charge of the collection; an appointment which provides another example of the openings for women provided by the USDA (see Fig. 24). Mrs Patterson held the post until 1923 (latterly as mycologist in the Division of Mycology and Plant Disease Survey headed by C. L. Shear) when she was succeeded by J. R. Weir (1923–7) and John A. Stevenson (1927–60) and on Stevenson's retirement there were some 650000 specimens in the herbarium. The current (1978) total exceeds three-quarters of a million. The herbarium is supplemented by various card indexes including a survey of the world literature of mycology and plant pathology which has more than a million entries. The National Fungus Collections also incorporate the comprehensive John A. Stevenson Mycological Library of 5000 volumes and 50000 reprints for which Dr Stevenson made an endowment to help meet future needs.

Canada. Canada lagged behind the United States in the attention paid to plant disease but this late start has long been overcome (for details see Conners, 1972), in large measure due to the sound foundation laid by the first Dominion Botanist, Hans T. Güssow (Fig. 73), a native of Breslau who after serving as assistant to William Carruthers, consulting botanist to the Royal Agricultural Society of England, was appointed to the post in 1909. A notable feature of the Canadian scene is the now world-famous Dominion Rust Laboratory at Winnipeg (founded in 1925) which specialises on cereal diseases.

Central and South America

Sidney F. Ashby (who subsequently succeeded Butler as director of the Commonwealth Mycological Institute) was appointed to the Department of Agriculture in Jamaica in 1906 where he became increasingly involved in plant disease problems. Later the American J. B. Rorer was appointed mycologist to the Board of Agriculture in Trinidad and this example was subsequently followed by other West Indian islands including Bermuda where the appointment of Laurence Ogilvie in 1923 resulted from a vacation visit of Professor Whetzel of Cornell University to study the plant diseases and fungi of the island, a census catalogue for which was published by Waterston in 1947.

In South America pioneering work was done in Surinam by Gerold Stahel who in 1915 elucidated the aetiology of witches' broom of cacao, named the pathogen *Marasmius perniciosus* (a name changed to *Crinipellis perniciosus*), and investigated the disease in detail (Stahel, 1919). The Biological Institute founded in São Paulo in 1928 has been the most influential centre for plant pathological work in Brazil where over a period of fifty years a firm foundation was laid by A. A. Bitancourt and his numerous collaborators (Bitancourt, 1978). Notable contributions were also made, among others, by A. P. Viégas at the agricultural institute at Campinas in the state of São Paulo and A. Chaves Batista at the University of Recife. In the Argentine many of the plant pathogenic fungi were first recorded or described by Carlos Spegazzini [1858–1926], a versatile Italian immigrant, and the first comprehensive regional phytopathological text was that by Fernandez Valiela (1942).

Asia

India. The first major study of a plant disease in the Indian subcontinent was that by Marshall Ward on coffee rust in Sri Lanka in the 1870s. At about the same time observations by Europeans on plant diseases of India began to be made by members of the Government Medical Service including D. D. Cunningham (who proposed the new genus *Mycoidea* for the parasitic alga, now known as *Cephaleuros virescens*, responsible for red rust of tea) and A. Barclay (still remembered for his studies on Indian rusts). In 1881–2, as a result of recommendations by the Famine Committee of 1880, Imperial and Provincial Departments of Agriculture were established for the amelioration of agriculture. This development drew attention to the need for research and an Imperial Agricultural Chemist was appointed in 1889. Posts for an Imperial Cryptogamic Botanist and an Imperial Entomologist were created shortly afterwards. The turning point for plant pathology was the appointment of E. J. Butler – of Irish extraction and a medical man by training – as Imperial Botanist in 1901. At first stationed at Dehra Dun, Butler moved in 1905, as the first Imperial Mycologist, to the impressive newly-built Agricultural Research Institute at Pusa in Bihar where during the next two decades he laid a firm foundation for the Indian phytopathological service. A disastrous earthquake in 1934 so badly damaged the Pusa Institute that it was decided to rebuild the Institute at New Delhi as the Indian Agricultural Research Institute which includes a flourishing Division of Mycology and Plant Pathology. The phytopathological work at New Delhi is currently supplemented by investigations at a number of substations and regional laboratories devoted to particular crops or special topics (see Raychaudhuri *et al.*, 1972).

The many important contributions to the plant pathology of Sri Lanka by Tom Petch between 1905 and 1924 will long be remembered (Petch, 1921, 1923).

Japan. In the Far East Japan has for the last half century been in the van of plant pathological research. Following the revolution of 1866, the Japanese government in its efforts to introduce western practices invited to Japan European and American professors in all branches of science and the arts. Plant pathology was mainly introduced through Tokyo and Sapporo in Hokkaido, the northernmost island of Japan. The first agricultural experiment station was founded at Nishigahara, Tokyo, in 1893 and this, in 1950, became the present National Institute of Agricultural Sciences while the Hokkaido Agricultural Experiment Station was established in 1901 at Sapporo where since 1876 there had been the agricultural college (the

Sapporo Nogakko) organised by president W. S. Clark and professor W. P. Brooks of the Massachusetts Agricultural College and where the latter lectured on plant pathology for 12 years. Both these stations have divisions for plant pathology and entomology.

At the turn of the century additional research stations were established for tea (1896) and horticulture (1902) and other regional stations have been started since, all of which give attention to disease in plants. There has been a Forest Experiment Station at Tokyo since 1905 (with nine substations in 1966) and an Institute for Plant Virus Research was founded at Chiba in 1964.

Australasia

Australia. In Australia the two leading phytopathological pioneers were D. McAlpine and N. A. Cobb. Daniel McAlpine (Fig. 71) – a Scottish immigrant trained under T. H. Huxley and W. Thistleton-Dyer at the Royal College of Science in London – became the first professional phytopathologist in the Dominion on his appointment as Vegetable Pathologist to the Department of Agriculture of the State of Victoria in 1890 at Melbourne from where he made major contributions to the knowledge of

Fig. 71. Daniel McAlpine (1849–1932).

diseases of citrus (1899), stone fruit (1902), potato (1911) as well as his well-known investigations on take-all of wheat (1904), bitter-pit of apple (1911–16), and his important monographs on the Australian rusts (1906) and smuts (1910). Nathanial Cobb, a United States citizen did good work from 1890 in a corresponding post for the New South Wales Department of Agriculture. He was the first to describe the bacterium of the gumming disease of sugarcane (as *Bacterium* [now *Xanthomonas*] *vasculorum*; Cobb, 1895) and coined the name 'bitter pit' for the physiological disorder of apples. In 1905 he became chief pathologist at the experiment station of the Hawaiian Planters' Association in Honolulu and later transferred to the USDA Bureau of Plant Industry at Washington, where his main interest was in nematology.

New Zealand. In New Zealand, Thomas A. Kirk (who emigrated from England in 1863) did useful work from 1893 to 1909 on plant diseases in his

Fig. 72. George Herriot Cunningham (1892–1962).

capacity as Government Biologist to the Department of Agriculture, as did his successor A. H. Cockayne. It was not, however, until 1928 that the New Zealand-born George H. Cunningham (Fig. 72), then aged 36, was appointed as the first mycologist in the Department of Agriculture. Cunningham came from a farming background and both before and after his appointment in 1917 as an assistant in the Department of Agriculture under Kirk and Cockayne his life was a struggle both for a post and the educational qualifications necessary for professional advancement which finally enabled him to make notable contributions – in six books (e.g. Cunningham, 1925, 1935) and more than 200 scientific papers – to both the plant pathology and mycology of New Zealand, investigations which brought him numerous honours including Fellowship of the Royal Society of London.

Africa

In Africa, between 1911 and 1949 René Maire, as professor of botany at the University of Algiers, made notable contributions to the knowledge of plant diseases and fungi of North Africa and before the First World War the Germans established the agricultural research station at Amani in German East Africa (later Tanganyika and now Tanzania) which in 1928 became the East African Research Institute to serve Kenya, Uganda, and Tanganyika. It was at Amani during the next twenty years that H. H. Storey made his important studies on a range of virus diseases and on virus transmission by insects. Today the station, as a component of the East African Agriculture and Forestry Research Organization (EAAFRO) based on Nairobi, plays no part in East African plant pathology.

The first phytopathologist to be appointed in British tropical Africa was C. O. Farquharson in Southern Nigeria in 1911 and in 1913 W. J. Dowson was posted as mycologist to British East Africa (now Kenya). The names of most of the former British African possessions are still associated with the names of one or more English pathologists who pioneered the development of plant pathology in the region during the inter-war years. Among others, R. H. Bunting and H. A. Dade are associated with the Gold Coast (Ghana), F. C. Deighton with Sierra Leone, T. D. Maitland, C. G. Hansford, and W. Small with Uganda, and J. C. F. Hopkins with Southern Rhodesia (Zimbabwe) while in 1922 E. F. S. Shepherd, trained at Macdonald College, Quebec was appointed botanist and mycologist to the Mauritius Department of Agriculture.

The British also helped to lay a foundation for plant pathology in Egypt by the investigations on diseases of cotton made by W. Lawrence Balls between 1904 and 1913 and for many years there was a British chief of the plant pathological service of the Anglo-Egyptian Sudan (now the Republic of the Sudan). This post was held from 1912–1936 by R. E. Massey.

In South Africa I. B. Pole Evans (trained by Marshall Ward) was appointed mycologist to the Transvaal Government in 1905 and became chief of the Division of Plant Pathology and Mycology of the Union Department of Agriculture in 1911. Notable among his successors was his former assistant Ethel M. Doidge.

By international co-operation

International co-operation for the control of pests and diseases was slow to develop and, as described in Chapter 10, has always been mainly directed towards the co-ordination of legislation regulating the movement of plants and plant products between countries.

One of the most useful outcomes of international co-operation was the initiation of the Imperial Mycological Bureau in 1920, under the directorship of E. J. Butler, by collaboration between the Dominions and the Colonies of the British Empire (Butler, 1921).[4] The origin of the Bureau can be traced back to 1913 when the Imperial Bureau of Entomology was established in London at the British Museum of Natural History to act as a clearing house for information on insect and other pests of plants throughout the British Empire and to abstract the world literature in this field. The success of the venture stimulated the Imperial War Conference of 1918 to recommend the creation of a complementary bureau for plant diseases. This recommendation was implemented two years later by installing the new Bureau at 17–19 Kew Green, two houses recently vacated by the Ministry of Agriculture's Plant Pathology Laboratory, and the publication of the first monthly part of the *Review of applied mycology* (since 1970 the *Review of plant pathology*) followed in January 1922. Subsequently the Bureau was incorporated (as the Imperial (later Commonwealth) Mycological Institute) into the Imperial (later Commonwealth) Agricultural Bureaux which offers a comprehensive abstracting service on all aspects of agriculture. The Institute moved to a purpose-built building at Kew in 1931 which enabled both the information and the mycological identification services to be expanded. Since 1959 bacterial identifications have also been undertaken.

Between 1924 and 1975 the Institute convened nine Conferences of Commonwealth Plant Pathologists, one duty of which was to arrange for an inspection of the work of the Institute and to make recommendations for its policy.

During the last few decades assistance has been given by the developed to the developing countries, both by individual countries and by international organisations such as the Food and Agriculture Organisation of the United Nations (FAO). This help has frequently taken the form of the secondment of specialist plant pathologists for relatively short-term investigations of particular problems. For example, for some years a small 'Pool of Plant Pathologists' was based at the Commonwealth Mycological Institute from where members of the Pool were seconded to survey the plant diseases of Malta and investigate a disease of pepper in Sabah and one of cloves in Zanzibar, to cite a few typical projects. Currently the Overseas Development Administration maintains two tropical plant pathologists as Liaison Officers at CMI.

There are also a number of world-wide or regional organisations covering aspects of the phytopathology of individual crops. These include the Cooperative Centre for Scientific Research Relative to Tobacco (CORESTA), the International Council for the Study of Viruses and Virus Diseases of the Grapevine (ICVG), the International Institute for Sugar-Beet Research (IIRB), and the European Association for Potato Research (EAPR).

By industry and private foundations

One strand of the involvement of industry in plant pathology is the initiative by farmers and growers. As long ago as 1873 the Royal Agricultural Society of England commissioned a report on potato blight from de Bary and in 1907 the Royal Horticultural Society opened a small laboratory at their Wisley Gardens with F. J. Chittenden as director. Ten years later this laboratory was enlarged[5] and W. J. Dowson appointed as the first specialist plant pathologist. In 1928 he was succeeded by D. E. Green who held the post for the next 35 years. Research was undertaken and free advice offered on plant pathological problems to both amateur and professional members of the Society. In the 1870s Marshall Ward was sent officially to Sri Lanka to investigate coffee rust; 25 years later when faced with canker of cacao the Ceylon Planters' Association in 1897 appointed J. B. Carruthers (son of William Carruthers, keeper of botany at the British Museum) as its own cryptogamist. This initiative by planters and growers has been quite widely followed.

In England, several major research stations – now largely maintained by public funds administered by the Agricultural Research Council – began as private ventures before the First World War. The Long Ashton Research Station was established in 1912 by the incorporation of the Cider Research Institute (founded 1903) with the Department of Agricultural and Horticultural Research of Bristol University.[6] The next year the fruit growers initiated the East Malling Research Station, where the investigations of H. H. Wormald on bacterial diseases and brown rot of fruit trees became famous (see Wormald, 1935), and in 1914 glasshouse growers in the Lea Valley north of London established the Lea Valley Research Station (better known as the Cheshunt Experimental and Research Station[7]) which in 1955 moved to Sussex to become the Glasshouse Crops Research Institute. This pattern has also been reflected by the tea planters in Ceylon, rubber planters in Malaya, sugar growers in the Philippines, and the peanut industry in North America. In Malaya the appointment of plant pathologists by British and French growers and by the Department of Agriculture led to duplication of research and much waste of effort so that in 1927 the planters and the government co-operated in founding the Rubber Research Institute of Malaya at Kuala Lumpur with James R. Weir (formerly of the USDA) as head of the department of plant pathology.

Between 1921 and 1976 much important work on the genetics and physiology of cotton and its pests and diseases was organised and funded by the Empire Cotton Growing Corporation (later Cotton Research Corporation) at a series of laboratories and research stations in cotton growing areas of the West Indies (Trinidad) and Africa (including South Africa, Rhodesia, Uganda, and the Sudan). Research on bacterial blight (*Xanthomonas malvacearum*) was a prominent feature of the plant pathological programme as recorded in the Annual Reports of the Corporation.

A second strand is provided by the initiative of chemical firms, engaged in the manufacture of fertilisers, fungicides, pesticides, and veterinary preparations, which established field stations at which trials of their products could be made and various branches of research, including aspects of plant pathology, undertaken. The German chemical firm of Bayer was early in this field and established a field station at Leverkusen while in England Imperial Chemical Industries Ltd in 1925 set up the field station at Jealott's Hill (under the directorship of Sir Frederick Keeble – formerly Sherardian professor of botany at the University of Oxford), which recently celebrated its Golden Jubilee (Peacock, 1978).

Another major contribution to knowledge of plant disease has been due to the benefactions of philanthropists. In 1843 the landowner John Bennett

Lawes, stimulated by Justus von Liebig's *Organic chemistry in its applications to agriculture and physiology*, 1840, in collaboration with the Scottish chemist J. H. Gilbert (who had studied under Liebig) laid out on his Rothamsted estate in Hertfordshire the world-famous fertiliser experiments on cereals and other crops; these marked the beginning of what was to become the Rothamsted Experimental Station, endowed by Lawes, although it was nearly 80 years before plant diseases were given special attention at Rothamsted. During the twentieth century two outstanding examples of the use of trust money to further phytopathological research were the contributions to plant virology made by workers at the Boyce Thompson Institute in New York and the Rockefeller Institute at Princeton, New Jersey, especially during the 1930s.

By plant pathologists

Since the turn of the century, plant pathologists have with increasing frequency organised themselves into societies. At the International Congress of Agriculture and Forestry held in Vienna in September 1890 an International Phytopathological Committee was established. The next year the Netherlands Section of this Committee sponsored the Nederlandse Plantziektenkundige Vereniging which was founded on 11 April 1891 by Professor Hugo de Vries and Dr J. Ritzema Bos, with Dr J. H. Krelage as its first president. On 1 April 1895 the first number of the *Tijdschrift over Plantenziekten* was issued by the Wille Commelin Scholten Laboratory of Amsterdam and the Kruidkundig Genootschap Dodonaea of Ghent with the financial support of the latter society and the Netherlands Society of Plant Pathology. Ritzema Bos and G. Staes (of the Hoogeschool, Ghent) were the first editors and the former continued to serve in this capacity for the next 27 years. The 75th anniversary of the society, which flourished, was marked by a special issue of the *Netherlands journal of plant pathology* which although a journal of international standing, and since 1963 written in English, only accepts contributions from members of the Society.

The American Phytopathological Society was the next to be founded. During the summer of 1908 C. L. Shear suggested to several of his colleagues in the Bureau of Plant Industry the desirability of an organisation of plant pathologists. As a result, a meeting of the pathologists of the Department of Agriculture was called for the 15 December 1908 when it was 'Resolved, That it is the sense of this meeting that an association of plant pathologists be organised'. A committee of three – C. L. Shear, Donald Reddick, and W. A. Orton – was appointed to arrange for the implementation of this decision which took place on the occasion of the meeting of the American Association for the Advancement of Science at Baltimore on 30 December 1908 with 54 persons present. Professor L. R. Jones of Wisconsin, was the first president, C. L. Shear, Secretary-Treasurer. The first regular meeting of the American Phytopathological Society was held at Harvard Medical School, Boston the next December when 41 papers were read (see Shear, 1910). By then 130 plant pathologists had accepted the invitation to become 'charter members' – an acceptance each had been asked to signify 'by remitting to the secretary-treasurer 50 cents by post-office money order, to be used to defray the expenses of the society'. The membership now exceeds 3000. Between 1915 and 1947 geographical Divisions (the Pacific, Southern, Canadian, New England, Potomac, and North Central) were formed within the society and in 1929 the Canadian Division became the independent Canadian Phytopathological Society. The American Phytopathological Society is now the leading national plant pathological society of the world and includes several hundred plant pathologists of other countries in its membership.

The only other strictly phytopathological society founded up to the end of the First World War was the Phytopathological Society of Japan (Nihon Shokubutsu Byōrigaku-Kai) (in 1916 with Professor M. Shirai as the first president), which, with nearly 2000 members, is today the second largest. Although during the inter-war years there was much interest in plant pathology few new societies were formed. The Indian Phytopathological Society, founded in 1947 met a need but it was not until the 1960s and 1970s that new phytopathological societies began to multiply. This trend was stimulated by the First International Congress of Plant Pathology held in London in 1968 and during the next decade the number of national phytopathological societies more than doubled.

From 1926 the most important international meetings of plant pathologists were those associated with the series of International Botanical Congresses begun in 1900. These Congresses included a special section for plant pathology – that of the Sixth Congress, held in Amsterdam in 1935, being a particularly wide and representative gathering of plant pathologists, see Fig. 73 – and it was at a meeting of those participating in the Plant Pathology Section of the Edinburgh Congress of 1964 that the decision was made to organise an independent international congress for plant pathology. As a result, the first, and very successful International Congress of Plant Pathology attended by 1216 pathologists from 71 countries, was held in London in July 1968 under the presidency of Sir Frederick Bawden. This congress initiated the International Society of Plant Pathologists – one

Fig. 73. Plant pathologists at the 6th International Botanical Congress, Amsterdam, 1935.

Top row (left to right): *H. M. Quanjer, J. Henderson Smith, E. J. Butler, G. D. Darnell-Smith, H. Güssow, I. Jørstad, E. Gram.*
Middle row: *Mrs N. L. Alcock, F. T. Brooks, G. H. Pethybridge, J. Johnson, O. Appel, Johanna Westerdijk, E. C. Stakman, E. Föex.*
Bottom row: *P. Murphy, N. E. Stevens, E. Riehm, E. van Slogteren, I. E. Melhus, G. P. Clinton, R. N. Salaman.*

function of which was to arrange future congresses – which subsequently secured recognition as the Section Phytopathology of the Division of Botany of the International Union of Biological Sciences (IUBS), the organisation within which most international biological congresses, federations, commissions, and other wholly biological groups function and are promoted and co-ordinated. The IUBS is the primary biological component of the International Council of Scientific Unions (ICSU) which is funded by subventions from UNESCO (United Nations Educational Scientific & Cultural Organisation) and by financial contributions from its adhering countries.

In the United Kingdom there was for long no phytopathological society the functions of which were shared mainly by the British Mycological Society (founded in 1896), the strong plant pathological interests of which were from 1926 delegated to a Plant Pathology Committee which organised paper-reading and field meetings on plant disease – especially fungal diseases, and the Association of Applied Biologists, founded in 1904, as the Association of Economic Biologists, which covered not only diseases in plants but also pests and other aspects of agriculture and economic biology and which during 1943–53 had a Pests and Diseases Committee. The impending International Congress of Plant Pathology emphasised the need for an organisation to represent British plant pathologists and this led the two societies in 1966 to inaugurate the Federation of British Plant Pathologists into which any member of either society could opt. This lead was followed by Irish plant pathologists in 1968. Subsequently both the Federation (now 900 strong) and the Society of Irish Plant Pathologists became associated societies of the International Society of Plant Pathology as did the other older societies and many new ones so that by the time of the Third Congress held in Munich in 1978 phytopathological societies of 33 countries (representing all five continents) were affiliated to the International Society, as were the Mediterranean Phytopathological Union (founded 1960) and the Associacion Latinamericana de Fitopatologia (1964).

Education for plant pathology

Formal teaching of plant pathology began in Europe and during the second half of the nineteenth century the rest of the world was dependent on European textbooks, particularly German. The most famous and influential teacher was Anton de Bary who after relinquishing medical practice became successively professor of botany at the universities of Freiburg,

Halle, and Strasbourg. Essentially a mycologist, de Bary maintained a life-long interest in fungal diseases of plants as exemplified by his classical *Untersuchungen über die Brandpilze* of 1853 (Fig. 13), based on investigations made while still a student at Berlin, which was followed by many other studies including that on potato blight (de Bary, 1861), proof of heteroecism in *Puccinia graminis* (de Bary, 1865–6), and, towards the end of his life, the role of pectolytic enzymes in fungal parasitism (see de Bary, 1886). The list of the many students who worked in his laboratory and became noted for their contributions to plant pathology is impressive. It includes Millardet (of Bordeaux mixture fame) and the Russian Michael Woronin (with whom de Bary collaborated) at Freiburg; Brefeld, Briosi, D. D. Cunningham, and the Pole Jaczewski at Halle; while of the many at Strasbourg Beijerinck and Wakker from the Netherlands, Jean Dufour from Switzerland, W. G. Farlow and W. R. Dudley from the United States, Matirolo from Italy, and Marshall Ward from England all, as research workers or teachers, made notable phytopathological contributions after returning to their own countries.

This pupil–teacher relationship has always been important. In a recent analysis of the supervisors under whom the first 70 presidents of the American Phytopathological Society worked for their doctorates Horsfall & Cowling (1977–8, 1: 24–7) found that all but three were 'descendants' of three men who had studied under de Bary: Farlow (of Harvard), Dudley (of Cornell), and Marshall Ward (of Cambridge, England) whose student E. M. Freeman introduced plant pathology to the University of Minnesota. The clustering of presidents was also striking. Nine had worked under L. R. Jones (University of Wisconsin), seven under Stakman (Minnesota), and seven under Whetzel (Cornell). In Great Britain during the inter-war years it seemed that every plant pathologist had studied under, or been influenced by, either F. T. Brooks (Fig. 73) at the University of Cambridge (who was one of Marshall Ward's students) or William Brown (Fig. 74) of the Imperial College of Science & Technology in London. The tradition of post-graduate plant pathological training continues to flourish at both these centres although it was not until 1964 that R. K. S. Wood at the Imperial College was appointed as the first professor of plant pathology in England. Cambridge has never created such an 'established' post, Brooks was reader in plant pathology, for a single tenure only.

The first chair ('Lehrstuhl') of plant pathology in the world was that created at the Royal Veterinary and Agricultural High School, Copenhagen to which Emil Rostrup was appointed in 1883 and where he was finally designated professor in 1902, five years before his death.

Fig. 74. William Brown (1888–1975).

Rostrup was succeeded by his assistant F. K. Ravn who died in 1920 at the age of 47. The next two holders of this chair, C. Ferdinandsen and Fabritius Buchwald, each served for a quarter of a century. Professor Buchwald still lives in retirement. E. E. Prillieux, professor of botany at the Institut National Agronomique from 1876 to 1898 and the author of a two-volume text on plant pathology (Prillieux, 1895–7) was largely responsible for initiating phytopathological teaching in France while P. Viala who held the chair of viticulture in the same institute for 44 years (1890–1934) is remembered by plant pathologists for his investigations on diseases of the grape vine (Viala, 1885). In Italy, too, attention was paid to phytopathology in several universities before the end of the nineteenth century (e.g. by Orazio Comes and Luigi Salvatore Savastano at Portici, Fridiano Cavara at Naples, and A. N. Berlese at Milan) and for 30 years from 1920 the investigations and teaching of the versatile Rafaele Ciferri made its mark at Pavia. Johanna Westerdijk, the first director of the Centraalbureau voor Schimmelcultures and director of the Wille Commelin Scholten phytopathological laboratory from 1907, later also held a chair of phytopathology in the universities of Utrecht (where she delivered her inaugural lecture in 1917) and Amsterdam and thus became the first woman to hold such a position and the first woman professor in the Netherlands.

Ernst Gäumann's tenure as professor of botany and director of the Institut für spezielle Botanik of the Eidgenössischen Technischen Hochschule at Zurich from 1917 until his death in 1963 was notable. Gäumann (Fig. 17) gained his doctorate under Eduard Fischer at the University of Bern in 1917 and before his appointment at the Technical High School had worked as a plant pathologist in Buitenzorg, Java (1919–22) and then as botanist at the Eidgenössischen Landwirtschaftlichen Versuchsanstalt at Zurich-Oerliken (1922–7). This plant pathological and mycological experience enabled his institute to gain a world-wide reputation for teaching and research, especially on epidemiological and physiological aspects of plant pathology, while Gäumann's own reputation was enhanced by a series of notable phytopathological texts (Gäumann & Fischer, 1929; Gäumann, 1946, 1959).

In 1914, Ernst Schaffnit was appointed head of the Pflanzenschutzstelle at Bonn-Poppelsdorf where in 1921 he became the first full professor of plant pathology in Germany (at the agricultural high school) and from 1927 director of the Institut für Pflanzenkrankheiten[8] – the first German phytopathological high school institute. Three years later he founded the *Phytopathologische Zeitschrift*.

Phytopathological teaching has always been somewhat neglected in the British Isles. The first chair of plant pathology was that in the Faculty of Agriculture of University College, Dublin to which Paul Murphy was appointed in 1927. Four years later a department of mycology and plant pathology was created at Queen's University, Belfast where a chair of plant pathology was established in 1945 for Professor A. E. Muskett with whose name plant pathological teaching and research in Northern Ireland during the next two decades will always be associated. In England plant pathological teaching, as already noted, has been particularly associated with the University of Cambridge and the Imperial College of Science & Technology of London University but phytopathology has also at times been given emphasis at a number of other university centres including Newcastle-upon-Tyne, Leeds, Manchester, and Exeter (the last being currently the only centre offering a MSc course exclusively in plant pathology*); at Bangor and Aberystwyth in Wales; and, in Scotland, at Glasgow, Edinburgh, Aberdeen, and Dundee.

The public lectures on plant diseases given at the British Museum in South Kensington under the auspices of the Institute of Agriculture, which

* Discontinued 1980.

UNIVERSITY OF BIRMINGHAM.

Department of Botany.

SKETCH SYLLABUS OF COURSE OF LECTURES ON

PLANT DISEASES,

With Special Reference to Cultivated Plants and the Destruction
of Timber by Fungi,

BY A. H. REGINALD BULLER, D.Sc., PH.D.

(Lecturer on Botany in the University.)

I.	Introductory.
II.	Parasitic Flowering Plants.
III.	The Potato Disease. The False Mildew of the Vine.
IV.	The Damping Off of Seedlings. The White Rust of *Cruciferæ*.
V.	Witches' Brooms and *Exoascus*. Mildews.
VI.	*Nectria cinnabarina*. Canker of Apple Trees, Beeches, &c.
VII.	Larch Canker. Ergot of Rye.
VIII.	Smut and Bunt of Cereals.
IX.	Rust Diseases.
X.	Poisonous and Edible Fungi. Fairy Rings.
XI.	Parasitic Toadstools. Tree-destroying Fungi.
XII.	The Destruction of Wood in Houses; Dry Rot.
XIII.	Bacterial Diseases of Plants. Gummosis.
XIV.	Slime Fungi. The Fingers and Toes Disease of Cabbage and Turnips.
XV.	Symbiosis. Root Tubercles of *Leguminosæ*. Mycorhiza. Lichens.
XVI.	Fungi which attack Animals.
XVII.	Vegetable Galls caused by Animals.
XVIII.	Plant Diseases due to Insects, Mites, and Worms.
XIX.	Malformations.
XX.	Effects of Cold, Heat, Lightning, and Wind. The Healing of Wounds.

The Lectures will be given in the Biological Lecture Theatre,
at 4 p.m., on Tuesdays during the Winter and Spring Terms,
commencing **Tuesday, October 13th.** They will be illustrated
by Specimens, Microscope, and Lantern, and may be accompanied
by Laboratory Work.

Laboratory—By arrangement. So far as practicable, living
material will be dealt with.

*FEES.—Lectures and Laboratory, £1 1s. per Term;
or Lectures alone, £1 1s. the Course.*

Fig. 75. Prospectus for lectures on plant diseases by A. H. R. Buller, 1903–4.

form the basis of Worthington G. Smith's little textbook of 1884, were an innovation. The first course of lectures and practical work on plant disease in a British university appears to have been that given at Cambridge in the autumn of 1893 by W. G. P. Ellis of St Catharine's College, university demonstrator in botany, while in 1903–4 A. H. R. Buller gave a series of 20 weekly lectures (the prospectus for which has survived, see Fig. 75) at the University of Birmingham. A foundation for the phytopathological tradition at Cambridge was laid by Marshall Ward who became professor of botany there in 1895 and after his death in 1906 by his student F. T. Brooks who 30 years later succeeded to the chair of botany. At the Imperial College plant pathological teaching was introduced before the First World War by V. H. Blackman whose general lectures were supplemented by others by E. S. Salmon of the Wye Agricultural College. Blackman was instrumental in the appointment of William Brown (mycologist) and Sydney G. Paine (bacteriologist) to the College staff and during the interwar years Brown developed an international reputation as a teacher and for his researches on the physiology of parasitism. Both Cambridge and the Imperial College trained many students for service overseas and from overseas. The College developed particularly strong ties with India and Egypt where it is still possible to detect the influence of the teaching of Brown's school.

In North America[9] T. J. Burrill (see Fig. 22), during the winter of 1869–70, in his first public address as assistant professor of natural history at the University of Illinois was one of the first to draw attention to the need for studies on diseases of plants. The next year, to use his own words, he 'introduced in the course of instruction a term's work called "Cryptogamic Botany"; this was most surely curious teaching but parasitic fungi was made part of the work. By 1873, however, we had gotten better into the subject...'. He based his course on M. C. Cooke's little *Rust, smut, mildew and mould*, 1865, a book in print for the next 40 years. During 1874–5 W. G. Farlow, shortly after his return from Germany and his appointment as assistant professor of botany at Harvard University, gave a course at the Bussey Institution at Jamaica Plains (where he was stationed) which included 'Rudiments of cryptogamic botany. Fungi, especially those injurious to vegetation' and the announcement of the course also stated 'Special investigations on diseases of plants will be pursued'; Farlow, like Burrill, being actively engaged on personal investigations of diverse diseases, including the onion smut (*Urocystis cepulae*) of which he published the first formal description (Farlow, 1877). It was beginnings such as these which provided the inspiration for similar developments elsewhere during the 1880s and after by teachers and research workers including E. A. Bessy at

Iowa Agricultural College (and later at the Universities of Nebraska and Michigan) and J. C. Arthur, botanist at the New York Experiment Station, Geneva. It was not, however, until 1907 that the first university department of plant pathology was established in the United States, at Cornell University under the dynamic Herman Hice Whetzel (see Fig. 76). Two years later a similar department was created at the University of Wisconsin, Madison under L. R. Jones. These two departments became world famous as have a number of others in the United States including those at the Universities of California (Berkeley), Illinois (Urbana), Minnesota (St Paul's), and Washington (Pullman).

Today there are teaching centres for phytopathology in most of the larger and many of the smaller countries. In Asia there are good facilities in India and China as there are in Japan where plant pathological teaching was introduced by the American W. P. Brooks (see above) as early as 1877 and it was at the Agricultural College of Tokyo Imperial University that the first chair of plant pathology was instituted in 1906 for Mitsutaro Shirai; a similar chair being created the next year at Sapporo for Kingo Miyabe, who had studied at Harvard University under Farlow. Australia and New Zealand are adequately catered for by the major universities and among

First Department of Plant Pathology in its Second Winter, 1908–1909. The entire department group is shown in the south laboratory of the top floor of Stone Hall, Cornell University. Back row—Gertrude Whetzel, H. H. Whetzel, C. N. Jensen, J. J. Taubenhaus, M. F. Barrus, Donald Reddick. Front row—Mrs. Lucy Whetzel, Agnes McAllister, Errett Wallace.

Fig. 76. *Whetzel's Plant Pathology Dept, 1908–9, (H.M. Fitzpatrick,* Mycologia **37**:*401, 1945.)*

South American countries Brazil and Argentina have both established important phytopathological teaching centres.

An important educational innovation was the establishment in 1967 by the Danish Government of the Institute of Seed Pathology for Developing Countries under the direction of Dr Neergaard which has supplemented its home-based teaching with workshops in overseas countries including the Philippines, Ghana, Turkey, and Argentina. Previously the most comprehensive training in this field was offered by Professor R. H. Porter's seed technology course at Iowa which included 'Recognition of the importance of seed-borne organisms in seed laboratory practice, the use of fungicides in seed testing and development of laboratory methods for measuring seed germination under adverse conditions' as set out in his review of 1949 (Porter, 1949).

Extension work

One vital aspect of phytopathological education is that classified as 'extension work', the education of farmers and growers in proved methods for the field recognition, treatment, and prevention of disease in their crops. Such information, like that used for the instruction of plant pathologists, has to be constantly updated as new and better practices are derived from research. The most widely used technique is by the printed word and most departments of agriculture in addition to bulletins aimed at the research worker and specialist adviser issue – frequently gratis – bulletins and leaflets of popular appeal. One famous series is that of the *Farmer's bulletins* (begun in 1889) which supplemented the more technical bulletins of the United States Department of Agriculture. In England and in Wales, in 1894 the Board (now Ministry) of Agriculture initiated the monthly *Journal of the Board of Agriculture* (since 1940 re-titled *Agriculture*), which included articles on plant diseases, and also a series of leaflets 'on subjects relating to agriculture and other matters of practical interest to farmers'; the first on a plant disease being Leaflet No. 23 (currently No. 271 in the new series of Advisory Leaflets begun in 1930) on potato blight. Many agricultural colleges and experiment stations have also produced popular bulletins and many of these are accurately and skilfully written in a manner both attractive to the grower and inoffensive to the plant pathologist (see Fig. 77). In some countries where literacy is low, information on control measures has been offered in pictorial form. During recent years, in addition to the verbal advice given by pathologists by lectures and in the field and by announcements in the daily press, advice and information regarding plant disease

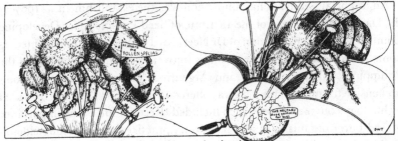

Insects carry the Blighters from diseased to healthy blossoms

THE POLLEN-gathering bee combs the pollen from the anthers, or pollen sacs, which are the uppermost organs in a blossom, and accumulates it in the so-called pollen baskets on its rear legs. Numerous pollen grains including those contaminated with the bacteria also become attached to the insect's body in the process and find easy access to the anthers or stigmas of flowers subsequently visited. The nectar-sucking bee likewise becomes contaminated when it pokes its proboscis, or sucking tube, into the nectary of an infected flower. It would be difficult, if not impossible, for a pollinating insect to visit an infected blossom without itself becoming contaminated.

The busy bee ranges far and free, and takes the Blighter with him

INSECTS spread the bacteria to other flowers. The busy insect moves from flower to flower until loaded and then returns to the hive leaving bacteria along the way. How long a contaminated insect may continue to inoculate blossoms is not known, but theoretically it could continue as long as two days or until all of the living pathogenic bacteria have been eliminated from its body. It should be mentioned here that although dry weather would ordinarily dry the nectar in pear blossoms and prevent infection

Fig. 77. E. M. Hildebrand, Fire blight and its control, **Cornell Extension Bulletin** **405:***12, 1939.*

regulations and disease forecasts have been increasingly frequently propagated by film strips, sound films, and by radio and television.

Plant pathological information

Books

The textbooks and other handbooks of any period designed for students or practitioners of plant pathology tend to reflect the prevailing state of the art and a study of their content and succession throws light on the development of the subject. To locate copies of standard texts and successive editions of popular texts is, however, often surprisingly difficult due to the tendency to discard outdated and little used technical books for their modern replacements.

General texts on plant diseases have usually been characterised by a taxonomic approach. In those which began to appear towards the end of the eighteenth century, as in the monographs of Zallinger (1773), Fabricius (1774), Plenck (1794), and Ré (1807), the diseases were taxonomised, following medical practice, into symptomatological categories particularised as genera and species and dignified by Latin names (see Table 1, Chapter 2). Since the acceptance of the concept of pathogenicity diseases have commonly been grouped according to the type of pathogen – fungus, bacterium, virus – and subdivided according to the taxonomy of the pathogens: fungal diseases, for example, into those caused by phycomycetes, ascomycetes, basidiomycetes, and fungi imperfecti. Although never issued as a book, the 173 articles by the English country clergyman Miles Joseph Berkeley published in the *Gardeners' chronicle* during 1854–7 are a bridge between the old and the new. Berkeley, who was incredibly industrious and a prolific writer of books, scientific papers (in which he described for the first time more than 5000 species of fungi including many plant pathogens), and semi-popular articles, adopted a generic classification of diseases, based on Ré's system, but recognised fungi as primary pathogens, while his approach was broadly based and covered plants in a state of health as well as nutritional and genetic disorders and the depredations of pests.

It was the publication in 1858 of *Die Krankheiten der Kulturgewächse, ihre Ursachen und ihre Verhütung* by Julius Kühn (see Fig. 15) that initiated the modern era. Other German textbooks soon followed including those by Ernst Hallier of Jena in 1868 (a work vitiated by what de Bary described as the author's 'pleomorphic extravagances' which involved bacteria in the life histories of fungi), Paul Sorauer (1874), Georg Winter (1878),

A. B. Frank (1880), and von Tubeuf (1895) (Fig. 78). Of these, the most influential has been the *Handbuch der Pflanzenkrankheiten* by Sorauer the 1st (1874) edition of which was in one volume, the 2nd (1886) in two, and the 3rd (1905–9) in three (including one on insect pests), while the current edition (the 6th), a multi-author multi-volume series, is the standard German phytopathological text.

Up to the 1890s English-speaking plant pathologists were dependent on German texts and in 1897 von Tubeuf's textbook appeared in English translation as later did parts of the 3rd edition of Sorauer's *Handbuch*. It was also in the closing years of the century that the first comprehensive texts were published in Japan, by Shirai (1893–4), and in France, by Prillieux (1895–7). In England little introductory semi-popular books by the botanical artist and amateur mycologist Worthington G. Smith and by Marshall Ward appeared in 1884 and 1889, respectively, but it was not until the last year of the century that the general text by George Massee, mycologist at the Royal Botanic Gardens, Kew, was published. This work, and a revised successor in 1910 (for both of which there were calls for further editions) were the standard British texts until the publication of *Plant diseases* by the Cambridge botanist F. T. Brooks in 1928. Similarly in the United States (where Lampson-Scribner wrote a little book on diseases of the grape vine and other plants in 1890) it was not until 1909 that the first comprehensive text was furnished by B. M. Duggar who set a new standard with his *Fungous diseases of plants*. This paved the way for F. D. Heald's famous textbook of 1926 and his less detailed *Introduction to plant pathology* in 1937. These were followed, among others, by similar volumes by Melhus & Kent (1939) and J. C. Walker (1950) which received wide circulation and increased the dependence of the English-speaking world on American textbooks.

During the first decade of the present century several general works appeared in France and plant pathological texts were early available in Italy while between the wars and since many regional treatments with emphasis on diseases of local importance have been published in a diversity of languages (see analysis of Bibliography).

Other books have covered more restricted topics. One, now unfashionable, by-way is teratology – the histological study of abnormalities whether caused by genetical defects, pathogens, or pests – which was the subject of monographs extending from that of Moquin-Tandon in 1841 to the two-volume *Principles of plant-teratology*, 1915–16 by Worsdell. E. J. Butler in 1929 reviewed aspects of the morbid anatomy of plants, with emphasis on tissue modification caused by the action of gall-inducing fungi and insects, in his presidential address to the Association of Economic [Applied] Biolog-

Fig. 78. Karl von Tubeuf (1862–1941).

ists (Butler, 1930). Another series, stemming from Marshall Ward's *Plant diseases* of 1901 has dealt with principles rather than with descriptions of specific diseases: for example, the publications by Klebahn (1912) in Germany, the Americans Owens (1928), N. E. & R. E. Stevens (1952), and Stakman & Harrar (1957), and S. A. J. Tarr (1972) in England. Others have had an epidemiological bias (e.g. Gäumann, 1946; Vanderplank, 1963). A third, and longer series, is of books restricted to the diseases of particular crops and today there are handbooks covering disease in most major crops. Cereal diseases were, not unexpectedly, the first to receive such treatment – by Francesco Ginanni in Italy as long ago as 1759 (although the author attributed all diseases to insects) and the Abbé Tessier (1783) in France. Forest and garden trees were first given book-length treatment by Schreger (1795) and several minor works preceded Robert Hartig's classics of 1874 and 1882. Today this area of plant pathology is well served by comprehensive texts.[10] Fruit trees also received early attention by Ehrenfels (1795), Sorauer (1882), and von Thümen (1879, 1887) as did the vine (by Viala, 1885, von Thümen and others). Later, both Taubenhaus (1920) and Bewley (1922) wrote on diseases of glasshouse crops. The first extended account of disease in tropical crops was that by Delacroix (1911) and in this field E. J. Butler's 1918 text on plant diseases with special reference to India is a classic.

Although innumerable bulletins or pamphlets have been published on individual diseases few plant diseases have received book-length treatment which is in marked contrast to medical mycology where most of the major mycoses of man have been monographed. The exceptions include black stem rust (*Puccinia graminis*) (Lehmann *et al.*, 1937) and brown rust (*P. recondita*) (Chester, 1946) of wheat, wheat bunt (*Tilletia caries*) (Holton & Heald, 1941), rice blast (*Pyricularia oryzae*) (Ou, 1965), and brown rot of fruit (*Sclerotinia* spp.) (Byrde & Willetts, 1977).

Other useful aids have been provided by the succession of host indexes published since the first major compilation made by Saccardo in vol. **13** (1898) of his *Sylloge fungorum* (1882–1972) which included not only parasites but also saprobes found associated with specific plants. Outstanding among these are Oudemann's massive five-volume *Enumeratio systematica fungorum*, 1919–24, for Europe – which is more comprehensive than might be supposed because for any exotic plant grown in Europe all the known fungal pathogens were compiled – and Seymour's *Host index of North American fungi*, 1929. Although only a pamphlet, J. A. Stevenson's USDA Bulletin 'Foreign plant diseases', 1926, proved most useful and was much thumbed as it listed the major diseases not recorded or established in the United States of the important (and many of the less important) economic plants of the world. A development from the host index was the regional listing of plant diseases. Not surprisingly, for North America, which has the highest incidence of plant pathologists, the census of its diseases is virtually complete and readily accessible in the 500-page USDA Agriculture Handbook 165, 'Index of plant diseases in the United States', 1960 and the complementary *Annotated index of plant diseases in Canada* by I. L. Conners, 1967. Among other countries for which the plant disease census has been comprehensive, particularly for fungal diseases, is India where Butler & Bisby's listing of 1931 is constantly being updated. In England and Wales the Ministry of Agriculture initiated disease surveys in 1917 as a war-time measure. These annual surveys have been periodically summarised, most recently by J. J. Baker (1972) for the years 1957–68, and they provided the main source for W. C. Moore's *British parasitic fungi*, 1959. Many less ambitious lists for all parts of the world have been published in book form (e.g. Waterston (1947), Bermuda; Tarr (1955), Republic of the Sudan) or in the periodical literature.

In addition to the contributions plant pathologists have made to knowledge of the geographical distribution of fungi they have made notable advances in mycological taxonomy by monographs on plant pathogenic fungi in general and on many important groups – particularly rusts (e.g. Arthur, 1934, 1929; Gäumann, 1959) and smuts (e.g. G. W. Fischer, 1953), but also to many other families and genera (for example, Salmon (1900) on the Erysiphaceae; Chupp (1954) on *Cercospora*).

Plant pathologists have also produced atlases to aid diagnosis and written books on bacteria and viruses and the diseases they cause, on fungicides, legislation, and phytopathological techniques, and on aspects of training phytopathologists, representative examples of which will be found listed in the introductory analysis to the Bibliography.

In conclusion, one modern trend may be noted. Most books up to the 1950s were written by single authors – even advanced texts intended for professional use. With the increasing size and technicality of the literature it has become increasingly difficult for any one individual to cope with its quantity and diversity or to have a specialist knowledge of more than a few fields of interest so that, as Lindau and Sydow's compilation of the literature has been taken over by organisations such as the Commonwealth Mycological Institute and Biological Abstracts Inc. and Sorauer's *Handbuch*, as already noted, has become a multi-author work, advanced treatises whether on general plant pathology (e.g. Horsfall & Dimond. 1959–60; Horsfall & Cowling, 1977–8) or special topics such as fungicides are now frequently produced by teams of writers under varying degrees of editorial guidance. This has both advantages and disadvantages. The information can be of a high technical standard and made available promptly but it is rarely possible to impose on such works consistency in style and presentation. Currently a popular variant of this approach is to publish as a volume the papers contributed to symposia (and sometimes also a summary of the discussion engendered) at international congresses or at specially convened meetings organised by scientific societies or by groups of workers with a common interest. One of the first of these in the field of plant pathology, and still one of the most successful, is the volume reporting the symposia on a variety of topics arranged to celebrate the fiftieth anniversary of the American Phytopathological Society in 1958 (Holton *et al.*, 1959).

Periodicals

Today, for the production of the *Review of plant pathology*, a monthly journal of abstracts of the world literature on plant pathology, the Commonwealth Mycological Institute routinely scans more than a thousand serial publications and occasional relevant articles appear in at least an equal number of others. Only a small number of these publications are entirely devoted to phytopathology but most major countries have at least one journal specialising in plant pathology. In 1885 three American mycologists – W. A. Kellerman, J. B. Ellis, and B. M. Everhart – initiated *The journal of mycology* which four years later was taken over by the Section of Vegetable Pathology of the USDA under the editorship of the chief of the Section, B. T. Galloway, and from Vol. 5 (1889) the title-page carried the subtitle 'Devoted to the study of fungi, especially in their relation to plant disease' so that the *Journal of mycology* could be claimed as the first phytopathological periodical. Shortly afterwards, in 1891, Paul Sorauer founded the *Zeitschrift für Pflanzenkrankheiten* (Fig. 79) which he edited (see Rademacher, 1966) and this was followed in Italy in 1892 by the *Rivista di patologia vegetale* (edited by A. N. Berlese) and in 1895 by the Dutch *Tijdschrift over plantenziekten*. These three periodicals are still leading phytopathological journals of today – the last under the title *Netherlands journal of plant pathology*. Many primary journals in a number of different languages were subsequently established (see Table 9). Although the leading journal, *Phytopathology* (started in 1911 with C. L. Shear and H. H. Whetzel as editors), is the official organ of the American Phytopathological Society (founded in 1908), society journals have not played such a dominant role in the primary publication of the results of research on plant disease as they have for mycology, bacteriology, and virology. State concern for plant losses caused by pests and diseases has resulted in many departments or ministries of agriculture issuing periodical publications on phytopathology and more recently international bodies such as the Food and Agricultural Organisation (FAO) of the United Nations have contributed to this field.

The first comprehensive secondary serial publication was the German annual *Bibliographie der Pflanzenschutz-literaturen* (begun in 1914) which attempts a complete listing of the titles of publications on diseases and pests of plants and their control. The leading abstracting service is still that offered by the monthly *Review of plant pathology* begun at Kew in 1922 by the Imperial Mycological Bureau (now the Commonwealth Mycological Institute). The *Distribution maps of plant disease*, the *Descriptions of pathogenic fungi and bacteria* (both initiated by S. P. Wiltshire (Fig. 80)), and the *Descriptions of plant viruses* published by the CMI (the last in collaboration with the

ZEITSCHRIFT für Pflanzenkrankheiten.

Organ für die Gesamtinteressen des Pflanzenschutzes.

Unter Mitwirkung
der
internationalen phytopathologischen Kommission
bestehend aus

Prof. Dr. **Alpine** (Melbourne), Dr. **F. Benecke** (Samarang — Java), Prof. Dr. **Briosi** (Pavia), Prof. Dr. **Maxime Cornu** (Paris), Prof. Dr. **Cuboni** (Rom), Prof. Dr. **Dafert** (Rio de Janeiro), Dr. **Dufour** (Lausanne), Prof. Dr. **Eriksson** (Stockholm), Prof. Dr. **Farlow** (Cambridge), Staatsrat Prof. Dr. **Fischer von Waldheim**, Excellenz (Warschau), Prof. Dr. **Frank** (Berlin), Prof. Dr. **Galloway** (Washington), Prof. Dr. **Gennadius** (Athen), Forstrat Prof. Dr. **Henschel** (Wien), Prof. Dr. **Humphrey** (Amherst — Massachusetts), Prof. Dr. **Johow** (Santiago — Chile), Prof. Dr. **O. Kirchner** (Hohenheim). Geh. Reg.-Rat Prof. Dr. **Kühn** (Halle), Prof. Dr. **Lagerheim** (Quito-Ecuador), Prof. Dr. **Ritter von Liebenberg** (Wien), Direktor **Mach** (St. Michele). Prof. Dr. **Masters** (London), Prof. Dr. **Mayor** (Herestrau — Rumänien), Prof. Dr. **Millardet** (Bordeaux), Prof. Dr. **Mac Owan** (Capetown), Prof. Dr. **O. Penzig** (Genua), Prof. Dr. **Charles Plowright** (Kings Lynn — England), Prof. Dr. **Prillieux** (Paris). Prof. Dr. **Rathay** (Klosterneuburg), Dozent Dr. **Ritzema-Bos** (Wageningen — Holland), Prof. **Rostrup** (Kopenhagen), Prof. Dr. **Saccardo** (Padua), Prof. Dr. **Solla** (Vallombrosa), Dr. **Paul Sorauer**, Schriftführer (Proskau), Prof. Dr. **Sorokin**, Wirkl. Staatsrat (Kasan), Freiherr von **Thümen** (Teplitz), Prof. Dr. **De Toni** (Venedig), Prof. Dr. **H. Trail** (Aberdeen — Schottland), Prof. Dr. **Treub** (Buitenzorg — Java), Direktor **Vermorel** (Villefranche), Prof. Dr. **Hugo de Vries** (Amsterdam), Prof. Dr. **Marshall Ward** (Coopers Hill — Surrey), Prof. Dr. **Woronin** (St. Petersburg), Prof. Dr **Zopf** (Halle)

herausgegeben von

Dr. Paul Sorauer.

Band I.
Jahrgang 1891

Stuttgart
VERLAG von EUGEN ULMER.

Fig. 79. Title page of first volume of the Zeitschrift für Pflanzenkrankheiten, *1891.*

Table 9. *Representative phytopathological periodicals and serial publications*

1891	*Zeitschrift für Pflanzenkrankheiten*	Germany
1892	*Rivista di patologia vegetale*	Italy
1895	*Tijdschrift over plantenziekten*	Netherlands
	(*Netherlands journal of plant pathology*, 1963–)	
1907	*Bolezni rastenij*	USSR
1911	*Phytopathology* (Amer. Phytopath. Soc.)	USA
1913	*Annales du service des épiphyties*	France
	(*Annales des épiphyties*, 1950–68;	
	Annales de phytopathologie, 1969–)	
1914	*Bibliographie der Pflanzenschutzliteratur*	Germany
1916	*Plant disease reporter* (USDA) (From 1980 as	USA
	Plant disease. Amer. Phytopath. Soc.)	
1919	*Phytoprotection*	Canada
1921	*Ochrana rostlin*	Czechoslovakia
	Annals of the Phytopathological Society of Japan	Japan
1922	*Review of applied mycology*	UK
	(*Review of plant pathology*, 1970–)	
1926	*Phytopathological classics* (Amer. Phytopath. Soc.)	USA
1929	*Proceedings of the Canadian Phytopathological Society*	Canada
1930	*Phytopathologische Zeitschrift*	Germany
1935	*Annales de l'Institut Phytopathologique Benaki*	Greece
1942	*CMI Distribution maps of plant diseases*	UK
1948	*Indian phytopathology*	India
1952	*Plant protection bulletin* (FAO)	Italy
	Plant pathology	UK
1956	*Phytopathological papers* (CMI)	UK
1961	*American phytopathological monographs*	USA
	(Amer. Phytopath. Soc.)	
1962	*Fitopathologica Mexicana*	Mexico
	Phytopathologia Mediterranea	Italy
	CMI Descriptions of pathogenic fungi and bacteria	UK
1963	*Iranian journal of plant pathology*	Iran
	Annual review of phytopathology	USA
1964	*Növényvédelem* (*Plant protection*)	Hungary
1965	*Philippine phytopathology*	Philippines
1966	*Acta phytopathologica academiae*	Hungary
1967	*Boletim da Sociedada Brasiliera de Fitopatologia*	Brazil
	Mikologiya i fitopatologiya	USSR
	Revista de la Sociedad Chilena de Fitopatologia	Chile
	Physiological plant pathology	UK
1970	*CMI/AAB Descriptions of plant viruses*	UK
1971	*Indian journal of mycology and plant pathology*	India
1972	*Phytoparasitica*	Israel

Fig. 80. Samuel Paul Wiltshire (1891–1967).

Association of Applied Biologists) have all proved notable secondary publications as, since 1963, has the *Annual review of phytopathology* in the Annual Reviews Inc. series.

The recent trend among specialist workers in many fields of biology and science in general to keep in touch with one another by 'newsletters' has not been very widely adopted by plant pathologists. The American Phytopathological Society has recently supplemented *Phytopathology* by printed *Proceedings* by which news of members and their activities are given publicity within the society and the International Society for Plant Pathology keeps its members and associated societies informed by the *International newsletter on plant pathology* as since 1973 do workers on Phytophthora by the mimeographed *Phytophthora newsletter*.

12

Recent trends and future prospects

For more than half a century the Commonwealth Mycological Institute has surveyed the world literature on plant disease in the *Review of plant pathology* and over the years certain statistical data have been extracted from annual volumes of the *Review*[1] for the information of successive Commonwealth Conferences on Plant Pathology – one duty of which was to consider the policy of the Institute. More recently McCallan has examined the changing pattern of plant pathology in North America by a subject analysis of the abstracts of papers read at the annual meetings of the American Phytopathological Society between 1918 and 1948 and from the fields of interest claimed by members in successive Directories of that society from 1953 to 1968. Horsfall & Cowling (1977–8, 1: 112) also used this latter method to contrast interest in fungicides with that in breeding for resistance while Maan & Zadoks (1977) have reported trends in phytopathological research in the Netherlands by sampling the contents of the *Netherlands journal of plant pathology* at five-yearly intervals from 1895 to 1973. Examination of these statistics reveals a number of trends which clearly reflect changes in the practice of plant pathology.

The literature

Size

The first statistic to consider is the size of the literature which is directly correlated with the number of plant pathologists. Up to 1910, approximately 12 thousand of the 42 thousand items compiled in Lindau & Sydow's *Thesaurus literaturae mycologicae* dealt with diseases of plants. Since 1922 the main evidence for the output of plant pathological publications has been provided by the survey of the literature by the Commonwealth Mycological Institute which has, however, never been exhaustive, always selective, because designed to meet the needs of professional plant pathologists, especially those engaged in research. Increasing literacy throughout the world and the need for higher agricultural output make inevitable the popularisation of much plant pathological writing. The results of research and information on practices best suited to local needs must, in order to supplement the advisory officer's verbal advice (either rendered directly, by films, or over the radio), be passed on by the printed word in growers' periodicals, local newspapers, leaflets, and other popular publications. Much of this literature is ephemeral and it is usually sufficient to draw the attention of plant pathologists in general to novelties and to examples of this genre of particular interest and excellence (see Fig. 77). Thus, selection of items which merit notice by abstract or title in the *Review* has always been made by editors who were formerly practising plant pathologists and in addition to excluding much popular writing they have selected only representative examples of the publications reporting academic research on aspects of disease in plants which shade off into pure science.

From Table 10 it is apparent that between 1925 and 1975 the size of the literature noticed annually has increased more than fourfold, an increase which certainly represents a smaller relative growth rate than that shown by the total literature. The number of authors has also increased – more than sevenfold – and it is interesting to note that during the past 25 years this increase has been at a greater rate than that of the literature. Reasons

Table 10. *Some statistics derived from the* Review of plant pathology (Review of applied mycology *up to 1970*), *1925–75*

Year	1925	1930	1939	1950	1958	1963	1975
Number of abstracts	1447	1626	1838	1773	3070	3414	5820
Number of authors	1022	1228	1665	1805	3309	4064	7235
Number of authors per journal article[a]	1.2	1.3	1.4	1.4	1.5	1.5	2.0
Number of pages per journal article[a]	12.4	12.4	14.1	9.9	7.2	6.4	7.5
Number of contributions per author[b]	1.5	1.4	1.5	1.3	1.4	1.5	1.8
Summary in a second language to journal articles[b]	3(1)[c]	3(1)	5(3)	9(7)	16(15)	25(22)	29(19)

[a]From a random sample of 100 journal articles.
[b]From a random sample of 100 entries in the Author Index.
[c]Number in English in parentheses.

Recent trends and future prospects

for this were touched on in Chapter 1. Here it is only necessary to emphasise that the attention paid to plant pathology is still on the increase.

How many plant pathologists there are today is uncertain. Horsfall & Cowling (1977–8:18) set the total for the world at 9000, a third of whom work in the United States. What is certain is that more workers are involved than ever before and generate a larger literature.

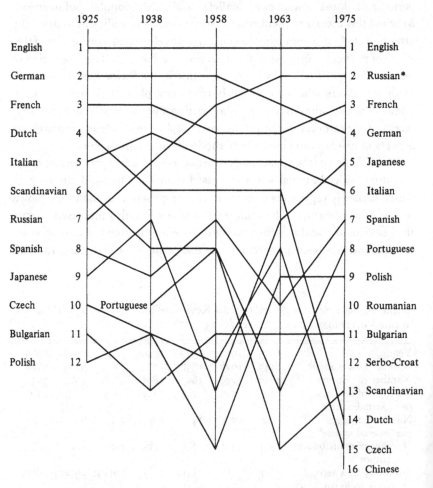

* Excluding Ukranian, Georgian, etc. If these languages are included 'Russian' in 1975 accounted for approximately 500 items, twice the number in French.

Fig. 81. The language of plant pathological literature, 1925–75.

Language

During the past 50 years the proportion of plant pathological literature written in English and noted in the *Review of plant pathology* has never fallen below 60 per cent, and for 1975 exceeded 70 per cent, while over the same period the practice of offering a summary in a second language has risen from 3 to 29 per cent and the frequency of English as the second language has more than doubled (see Table 10). English has thus retained its supremacy but in 1963 Russian overtook German, which had held second place since 1925, and in 1975 Russian was still well in the lead. Also in 1975, French exceeded German for the first time, by approximately 15 per cent.

Fig. 81 which summarises graphically the relative positions over the half century of the principal languages calls for little comment. In 1958 Poland and Japan had still not recovered from the effects of war while the fall in Scandinavian languages and Dutch reflects the excellent English in which workers of these regions write. The recent rise in the placing of Portuguese is largely due to plant pathological activity in Brazil.

The correlation between the distribution of the languages in which the literature is written and the geographical distribution of phytopathological investigation is naturally less close than that between the size of the literature and the number of investigators. Currently by far the greatest proportion of plant pathological work is undertaken in countries (such as the United States and Canada, the UK, Australia and New Zealand) where the first language is English and the USA leads the world in the amount of attention paid to disease in plants. There is a notable official interest in plant pathology in the Indian subcontinent which helps to increase the English language proportion as do the many contributions written by those whose mother tongue is a minority language such as Dutch or Finnish. Further, most African studies are reported in English with French as the runner up.

Content

Fungi, the first pathogens of plants to be recognised, are still the most widespread and economically important. Most of the publications on plant pathology up to 1910 were on fungal disease (less than 1 per cent on bacterial diseases) and this pattern has persisted. For the past 50 years approximately 60 per cent of the literature abstracted in the *Review of plant pathology* has been on fungus diseases or mycological aspects of plant pathology. In North America, however, according to McCallan's data, interest in mycology and fungus diseases has been in steady relative decline since 1918 while interest in virology has shown an equally steady increase, as has the world literature on plant virology from approximately 6 per cent in 1925 to 20 per cent in 1963. The literature on bacterial diseases has fluctuated around 5–6 per cent over the years but there seems to have been a recent increased interest in this field by North American phytopatholog-ists. That on non-parasitic disorders has tended to fall and in 1963 stood at 3.5 per cent.

Lindau & Sydow's analysis in vol. 4 of the *Thesaurus* shows that approxi-mately one third of all the phytopathological literature they compiled up to 1910 was on diseases of fruit – more than half on diseases of the grape vine – while that on vegetables and root crops came second at approximately one fifth – with potato diseases accounting for just under half of this category. Cereal diseases at 10 per cent came next followed by diseases of tropical crops, trees, and ornamental plants. More recently estimates of the atten-tion paid to diseases of 63 major crops were made by measuring the lengths of the index entries in representative volumes of the *Review of plant pathology* between 1925 and 1958 when it was found that the relative attention paid to different types of crops had remained more or less constant for the period. Diseases of vegetables accounted for about a third of the literature with tropical crops a close runner up. Cereals and fruit constituted another pair, each accounting for about a sixth of the literature. While the attention paid to ornamental plants in 1958 was standing at only 3 per cent, this was twice that of 1925; a reflection perhaps of the increasing affluence of western society. The indications are that this pattern is not greatly different today.

With regard to individual crops the survey indicated an increased interest in diseases of rice and decreased attention to those of the grape vine (which has long lost its status as the most studied host) and rubber, changes considered to be real trends readily associated with economic factors such as the need to increase food production and the introduction of synthetic substitutes for rubber and perhaps also with the evolution of satisfactory control measures for some of the much studied diseases.

These results may be contrasted with McCallan's findings for North America where cereals provide the dominant crop interest. During 1918–68 interest in ornamentals rose until 1958 and then declined while a marked feature is the increased interest in diseases of forest and shade trees during the last 25 years. Other growth areas have been soil-borne pathogens and diseases, epidemiology, and, possibly, teaching and extension work.

Two of the main components of plant disease control have shown con-trasting trends according to both McCallan and Horsfall & Cowling. Interest in fungicides within the American Phytopathological Society increased after the First World War to the end of the Second and then

Fig. 4. Distribution of articles in every fifth volume of the Netherlands Journal of Plant Pathology over the classes field, greenhouse, and laboratory.

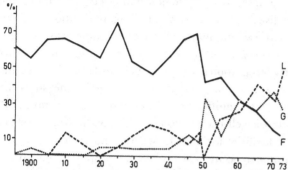

F = field/*veld;* G = greenhouse/*kas;* L = laboratory/*laboratorium.*

Fig. 82. Trends in research reported in the Netherlands Journal of Plant Pathology, *1895–1973. (Maan & Zadoks, 1977.)*

declined, that in breeding plants for resistance, in 1918 below that for fungicides, increased up to 1953, and then declined only to rise again after 1958 so that by 1975 it exceeded interest in fungicides. Horsfall and Cowling attribute this recent reversal, in part at least, to the effect of Rachel Carson's influential book *Silent spring,* published in 1962, in spite of the fact that it is the indiscriminate use of insecticides which has had the most detrimental effects on the environment.

A significant modern shift in emphasis in plant pathological research is that from observation to experiment, from the field to the greenhouse and the laboratory. This trend, which has been graphically expressed by Maan & Zadoks (1977) (see Fig. 82), is in part due to the experimental analysis of field problems under controlled conditions but principally due to a major shift towards invoking physiology and biochemistry as aids to the elucidation of host–pathogen relationships. For example, McCallan's figures show a fivefold increase in the interest of Americans in the 'physiology of parasitism' between 1953 and 1968 and in 1967 British plant pathologists started a new journal – *Physiological plant pathology* – to report the results of research in this field.

Future prospects

Both the short-term and long-term future for plant pathology seems assured. Crop losses from disease are still unacceptably high at a time of world food shortage and increasing population. The introduction of new crops generates new disease problems and changes in agricultural and horticultural practices frequently cause major changes in the pattern and relative importance of diseases already familiar. Strict precautions must be enforced to ensure that certain major diseases of some economically important crops remain geographically localised, and constant vigilance maintained for the early recognition of the introduction of such diseases into new areas which inevitably occurs because no preventive measures are infallible under the present conditions of world trade and air travel. All these activities necessitate a well-trained body of plant pathologists in every country with adequate facilities for both long term and *ad hoc* research.

Epidemics will continue to occur as a result of the conjunction of abnormal weather conditions with intensive farming and horticultural practice but the use of computers and a better understanding of the dynamics of disease should allow epidemics to be forecast with greater accuracy and even if they cannot always be prevented it should be possible to mitigate the social disasters which have in the past followed epidemic outbreaks of plant disease.

The interaction of the pathogen and host will continue to be an attractive topic for academic research. Plant breeders will continue to evolve new cultivars combining enhanced disease resistance with other desirable characters and better fungicides will be discovered and improved techniques for their use devised. Such developments should contribute to the better control of plant disease which is basically the ecological problem of how to adjust the host–pathogen–environment equation in favour of plant health.

NOTES ON THE TEXT

1. Historical patterns of plant pathology

1. Phytomedicine (from the German *Phytomedizin*) is occasionally met with while 'plant protection' and the French *phytiatrie* (*fitoiatra* in Italian) and *phytopharmacie* cover the control of pests, pathogenic organisms, and weeds. There is a German society Vereinigung Deutschen Pflanzenärzte e.V. but 'plant doctor' is hardly acceptable in England though more so in the United States. 'Animal health' has become established as a useful designation for university chairs, national and international bodies, and other institutions concerned with disorders of animals. 'Plant health', though less frequently met with, could have had an equally useful potential but during recent years its scope has become restricted to administrative methods of disease prevention and control.

2. Modern interest in phytopathological terminology began with H. H. Whetzel (1929). The American Phytopathological Society published a series of definitions of some widely used general terms in *Phytopathology* **30**:361–8, 1940 (which is supplemented by definitions in English, French, and German by G. Wilbrink) and subsequently a list of fungicide terms (*ibid.* **33**:624, 1943). Similar sets of definitions were offered by the Plant Pathology Committee of the British Mycological Society in the *Trans. Br. mycol. Soc.* **33**: 154–60, 1950 and the Nederlandse Plantezeiktenkundige Vereeniging in *Netherl. J. Pl. Path.* **74**:65–84, 1968 while R. A. Robinson (*Rev. appl. Mycol.* **48**:593–606, 1969) has reviewed disease resistance terminology. The latest, and most comprehensive, list is the Federation of British Plant Pathologists' *A guide to the use of terms in plant pathology*, 55 pp., 1973 (*Phytopath. Pap.* **17**). See also J. A. Sarejanni, Essai sur le concept de la maladie en pathologie végétale, *Ann. Inst. Phytopath. Benaki* **5**(2):88–127, 1951.

3. Cf. C. Harington, The debt of science to medicine, *Proc. roy. Soc. Med.* **57**:1–8, 1964.

4. In Horsfall & Dimond (1959–60), **1**:3–6, 1959; Horsfall & Cowling (1977–8), **1**:5–7, 1977.

5. Whetzel's 'Ungerian period' is roughly equivalent to Wehnelt's 'romantische Pflanzenpathologie' which was conditioned by the prevailing *Naturphilosophie*, his 'Kühnian' and 'Millardetian' periods to Wehnelt's 'naturwissenschaftliche Periode' (Wehnelt, 1943).

6. The lectures at the British Museum, South Kensington, on which Worthington G. Smith's (1884) textbook was based were preceded by a course of 20 lectures on Vegetable Physiology in relation to Farm Crops.

7. So much so that plant pathologists were frequently designated 'mycologists' and 'mycology' at times equated with plant pathology. As late as 1931 my own first post was that of an 'assistant mycologist' to investigate virus diseases of the tomato and

cucumber and in Great Britain it was not until the government agricultural advisory service was re-organised in 1946 that the 'provincial mycologists' became known as plant pathologists.

8. The Rev. Henry Bryant, the eighteenth-century vicar of Langham in Norfolk, attributed his motive to study the seed treatment of wheat against bunt to 'that of amusement, or the having something to do, ... for unless the mind has something to take hold of, and force us into action, that *Taedium Vitae*, so natural to Englishmen, will insensibly creep upon us, and we shall complain that we are wretched because we have done nothing to make us happy' (Bryant (1784):15–16).

2. Beginnings: problems of aetiology up to 1858

1. This section may be supplemented by the well-documented review by Orlob (1973) the objective of which was 'to compile all available records on the disorders and diseases of plants, that appeared from the beginning of history to about AD 1500'. I gratefully acknowledge my own indebtedness to the late Professor Orlob's study.
2. Hort trans. 1:401.
3. *Ibid.* 1:393.
4. *Ibid.* 1:393.
5. *Ibid.* 1:391.
6. *Ibid.* 2:203.
7. *Ibid.* 2:201.
8. *Ibid.* 1:395.
9. *Ibid.* 2:203.
10. *Ibid.* 1:395.
11. *Ibid.* 1:397.
12. I am grateful to Ian R. D. Mathewson, Dept. of Classics, University of Exeter, for help with this passage and for other translations from *De causis plantarum*.
13. Book XVII, ch. 37 (diseases of trees), 39 (treatment of diseases of trees); XVIII, ch. 44 (diseases of grain), 45 (remedies for diseases of grain); XIX, ch. 57 (maladies of garden plants).
14. Pliny himself notes (Bk XVII, ch. 37): 'Sideration ..., properly so called, is a certain heat and dryness that prevails at the rising of the Dog-star, and owing to which grafts and young trees pine away and die, the fig and the vine more particularly.'
15. Bostock & Riley trans. 4:199.
16. *Ibid.* 4:58–9.
17. *Ibid.* 4:58.
18. *Ibid.* 4:99.
19. Blasting, the Hebrew *shiddāphôn*, drying up, scorching; mildew, Hebrew *yērākon*, yellowness, pallor.
20. ... the fould fiend Flibbertigibbet ... mildews the white wheat (*Lear* III, iv).
21. H. Helbaek, Grauballemandens sidste Maltid, *Arbog. Arkaeol. Selskab* 1958: 83–116; while R. B. Stewart & W. Robinson (*Mycologia* **60**:701–4, 1968) claim to have identified *Puccinia graminis* and *Helminthosporium graminis* from prehistoric potsherds.
22. See Jessen (1864:71). (Engl. transl. by Orlob (1973):195).
23. Cited by Orlob (1964):197–8, Fig. 2.
24. Raychaudhuri & Kaw (1964):85–6.
25. R. Bradley, *A survey of the ancient husbandry and gardening* ..., 1725:325.
26. Raychaudhuri & Kaw (1964):93.
27. On the basis of this experimental design, Dr Garrett has pointed out to me that Tillet was almost certainly the first in the history of agronomy to design a factorial experiment.

28. In the British Museum (Natural History) there is a series of beautiful and detailed drawings (and manuscript) by Francis Bauer (Franz Andreas Bauer (1758–1840)) on 'Diseases in Corn' including wheat bunt and loose smut, black rust, etc.; also ergot of rye.

Diagnosis: the pathogenic agents

1. See Braun (1933):28.
2. See F. L. Wellman, *Ann. Rev. Phytopath.* **2**:43–56, 1964 (neotropical phanerogams); F. G. Hawksworth & D. Weins, *ibid.* **8**:187–208, 1970 (dwarf mistletoes).
3. See J. J. Joubert & F. H. J. Rijkenberg, Parasitic green algae, *Ann. Rev. Phytopath.* **9**:45–64, 1971.

3. Fungi

1. A terminology introduced by G. L. Hennebert & Luella K. Wersub, *Mycotaxon* **6**: 206–11, 1977.
2. *Mycol. Pap.* **71, 112, 124, 133, 137, 140**, 1959–76.
3. See M. A. & H. G. Ehrlich, *Ann. Rev. Phytopath.* **9**:155–84, 1971.
4. See T. Kosuge, *Ann. Rev. Phytopath.* **7**:195–22, 1969.
5. See the reviews by W. Brown, *Bot. Rev.* **5**:236–81, 1936; *Ann. Rev. Phytopath.* **3**:1–8, 1965; Wood (1967).

4. Bacteria, including actinomycetes

1. K. F. Baker (1971) has tabulated a number of early records. See also Kennedy *et al.* (1979).
2. Transl. by Estey (1974):549.
3. Transl. by Estey (1974):550.
4. E. F. Smith – A. Fischer controversy 1899–1901. (E.F.S.) Are there bacterial diseases of plants?, *Zbl. Bakt.* Abt. II **5**:271–78, 1899; (A.F.) Die Bakterienkrankheiten der Pflanzen, *ibid.* **5**:279–87, 1899; (E.F.S.) Dr Alfred Fischer in the role of pathologist, *ibid.* **5**:810–17, 1899; (E.F.S.) Entgegnung auf Alfred Fischer's 'Antwert' in Betreff der Existenz von durch Bakterien verursachten Pflanzenkrankheiten, *ibid.* **7**:88–100, 128–39, 190–9, 1901.
5. For a recent review of this field see N. W. Schrad, *Ann. Rev. Phytopath.* **17**:123–47, 1979 (177 refs.).

5. Viruses and organisms confused with viruses

1. Clusius, *Rariorum aliquot stirpium per Hispanias observatarum historia ...*, 1596 (Antwerp): 510–15.
2. McKay & Warner (1933): 208–9 reprinted Blagrave's description of 'An approved way to make any tulip of what colour you please, never before now printed'.
3. See 'An abridgement of several letters published by the Agriculture Society of Manchester, in consequence of a premium offered for discovering by actual experiment, the cause of Curled Disease in potatoes' in *Letters & Papers of Bath Society for the encouragement of Agriculture, etc.* **1**:237–59, 1780 (also later reprints) from which (pp. 258–9) the following 'general propositions' were drawn:

    First; that some kinds of potatoes are (*caeteris paribus*) much more liable to be affected by the disease than the rest; and that the *old red*, the *golden-dun*, and the *long-dun*, are the most free from it.

Secondly; that the disease is occasioned by one or more of the following causes, either singly or combined; 1st, by frost, either before or after the sets are planted; 2d, from planting sets cut out of large unripe potatoes; 3d, from planting too near the surface, and in old worn-out ground; 4th, from the first shoots of the sets being broken off before planting, by which means there is an incapacity in the *planta seminalis* to send forth others sufficiently vigorous to expand so fully as they ought.

Thirdly; that the most successful methods of preventing the disease are, cutting the sets from smooth middle-sized potatoes, that were fully ripe and had been kept dry after they were taken out of the ground; and without rubbing off their first shoots, planting them pretty deep in fresh earth, with a mixture of quick-lime, or on lime-stone land.

Later, James Anderson (of Cotfield, nr Leith, in Scotland) writing in the same periodical (4:85–92, 1788) on 'the disease called the curl in potatoes' stated (p. 86) 'The only thing that seems to be positively certain with regard to this disorder is, that it was scarcely, if at all known till very lately; and in particular that it was not known in the northern parts of this island till a very few years ago ... there is great reason to believe it was introduced by seed potatoes imported from the south country.' He conjectured (p. 88) 'that the disease depends entirely on the nature of the seed' and that it was 'highly probable that the curl in potatoes, like some hereditary diseases among animals, if once introduced, vitiates the prolific stamina, so as to be perpetuated as long as the infected breed continues to produce others'.

Atanasoff (1922) cites additional early references to curl.

4. *Rep. Conn. agric. Exp. Stn 1907/1908*: 857–8, pl. 66, fig. 6, 1909.
5. For example, three variants of the Latin binomial nomenclature method (with the names for tobacco mosaic virus) were those of W. D. Valleau (*Phytopathology* **30**: 820–30, 1940 [*Musivum tabaci*]), H. S. Fawcett (*Science* **92**:559–61, 1940 [*Nicotianaevir commune*]), and H. H. Thornberry (*Phytopathology* **31**:23, 1941 [*Phytovirus nicomosaicum*]).
6. First by G. A. Kausche, E. Pfankuch & H. Ruska, *Naturwissenschaften* **27**:292–9, 1939.
7. Takahashi & Rawlins (1932) concluded by double refraction studies 'that the virus of tobacco mosaic virus, or some substance regularly associated with it, is probably composed of rod-shaped particles'.
8. Hansen (1970), who had a long-standing interest in the classification of viruses, put forward a rather similar scheme but with the code names for 'genera' Latinised.
9. R. E. Davis & R. F. Whitcomb, *Ann. Rev. Phytopath.* **9**:119–54, 1971; R. Hull, *Rev. Pl. Path.* **50**:121–30, 1971.

6. A note on non-parasitic disorders

1. See R. W. G. Dennis & D. G. O'Brien, Boron in agriculture, *Res. Bull. W. Scotl. agric. Coll. (Pl. Husband. Dept)* **5**, 98 pp., 1937. Also, Winifred E. Brenchley, *Bot. Rev.* **2**: 173–96, 1936; **13**:169–93, 1947.
2. See M. Treshow, *Ann. Rev. Phytopath.* **9**:21–44, 1971.
3. See S. Rich, *ibid.* **2**:253–66, 1964.
4. See W. W. Heck, *ibid.* **6**:165–88, 1968.

7. Chemical control

1. The composition of Bordeaux mixture has been commonly expressed by the ratio of copper sulphate and lime in pounds to water in gallons (usually 100 gal.). Popular mixtures have been 8–8–100, 6–20–100.

2. *J. S.E. agric. Coll. Wye 1934*, **33**:39, 1934.
3. See Le cinquantenaire de la bouillie Bordelaise, *Rev. Path. vég. Ent. agric.* **22**, suppl.: 1–72, 1935.
4. *Phytopathology* **33**:627–32 (slide-germination method of evaluating protectant fungicides), 633–4 (standard laboratory Bordeaux mixture), 1943; **37**:354–6 (test tube dilution technique), 1947.
5. *Proc. roy. Soc. Lond.* **B101**:483–514, 1927.
6. For some diseases, such as Sigatoka (*Mycosphaerella musicola*) of banana, oil alone gives effective control; see L. Calpuzos, *Ann. Rev. Phytopath.* **4**:369–90, 1966.
7. Edn. **3**:238, 249.
8. W. Marshall in *Experiments and observations concerning agriculture and the weather*, 1779 tabulates the results of 67 experiments which showed no advantage for seed treatment including lime-water brine (expts v, vii, xix, lvi).
9. *Phytopath. Classics* **6**:80–1.
10. *Ibid.* **6**:82.
11. *Ibid.* **8**:10.

9. The epidemiological approach

1. 'Epidemic' was introduced for disease in man but has been widely used for both animals and plants. Unger (1833:329) coined *Epizootie* and *Epiphytozie* for epidemics among animals and plants, respectively, and the latter was popularised by Whetzel as 'epiphytotic'. Current usage is to accept epidemic in the broad sense. Whetzel (1929) has also been followed by some in his incorrect use of 'epiphytology' (the study of epiphytes) for 'epiphytotiology' but here again 'epidemiology' is standard practice.
2. The reviews by C. E. Foister on the relation of weather to plant disease (*Conf. Empire Meteorologists 1929* (Agric. Sect.) II. *Papers and discussions*: 168–215 (305 refs.), 1929; *Bot. Rev.* **1**:497–516 (129 refs.), 1935; **12**:548–91 (267 refs. after 1935), 1946) are a valuable introduction to this field.
3. For example by John Worlidge in *Systema agriculturae, being the mystery of husbandry discovered and layd open*, 1669. (Chap. X. Of common and known external injuries, inconveniences, enemies and diseases incidental to, and usually afflicting the husbandman; and their prevention and remedies.) London. '... mildews occurred in a dry summer and smuts come from wetness and fatness of the land.'
4. See F. W. Went, The Earhart Plant Research Laboratory, *Chron. Bot.* **12**:89–108, 1950.
5. Further details and references to other methods are given in the reviews by C. E. Yarwood and P. R. Miller in Holton *et al.* (1959; chap. 49, 50) and P. E. Waggoner, *Ann. Rev. Phytopath.* **3**:103–26, 1965.
6. In the early literature the take-all pathogen is cited as *Ophiobolus graminis*. Currently three varieties of *Gaeumannomyces graminis* are recognised: var. *tritici* on wheat and barley, var. *avenae* on oats, and the type variety var. *graminis* which is not usually pathogenic for wheat, barley or oats but which causes a basal sheath rot of rice. The varieties are not distinguished in this history.
7. Including S. D. Garrett, *Biol. Rev.* **9**:351–61, 1934 (cereal foot-rot fungi); **13**:159–85, 1938 (root infecting fungi); **25**:220–54, 1950 (ecology of root-inhibiting fungi); *New Phytol.* **50**:149–66, 1951 (ecological groups of soil fungi); *Ann. appl. Biol.* **42**:211–19, 1955 (a century of root disease investigation); Holton *et al.* (1959): 309–16 (biology and ecology of root disease fungi); Horsfall & Dimond (1959–60) **3**:23–56 (inoculum potential); Baker & Snyder (1965): 4–17 (biological control of soil-borne pathogens). See also root rots of cereals (P. M. Simmonds, *Bot. Rev.* **7**:308–32, 1941; II, **19**:131–46, 1953; III, B. J. Sallans, **31**:305–36, 1965) and certain non-cereal crops (G. H. Berkeley, *ibid.* **10**:67–123, 1944).

8. English translation by Richard Bradley, 1725 (London).
9. A. H. R. Buller, *Essays on wheat*, 1919. New York. (Chap. 3. The discovery and introduction of Marquis wheat.)
10. *Ann. appl. Biol.* **33**:336–7, 1946; Gäumann (1946, Engl. transl.): vii–viii.
11. *Trans. Br. mycol. Soc.* **48**:159, 1965.

10. Legislation and quarantine

1. The text of the Act is reprinted by Plowright (1889): 302–4.
2. *J. Bd. Agric.* **25**:304–5, 1908.
3. The EEC Wart Disease Directive introduced the concept of 'Wart Disease Safety Zones' in which only immune varieties could be grown. In 1973 an Order implemented this Directive for England and Wales but an amended Order had to be issued the next year to allow the planting of non-immune cultivars in gardens and allotments because of the outcry from gardeners and allotment holders at the ban on the use of King Edward potato in safety zones.
4. The texts of all these Acts are reprinted by Smith (1888–94) *Bull.* **9**:198–208, 1888.
5. 'An Act to Define and enlarge the Duties and Powers of the Board of State Viticultural Commissioners and to Authorise the Appointment of Certain Officers to Protect the Interests of Horticulture and Agriculture' (H. S. Smith *et al.*, *Bull. Calif. agric. Exp. Stn* **553**:109, 1933).

11. Organisation for plant pathology

1. The main sources on which the regional accounts in this section are based are those cited in the analysis of the Bibliography (*Regional phytopathology*).
2. The three detailed reviews by A. C. True (A history of agricultural extension work in the United States, 1785–1923, *Misc. Publ. USDA* **15**, 220 pp., 1928; A history of agricultural education in the United States, 1785–1929, *ibid.* **36**, 436 pp., 1929, and A history of agricultural experimentation and research in the USA, 1607–1925, *ibid.* **251**, 321 pp., 1937) provide many facts and much valuable background information.
3. *Inventory of agricultural research FY1977*, 2, Table 2D (Ser. 200), 1979, an extract from which was kindly made available by Dr Paul L. Lentz. The same table gives the number of 'scientist/years' devoted to plant disease control for 1977 as 730.2, the number of individual scientists involved being greatly in excess of this figure.
4. See also G. C. Ainsworth, C.M.I., 1920–80, *Rev. Pl. Path.* **59**:249–55, 1980.
5. See *J. roy. hort. Soc.* **42**:115–21, 1916–17; also H. R. Fletcher, *The story of the Royal Horticultural Society 1804–1968*, 1969 (London).
6. See T. Wallace & R. W. Marsh *Science and fruit. Commemorating the Jubilee of the Long Ashton Research Station 1903–1953*, 1953 (Univ. Bristol).
7. See W. F. Bewley, *Scient. Hort.* **4**:114–25, 1936; *The history and work of the Experimental & Research Station Cheshunt*, 35 pp., 1935 (Cheshunt).
8. Institut für Pflanzenkrankheiten, *Festschrift der Landw. Hochschule* 1930, 14 pp., Bonn Univ.
9. According to Horsfall (*Phytopathology* **59**:1760, 1969) the first university lecture on plant pathology in the United States was one given by C. E. Goodrich at Yale in 1860 during a 'course of lectures on agricultural subjects'.
10. For lists see *Rev. appl. Mycol.* **46**:113–18, 1967; CMI (1968) 111–18, 1968.

12. Recent trends and future prospects

1. See *Rep. 6th Commonw. Mycol. Conf. 1960*: 17–22, 1961; also *ibid. 7th Conf. 1964*: 19–20, 1964; *8th Conf. 1968*: 20, 1969; *10th Conf. 1975*: 31, 1975. See also G. C. Ainsworth *Mycologia* **55**:65–72, 1963 (mycological information); *Riv. Pat. veget.* Ser. 3, **4**:309–17, 1964 (plant pathological information); *Sabouraudia* **5**:81–6, 1966 (medical and veterinary mycological information).

BIOGRAPHICAL REFERENCES

This list of biographical references to deceased authors mentioned in the text, though incomplete both for authors and references, includes most of the items scanned during the preparation of this history. As far as possible the references cited are those likely to be most easily accessible to plant pathologists. Additional references will frequently be found in J. H. Barnart, *Biographical notes upon botanists*, 3 vols, 1965 (Boston, Mass., Hunt Botanical Library).

Abbreviations

Desmond   R. Desmond, *Dictionary of British & Irish botanists and horticulturalists*, 1977 (London)
DNB   *Dictionary of National Biography*, compact edition, 2 vols., 1975 (Oxford)
Myc.   *Mycologia*
Phytop.   *Phytopathology*
TBMS   *Transactions of the British Mycological Society*
b   bibliography
p   portrait

Adanson, Michel (1727–1806). H. M. Lawrence, *Adanson*, 2 vols., 1963 (Pittsburgh: Hunt Bot. Lib.); Whetzel (1918): 119.
Aderhold, Rudolph Ferdinand Theodor (1865–1907). *Ber. dtsch. bot. Ges.* **25**: (47)–(56), b, 1907.
Alcock, Mrs Nora Lilian Leopard (c. 1875–1972). *Bull. Br. mycol. Soc.* **6**: 81, 1972; Desmond: 6.
Allard, Harry Ardell (1880–1964). *Phytop.* **54**: 125–6, p, 1964.
Anderson, Harry Warren (1885–1971). *Phytop.* **63**: 907, p, 1974.
Appel, Friedrich Carl Louis Otto (1867–1952). *Z. Pflanzenkr.* **59**: 117–18, p (85th birthday), 417, p (death), 1952.
Arthur, Joseph Charles (1850–1942). *Phytop.* **32**: 833–44, p, b, 1942; *Myc.* **34**: 601–5, p, 1942; *Ann. Rev. Phytopath.* **16**: 19–30, 1978.
Ashby, Sidney Francis (1874–1954). *Chron. Bot.* **2**: 184, p, 1936; *Nature* **173**: 802–3, 1954.
Atkinson, George Francis (1854–1918). *Science* **49**: 371–2, 1919; *Amer. J. Bot.* **6**: 301–8, p, b, 1919.
Austen, Ralph (?–1676). *DNB* **1**: 61–2.
Bancroft, Claude Keith (1885–1919). *Kew Bull. 1919*: 86.
Banks, Sir Joseph (1743–1820). *DNB* **1**: 88; H. C. Cameron, *Sir Joseph Banks*, 1952 (London).
de Bary, Heinrich Anton (1831–88). *Ber. dtsch. bot. Ges.* **6**: viii–xxvi, b, 1888; *Myc.* **70**: 222–52, p, 1978.

Bauer, Edwin (1876–1933). *Phytopath. Classics* **7**: 53–4, p, 1942.

Bawden, Frederick Charles (1908–72). *Biogr. Mem. Fel. roy. Soc.* **19**: 19–63, p, b, 1973; *Ann. Rev. Phytopath.* **8**: 1–12, p, 1970.

Beijerinck, Martinus Willem (1852–1931). *Soil Sci.* **31**: 285–6, p, 1931; *Phytopath. Classics* **7**: 31–2, p, 1942; G. van Iterson *et al.*, *Martinus Willem Beijerinck, his life and work*, 1940 (The Hague).

Berkeley, Miles Joseph (1803–89). *DNB* **2**: 2374; F. W. Oliver, *Makers of British botany*, 1913 (Cambridge): 225–32, p; *Phytopath. Classics* **8**: 5–12, p, 1948; Desmond: 60.

Berlese, Agusto Napoleone (1864–1903). *Ann. mycol. Berlin* **1**: 178–80, 1903; *Riv. Pat. veg.* **10**: 347–94, p, b, 1904; Lazzari (1973): 288–90, p.

Bewley, William Fleming (1891–1976). *Ann. Rep. Glasshouse Crops Res. Stn 1976*: 12–13, p, 1977.

Biffen, Roland Harry (1874–1949). *TBMS* **33**: 166–8, p, 1950; *Obit. Not. Fel. roy. Soc.* **6**: 9–24, p, b, 1950; Desmond: 63.

Bisby, Guy Richard (1889–1958). *Nature* **182**: 987, 1959; *TBMS* **42**: 129–31, p, 1959; *Phytop.* **49**: 323, p, 1959.

Blackman, Vernon Herbert (1872–1967). *Ann. Bot. Lond.* **32**: 233–5, p, 1968.

Bock, Jerome (1498–1544). Agnes Arber, *Herbals* (edn 2), 1938: 55–61, p.

Bolley, Henry Luke (1865–1956). *Amer. Men of Sci.* **7**: 173, 1944.

Bradley, R. (?–1732). *DNB* **1**: 206.

Brefeld, Oscar (1839–1925). *Nature* **116**: 369, 1925; *Norske Vidensk.-Akad. Årbok* 83–6, 1926.

Brierley, William Broadhurst (1889–1963). *Nature* **198**: 133, 1963; *Ann. appl. Biol.* **51**: 509–10, 1963; Desmond: 88.

Brooks, Frederick Tom (1882–1952). *TBMS* **36**: 177–9, p, 1953; *Obit. Not. Fel. roy. Soc.* **8**: 341–54, p, b, 1953; Desmond: 92.

Brown, Nellie Adalasea (1877–1956). *Wash. Evening Star* 15 Sept. 1956.

Brown, William (1888–1975). *Biogr. Mem. Fel. roy. Soc.* **21**: 155–74, p, b, 1975.

Bryant, Henry (1721–99). *DNB* **1**: 243.

Buller, Arthur Henry Reginald (1874–1944). *Nature* **154**: 173, 1944; *Obit. Not. Fel. roy. Soc.* **5**: 51–9, p, b, 1945.

Bunting, Robert Hugh (1879–1966). *Proc. Linn. Soc. Lond.* **178**: 89, 1967; Desmond: 105.

Burrill, Thomas Johnathan (1839–1916). *J. Bact.* **1**: 269–71, p, 1916; *Bot. Gaz.* **42**: 153–5, 1916; *Phytop.* 1–4, p, 1918; Rogers (1952): 99.

Butler, Edwin John (1874–1943). *Obit. Not. Fel. roy. Scc.* **4**: 455–74, p, b, 1943; *Phytop.* **34**: 149–50, p, 1944; Desmond: 110.

Butler, Ormond Rourke (1877–1940). *Phytop.* **32**: 447–50, p, b, 1942.

Carleton, Mark Alfred (1866–1925). *Phytop.* **19**: 321–5, p, b, 1929; Rogers (1952): 204.

Carsner, Eubanks (1891–1979). *Phytop.* **70**: 174, p, 1980.

Cavara, Fridiano (1857–1929). *Atti Ist. Bot. Lab. Critt. Pavia*, Ser. iv, **1**: i–xvi, p, b, 1929; Lazzi (1973): 277–8.

Charles, Vera Katherine (1877–1954). *Myc.* **47**: 263–5, b, 1955.

Cheremisinov, Nikifor Adrianovich (1907–76). *Mikol. Fitopath.* **11**: 524–9, b, 1977. [Russ.]

Chupp, Charles David (1886–1967). *Phytop.* **58**: 1200, p, 1968.

Ciferri, Rafaele (1897–1964). *Atti Ist. Bot. Univ. Pavia*, Ser. v, **21** supplemento, 55 pp., b, 1964; *Myc.* **57**: 198–201, p, 1965.

Clinton, George Perkins (1867–1937). *Phytop.* **28**: 379–87, p, b, 1938; *Myc.* **30**: 481–93, p, 1938.

Cobb, Nathan Augustus (1859–1932). *Asa Gray Bull.* NS **3**: 205–72, p, b, 1957.

Comes, Orazio (1848–1923). Whetzel (1918): 96; Lazzari (1973): 242–4, p.

Cook, Harold Thurston (1903–75). *Phytop.* **66**: 933–4, p, 1976.

Cook, Melville Thurston (1869–1952). *Phytop.* **43**: 591, p, 1953.

Cooke, Mordecai Cubitt (1825–1914). *TBMS* **5**: 169–85, 1915; *Phytop.* **6**: 1–4, p, 1916; Desmond: 146.

Cordley, Arthur Burton (1864–1936). *Exp. Stn Rec.* **76**: 286, 1937.

Cotton, Arthur Disbrowe (1879–1962). *Nature* **197**: 951, 1963; *TBMS* **47**: 141–2, p, 1964; Desmond: 151.

Cunningham, David Douglas (1843–1914). *Proc. roy. Soc.* **B89**: xv–xx, 1917.

Cunningham, Gordon Herriot (1892–1962). *Nature* **197**: 17–18, 1963; *Biogr. Mem. roy. Soc.* **10**: 15–37, p, b, 1964.

Dade, Harry Arthur (1895–1978). *Bull. Br. mycol. Soc.* **13**: 74–5, 1979.

Darwin, Erasmus (1731–1802). H. Pearson, *Doctor Darwin*, 1930 (London); D. King-Hele, *Erasmus Darwin*, 1963 (London).

Davaine, Casimir-Joseph (1812–82). J. Théodorides, *Un grand médecin et biologiste Casimir-Joseph Davaine (1812–1882)*, 1968 (Oxford; Analecta Medico-Historico 4).

Delacroix, Edouard Georges (1858–1907). *Bull. Soc. mycol. France* **24**: 48–67, p, 1908.

Dickson, James Geere (1891–1962). *Phytop.* **52**: 1093–4, p, 1962; *Myc.* **55**: 537–9, 1963.

Dimond, Albert Eugene (1914–72). *Phytop.* **63**: 657, p, 1973.

Doidge, Ethel Mary (1887–1965). *Bothalia* **9**: 251–3, p, b, 1967.

Doolittle, Sears Polydore (1890–1961). *Amer. Men Sci.* (edn 4) **5**.

Dowson, Walter John (1887–1963). *Nature* **200**: 630–1, 1963.

Dufour, Léon (1862–1942). *Bull. Soc. mycol. France* **58**: 200–28, p, b, 1942.

Duggar, Benjamin Minge (1872–1956). *Nature* **178**: 834–5, 1956; *Phytop.* **47**: 379–80, p, 1957; *Biogr. Mem. nat. Acad. Sci. Wash.* **32**: 113–31, 1958.

Duhamel du Monceau, Henri Louis (1700–81). *Pl. Physiol.* **8**: 163–6, p, 1933.

Elliott, Charlotte (1883–1974). *Phytop.* **66**: 237, p, 1976.

Eriksson, Jakob (1848–1931). *Nature* **127**: 945–6, 1931.

Fabricius, Johan Christian (1745–1808). Lind (1913): 19–20, p; *Phytopath. Classics* **1**: 7–9, p, 1926; Buchwald (1967): 192–3.

Fairchild, David Grandison (1869–1954). Rogers (1952): 174; *Encyclop. Britannica* 1969, **9**: 35; *ibid.*, 5th edn, **IV**: 30, p, 1977.

Farlow, William Gilson (1844–1919). *Phytop.* **10**: 1–8, p, 1920.

Fawcett, Howard Samuel (1877–1948). *Phytop.* **39**: 865–8, p, 1949.

Ferdinandsen, Carl Christian Frederik (1879–1944). *Friesia* **3**: 83–93, p, 1945.

Föex, Étienne Edmond (1876–1944). *Ann. Épiphyties* **11**: 1–9, p, b, 1945.

Folsom, Donald (1891–1973). *Phytop.* **64**: 575, p, 1974.

Fontana, Felice (1730–1805). *Phytopath. Classics* **2**: 4, p, 1932; *Physis* **18**: 185–97, 1976.

Forsyth, William (1737–1804). *DNB* **1**: 721.

Frank, Albert Bernhard (1839–1900). *Ber. dtsch. bot. Ges.* **19**: (10)–(36), b, 1902.

Freeman, Edmund Monroe (1875–1954). *Amer. Men of Sci.* **7**: 600, 1944.

Galloway, Beverly Thomas (1863–1938). *Science* **88**: 6, 1938; Whetzel (1918): 103, p.

Gardner, Max William (1890–1979). *Phytop.* **70**: 79, p, 1980.

Gäumann, Ernst Albert (1893–1963). *Ann. appl. Biol.* **53**: 345–7, p, 1964; *Myc.* **57**: 1–5, p, 1964; *Verh. Schweiz. Naturf. Ges. 1963*: 194–206, p, b, 1963.

Grainger, J. (1904–78). *Phytop.* **69**: 4, p, 1979.

Gram, Ernst (1891–1964). *Phytop.* **55**: 373–4, p, 1965; Buchwald (1967): 193.

Green, Donald Edwin (1898–1968). *Gdnrs' Chron.* **165**: 35, p, 1969.

Güssow, Hans Theodor (1879–1961). *Phytop.* **51**: 739, p, 1961.

Hales, Stephen (1677–1761). *DNB* **1**: 867; F. W. Oliver, *Makers of British botany*, 1913: 65–83, p; A. E. Clark-Kennedy, *Stephen Hales*, 1929 (Cambridge).

Halsted, Byron David (1852–1918). *Phytop.* **9**: 1–6, p, 1919.

Hart, Helen (1900–71). *Phytop.* **61**: 1151, p, 1961. [First, and to date only, woman president of the Amer. Phytopath. Soc.]

Hartig, Heinrich Julius Adolph Robert (1839–1901). *Ber. dtsch. bot. Ges.* **20**: (8)–(28), b, 1901; *Phytop.* **5**: 1–4, p, 1915; *Phytopath. Classics* **12**: iv–xiv, p, b, 1975.

Heald, Frederick De Forest (1872–1954). *Science* **121**: 279–80, p, 1955; *Phytop.* **45**: 409, p, 1955

Hesler, Lexemuel Ray (1888–1977). *Phytop.* **68**: 1251, p, 1978; *Myc.* **70**: 757–65, p, b, 1978.

Hoggan, Ismé Aldyth (1899–1936). *Phytop.* **27**: 1029–32, p, b, 1937.

Hooke, Robert (1635–1703). *DNB* **1**: 999; Margaret 'Espinasse, *Robert Hooke*, 1956 (London).

Humphrey, Harry Baker (1873–1955). *Phytop.* **47**: 247–8, p, 1957.

Ivanovski, Dmitri Iosifovich (1864–1920). *Bact. Rev.* **36**: 135–45, p, b, 1972; *Phytopath. Classics* **7**: 25–6, p, 1942; Waterson & Wilkinson (1978): 193.

Jaczewski, Arthur (1863–1932). *Phytop.* **23**: 111–16, p, 1933.

Jensen, Jens Ludvig (1836–1904). Lind (1913): 31, p; *Phytop.* **7**: 1–4, p, 1917.

Johnson, James (1886–1952). *Phytop.* **44**: 335–6, p, 1954.

Johnson, Thorvaldur (1897–1979). *Phytop.* **70**: 173, p, 1980.

Jones, Leon Killy (1895–1966). *Phytopath. News* **1**(1): 3, 1967.

Jones, Lewis Ralph (1864–1945). *Phytop.* **36**: 1–17, p, b, 1946; *Ann. Rev. Phytopath.* **17**: 13–20, p, 1979.

Jørstad, Ivar (1887–1967). *Friesia* **8**: 113–16, p, 1967; *Myc.* **63**: 697–700, p, 1971.

Keitt, George Wannamaker (1889–1969). *Phytop.* **60**: 1155, p, 1970.

Kellerman, William Ashbrook (1850–1908). *J. Mycol.* **14**: 49–63, p, b, 1905.

Kendrick, James Blair (1893–1962). *Phytop.* **53**: 373, p, 1963.

Kirchner, Emil Otto Oskar (1851–1925). *Ber. dtsch. bot. Ges.* **43**: (47)–(59), p, b, 1925; Whetzel (1918): 77–9, p,.

Knight, Thomas Andrew (1759–1838). *DNB* **1**: 1144.

Kühn, Julius Gotthelf (1825–1910). Whetzel (1918): 122; *Ann. Rev. Phytopath.* **16**: 343–58, p, 1978.

Kunkel, Louis Otto (1884–1960). *Phytop.* **50**: 777–8, p,1960; *Biogr. Mem. nat. Acad. Sci.* **38**: 145–60, p, b, 1965.

Lamson-Scribner, Frank (1851–1938). *Science* **88**: 101–2, 1938.

Large, Ernest Charles (1902–76). *TBMS* **69**: 167–70, 1977.

Loudon, John Claudius (1783–1843). *DBN* **1**: 1245; Desmond: 394.

Loughnane, James B. (1905–70). *BMS Bull.* **5**(1): 29, 1971; *Internat. Newsletter Pl. Path.* **1**(2), p, 1972.

McAlpine, Daniel (1849–1932). Fish (1970): 14–15.

McKay, Robert (1889–1964). *Nature* **203**: 124, 1964; Muskett (1976): 404–5.

McKinney, Harold Hall (1889–1976). *Phytop.* **67**: 429, p, 1977.

Mains, Edwin Butterfield (1890–1968). *Myc.* **61**: 449–51, p, 1969.

Maire, René (1878–1949). *Bull. Soc. Hist. nat. Afr. nord* **41**: 65–113, p, 1952; *Bull. Soc. mycol. France* **69**: 7–49, p, 1953.

Malpighi, Marcello (1628–94). C. Singer, *A short history of biology*, 1931 (Oxford): 151–5.

Mangin, Louis (1852–1937). *Bull. Soc. mycol. Fr.* **54**: 11–22, p, 1938.

Markham, Roy (1916–79). *The Times* 21 Nov. 1979; *Nature* **285**: 57, 1980.

Massee, George Edward (1850–1917). *TBMS* **5**: 469–73, 1917; Desmond: 426.

Matsumoto, Takashi (1891–1968). *Phytop.* **58**: 1325, p, 1968.

Maublanc, André (1880–1958). *Rev. Mycol. Paris* **23**: 368–9, 1958; *Bull. Soc. mycol. France* **74**: xxii–xxvi, 1958.

Mayer, Adolf (1843–1942). *Phytopath. Classics* **7**: 9–10, p, 1942; Waterson & Wilkinson (1978): 198.

Melhus, Irving E. (1881–1969). *Phytop.* **62**: 391, p, 1972.

Meyen, Franz Julius Ferdinand (1804–40). *Bot. Z.* **2**: 793–4, 1844.

Micheli, Pier Antonie (1679–1737). Ainsworth (1976): 50–5.

Millardet, Pierre-Marie-Alexis (1838–1902). *Ber. dtsch. bor. Ges.* **22**: (10)–(14), 1904; *Phytop.* **4**: 1–4, p, 1914; *Phytopath. Classics* **3**: 4, p, 1933.

Montagne, Jean Pierre François Camille (1784–1866). *Bull. Soc. bot. Fr.* **12**: 277–84, b, 1865; *Ann. Sci. nat. Paris* Sér. 10, **16**: xl–xli, p, 18.

Moore, Walter Cecil (1900–67). *Ann. appl. Biol.* **61**: 167, p, 1968; *TBMS* **52**: 353–4, p, 1969.
Mortimer, John (1656?–1736). *DNB* **1**: 1428.
Mundkur, Balchandra Bhavanishankar (1896–1952). *Indian Phytopath.* **5**: 1–7, p, b, 1953.
Murphy, Paul Aloysius (1887–1938). Muskett (1976): 403.
Naumov, Nikolaï Aleksandrovich (1888–1959). *Bot. ZhSSSR* **44**: 1770–8, p, b, 1960.
Nishida, Tōji (1874–1927). *Phytop.* **19**: 881–3, p, b, 1929.
Orlob, Gert Bernard (1926–75). *Can. phytopath. Soc. News* **28**: 6, 1975.
Ørsted, Anders Sandøe (1816–72). Lind (1913): 17–18, p; *Trans. bot. Soc. Edinb.* **11**: 426–35, 1872.
Orton, Clayton Roberts (1885–1955). *Phytop.* **46**: 243, p, 1956.
Orton, William Allen (1877–1930). *Phytop.* **21**: 1–11, p, b, 1931.
Oudemans, Cornelius Antoon Jan Abraham (1825–1906). *Ber. dtsch. bot. Ges.* **26a**: (12)–(33), b, 1908.
Owens, Charles Elmer (1877–1957). *Phytop.* **48**: 291, p, 1958.
Pammel, Louis Hermann (1862–1931). *Phytop.* **22**: 669–74, p, b, 1932.
Patterson, Flora Wambaugh (1847–1928). *Phytop.* **18**: 877–9, b, 1928; *Myc.* **21**: 1–4, p, b, 1929.
Petch, Tom (1870–1948). *Nature* **163**: 202–3, 1949; *TBMS* **67**: 179–81, p, 1976.
Pethybridge, George Herbert (1871–1948). *Nature* **161**: 1002, 1948; *TBMS* **33**: 162–5, p, b, 1950; Muskett (1976): 401–2; Desmond: 491.
Piemeisel, Frank Joseph (1891–1925). *Phytop.* **17**: 135–6, p, b, 1927.
Pierce, Newton Barris (1856–1916). *Phytop.* **7**: 143–4, 1917.
Plowright, Charles Bagge (1849–1910). *TBMS* **3**: 231–2, 1910; Desmond: 498.
Pole Evans, Illtyd Buller (1879–1968). *Bothalia* **10**: 131–5, p, 1971.
Potter, Michael Cressé (1858–1948). *Who was who* **4** (1941–50): 929–30, 1952; Desmond: 501.
Prévost, Isaac-Bénédict (1755–1819). *Phytopath. Classics* **6**: 7–10, 1939; *Phytop.* **46**: 1–5, 1956.
Prillieux, Edouard Ernest (1829–1915). *Bull. Soc. mycol. Fr.* **32**: 7–16, p, b, 1916.
Quanjer, Hendrick Marius (1879–1961). *Netherl. J. Pl. Path.* **67**: 36–8, p, 1961.
Ravn, Frederik Kølpin (1873–1920). Lind (1913): 37; *Phytop.* **11**: 1–5, p, 1921; *Kgl. Vet. Landb. Årsskr 1976*: 1–27, p, 1975.
Rawlins, Thomas Ellsworth (1895–1972). *Phytop.* **64**: 573, p, 1974.
Ré, Filipo (1763–1817). *Malpighia* **20**: 273–4, 1906.
Reddick, Donald (1883–1955). *Phytop.* **46**: 299, p, b, 1955.
Ritzema Bos, Jan (1950–1928). *Ann. appl. Biol.* **16**: 483–5, p, 1929.
Rostrup, Emil (1831–1907). Lind (1913): 1–8, p; *Ber. dtsch. bot. Ges.* **26a**: (47)–(55), 1909.
Saccardo, Pier Andrea (1845–1920). *Nuovo Giorn. bot. ital.* NS **27**: 39–74, p, b, 1920.
Salaman, Redcliffe Nathan (1874–1955). *Lancet* **268**: 1333–4, 1955; *Biogr. Mem. Fel. roy. Soc.* **1**: 239–45, p, b, 1955.
Salmon, Ernest Stanley (1871–1959). *Nature* **184**: 1188, 1959.
Săvulescu, Alice (1905–1970). *Rev. roum. Biol. Sér. bot.* **15**: 139–41, p, 1970.
Săvulescu, Trajan (1889–1963). *Sydowia* **18**: 1–10, p, b, 1965.
Schaffnit, Johannes Ernst Martin Christian Otto (1878–1964). *Z. PflKrankh.* **55**: 257–60, 1948 (70th birthday); *Proc. Can. phytopath. Soc. 36th Session*, 1970: 34.
Schilbersky, Károly (1863–1935). *Acta Phytopath.* **1**: 5–10, 1966.
Schultz, Eugene Schultz (1884–1969). *Phytop.* **60**: 745, p, 1970.
Scott, Reynold (1538?–99). *DNB* **2**: 1873.
Selby, Augustine Dawson (1859–1924). *Phytop.* **15**: 1–10, p, b, 1925.
Seymour, Arthur Bliss (1859–1933). *Myc.* **26**: 279–90, p, b, 1934.
Shear, Cornelius Lott (1865–1956). *Phytop.* **47**: 321–2, p, 1957; *Myc.* **49**: 283–97, p, b, 1957; *Taxon* **6**: 7–10, p, b, 1957; *Sydowia* **11**: 1–17, p, b, 1958.

Smith, Erwin Frink (1854–1927). *Phytop.* **17**: 675–88, p, b, 1927; **18**: 475, 1928; Rogers (1952).
Smith, Worthington George (1835–1917). *Beds. hist. Record Soc.* **57**: 141–79, b, 1978.
Sorauer, Paul Carl Moritz (1838–1916). *Nature* **96**: 600, 1916; *Ber. dtsch. bot. Ges.* **34**: (50)–(57), p, 1916.
Spaulding, Perley (1878–1960). *Phytop.* **51**: 209–10, 1961.
Spegazzini, Carlos (1858–1926). *Nature* **118**: 704, 1926.
Sprague, Roderick (1901–62). *Myc.* **54**: 587–92, p, b, 1962.
Stakman, Elvin Charles (1885–1979). *Phytop.* **69**: 195, p, 1979.
Stanley, Wendell Meredith (1904–71). *Nature* **233**: 149–50, p, 1971.
Stevens, Frank Lincoln (1871–1934). *Myc.* **27**: 1–5, p, 1935.
Stevens, Neil Everett (1887–1949). *Phytop.* **40**: 413–18, b, 1950; *Myc.* **42**: 333–41, p, b, 1950.
Stewart, Fred Carleton (1868–1946). *Phytop.* **37**: 687–97, p, b, 1947.
Storey, Harold Haydon (1894–1969). *Ann. appl. Biol.* **64**: 188, p, 1969; *Biogr. Mem. Fel. roy. Soc.* **15**: 239–46, p, b, 1969; *Nature* **222**: 905, 1969.
Stoughton, Raymond Henry (1903–79). *The Times* 23 Nov. 1979.
Swingle, Deane Bret (1879–1944). *Phytop.* **34**: 769–71, p, b, 1944.
Swingle, Walter Tennyson (1871–1952). *Science* **118**: 288–9, 1953.
Targioni-Tozzetti, Giovanni (1712–83). *Phytopath. Classics* **9**: vii–xi, p, 1952.
Taubenhaus, Jacob Joseph (1884–1937). *Phytop.* **28**: 525–30, p, b, 1938.
Tehon, Leo Roy (1895–1954). *Phytop.* **45**: 115, p, 1955; *Myc.* **47**: 597–601, p, b, 1955.
Thaxter, Roland (1858–1932). *Phytop.* **23**: 565–71, p, 1933; *Myc.* **25**: 69–89, p, b, 1933; *Biogr. Mem. Nat. Acad. Sci. Wash.* **17**: 55–68, p, b, 1936; *Ann. Rev. Phytopath.* **17**: 29–35, 1979.
Thümen, Felix Karl Albert Joachim von (1839–92). *Ber. dtsch. bot. Ges.* **11**: (28)–(30), 1893; *Hedwigia* **32**: 247–8, 1893.
Tillet, Mathieu (1714–91). Wehnelt (1937).
Tournefort, Joseph Pitton de (1656–1708). Whetzel (1918): 124.
Tubeuf, Carl Freiherr von (1862–1941). *Ber. dtsch. bot. Ges.* **59**: (109)–(127), p, b, 1941.
Tull, Jethro (1674–1741). *DNB* **2**: 2119.
Unger, Franz Joseph Andreas Nicolaus (1800–70). *J. Bot.* **8**: 192–203, p, b, 1870; *Bot. Z.* **28**: 241–64, b, 1870.
Valleau, William Dorney (1891–1974). *Phytop.* **67**: 953, p, 1977.
Van Slogteren, Egbertus (1888–1968). *Netherl. J. Pl. Path.* **74**: [181–2], p, 1968.
Voglino, Pietro (1864–1933). *Riv. Pat. veg.* **24**: 1–10, p, 1934.
Waite, Merton Benway (1865–1945). *Phytop.* **36**: 175–9, p, b, 1946.
Wakker, Jan Hendrik (1859–1927). *Netherl. J. Pl. Path.* **72**: 38–47, p, b, 1966.
Ward, Harry Marshall (1854–1906). *DNB* **2**: 2948; *Phytop.* **3**: 1–2, p, 1913; *Proc. roy. Soc.* **B83**: i–xiv, p, 1911; F. W. Oliver, *Makers of British botany*, 1913 (Cambridge): 261–79, p; Desmond: 639.
Weigmann, Arend Joachim Friedrich (1771–1853). Whetzel (1918): 125.
Weir, James Robert (1881–1943). *Phytop.* **36**: 487–92, p, b, 1946.
Westerdijk, Johanna (1883–1961). *J. gen. Microbiol.* **32**: 1–9, p, b, 1963; *Ann. appl. Biol.* **50**: 372, 1962.
Weston, W. A. R. Dillon (1899–1953). *Nature* **172**: 480, 1953.
Whetzel, Herbert Hice (1877–1944). *Phytop.* **35**: 659–70, p, b, 1945; *Myc.* **37**: 393–413, p, b, 1945.
Wiltshire, Samuel Paul (1891–1967). *Nature* **215**: 221, 1967; *TBMS* **50**: 513–14, 1967.
Winter, Heinrich Georg (1848–87). *Ber. dtsch. bot. Ges.* **5**: 1–liv, 1887; *Hedwigia* **26**: 185–91, p, 1887; *Rev. mycol. Paris* **9**: 185–8, 1887.
Wollenweber, Hans Wilhelm (1879–1949). *Phytop.* **40**: 119, p, b, 1950.

Woods, Albert Fred (1866–1948). *J. Wash. Acad. Sci.* **39**: 313, 1949; Rogers (1952): 274.

Woolman, Horace Mann (1853–1932). *Phytop.* **23**: 931–3, p, b, 1933.

Wormald, Harry (1879–1955). *TBMS* **39**: 289–90, p, 1956; *Nature* **177**: 649, 1956; Desmond: 677.

Woronin, Michael Stephanovitch (1838–1903). *Ber. dtsch. bot. Ges.* **21**: (35)–(47), 1903; *Phytop.* **2**: 1–4, p, 1912; *Phytopath. Classics* **4**: 5–8, p, 1934; Dunin (1961).

Zallinger, Johann Baptista (1731–85). Whetzel (1918): 126.

BIBLIOGRAPHY

In addition to most of the literature cited in the text, this bibliography includes references to the principal publications on the history of plant pathology (*) together with a representative selection of phytopathological textbooks and handbooks.

History

General phytopathology. Pritzel (1871, bibliog.), Savastano (1890, Greek & Roman), Lindau & Sydow (1908–17, bibliog.), Sorauer (1914), Hecke (1915), Whetzel (1918), Meyer (1930), K. Braun (1933), Stevens (1934), Hino (1938), Large (1940), Reed (1942), Wehnelt (1943), Walker (1950), Keitt (1950), Orlob (1964, aetiology; 1971, Middle Ages; 1973, Ancient & Medieval), H. Braun (1965), Carefoot & Sprott (1969), Ordish (1976), Zadoks & Koster (1976, epidemiology), Ubrizsy (1977), Noble (1979, seed pathology).

Phytopathological chronologies. Smith (1929), Mayer (1959), Parris (1968).

Regional phytopathology. Africa, East (Bock, 1970); South (du Plessis, 1956); America, Central and South (Nolla & Fernandez Valiela, 1976); see also USA; Austria (Beran, 1951); Australia (Fish, 1970; Stubbs, 1971); Brazil (Puttemann, 1940; Bitancourt, 1978); Canada (Estey, 1970; Conners, 1972); China (Kelman & Cook, 1977); Denmark (Lind, 1913; Buchwald, 1967); France (Viennot-Bourgin, 1954; Moreau, 1926); Germany (Morstatt, 1921; Schlumberger, 1949; Richter *et al.*, 1955; Rademacher, 1966); Ghana (Bunting & Dade, 1925); Great Britain (Fryer & Pethybridge, 1924; W. C. Moore, 1942; Rae, 1955; Anon., 1968; Ainsworth, 1969; F. J. Moore, 1979); Hungary (Kiraly, 1972); India (Raychaudhuri, 1967, 1972); Ireland (Muskett, 1976); Italy (Lazzari, 1973); Jamaica (Turner & Henry, 1977); Japan (Hino, 1938; Akai, 1974); Malaysia (Ting, 1975); Malawi (Angus, 1963); Mexico (Nolla & Fernandez Valiela, 1976); Netherlands (Sirks, 1935; *Netherl. J. Pl. Path.*, 1966; Kerling, 1969); New Zealand (Chamberlain, 1969); Pakistan (Kausar, 1960); Rumania (Sorauer, 1891); Sumatra (Hartley & Rands, 1924); Switzerland (Sorauer, 1893); UK, see Great Britain, Ireland; USA (N. E. Stevens, 1933; Stevenson, 1951, 1954, 1959; Rogers, 1952; McCallan, 1959; Walker, 1975; Ellis, 1976); USSR (Jaczewski, 1903; Klemm, 1941).

Mycology. Ainsworth (1976), Arthur (1929, rusts), Woolman & Humphrey (1924) and Holton & Heald (1941, wheat bunt); Fischer & Holton (1957, smuts).

Bacteriology. Smith (1905–14, **2**), Bulloch (1938), Grainger (1958), Kennedy *et al.* (1979).

Virology. McKay & Warner (1933, tulip breaking), M. T. Cook (1938), Pethybridge (1939, potato leaf roll), Corbett & Sisler (1964), Bawden (1966), Markham (1977), Waterson & Wilkinson (1978).

Fungicides. Johnson (1935, copper fungicides), Butress & Dennis (1947, seed treatment), McCallan (1967).

General botany. Jessen (1864), Sachs (1890), Reed (1942)

Representative phytopathological textbooks and handbooks

General texts. Argentina (Fernandez Valiela, 1942); Austria (Unger, 1833; Thümen, 1886); Belgium (Marchal, 1925); Bulgaria (Atanasoff, 1934; Khristov, 1959); Czechoslovakia (Baudys *et al.*, 1958–62); Denmark (Ørsted, 1863, 1865; Rostrup, 1902; Ravn, 1914; Gram & Weber, 1944); Fiji (Graham, 1971); France (Arbois de Jubainville, 1878; Prillieux, 1895–7; Delacroix, 1902; Ducomet, 1908; Delacroix & Maublanc, 1908–9; Bourcart, 1910; Mangin, 1914–21; Nicolle & Magrou, 1922; Magrou, 1946); Germany (Wiegmann, 1839; Mayen, 1841; Kühn, 1858; Hallier, 1868; Sorauer, 1874; Winter, 1878; Frank, 1880; Wolf, 1887; Kirchner, 1890; Frank & Sorauer, 1892; Tubeuf, 1895; Klebahn, 1912); Ghana (Bunting & Dade, 1925); India (Butler, 1918; Mundkur, 1949; Kamat, 1953, 1956); Italy (Ré, 1807; Comes, 1882, 1891; Peglion, 1899; Ferraris, 1913; Voglino, 1924; Ciferri, 1941); Japan (Shirai, 1893–4; Ideta, 1904; Hara, 1930, 1948; Nakata, 1960); Lesser Antilles (Nowell, 1923); Madagascar (Bouriquet, 1946); Norway (Brunchorst, 1887; Ramsfeljell & Jeddalen, 1962); Poland (Garbowski, 1964); Rumania (Săvulescu & Săvulescu, 1959; Săvulescu *et al.*, 1960); Senegal (Bouhot & Mallamaire, 1965); Spain (Del Canizo Gomez & Gonzales de Andres, 1955); Sweden (Eriksson, 1910); Switzerland (Regel, 1854; Faes, 1909; Gäumann & Fischer, 1929; Gäumann, 1946); Tropics (Delacroix, 1911; M. T. Cooke, 1913; Roger, 1951–4; Wellman, 1972); UK (Smith, 1884; Ward, 1889, 1901; M. C. Cooke, 1906; Massee, 1899, 1910; Brooks, 1928; Weston & Taylor, 1948; Bawden, 1948; Butler & Jones, 1949; Wood, 1967; Wheeler, 1969; Tarr, 1972); USA (Lamson-Scribner, 1890; Weed, 1894; Freeman, 1905; Duggar, 1909; Stevens & Hall, 1910; Whetzel *et al.*, 1916; Heald, 1926, 1937; Owens, 1928; Melhus & Kent, 1939; Leach, 1940 (insect transmission); Chester, 1942; Walker, 1950; Westcott, 1950; Stevens & Stevens, 1952; Stefferud, 1953; Fulton *et al.*, 1955; Stakman & Harrar, 1957; Holton *et al.*, 1959; Horsfall & Dimond, 1959–60; Carter, 1962 (insect transmission); Horsfall & Cowling, 1977–8); USSR (Naumov, 1926, 1940; Sigrianskii, 1936; Dement'eva, 1970; Cheremisinov, 1973); West Indies (Bancroft, 1910).

Atlases. Sorauer (1887–93), Delacroix (1901), Kirchner & Boltshauser (1897–1902), Dressel (1924).

Phytopathological methods. Whetzel *et al.* (1916), Rawlins (1933), Riker & Riker (1936), CMI (1968).

Host indexes and regional lists of diseases. Saccardo (1882–1972, 13), Oudemans (1919–24), Stevenson (1926), and Seymour (1929) are important host indexes.

Lists include: Butler & Bisby, 1931 (India); Waterston, 1947 (Bermuda); Tarr, 1955 (Sudan); Moore, 1959; Baker, 1972 (England & Wales); USDA, 1960 (USA); Foister, 1961 (Scotland); Conners, 1967 (Canada).

Regional lists of fungi and diseases are compiled in *Rev. Pl. Path.* **47**: 553–8, 1968; **58**: 305–8, 1979 (Africa); **49**: 103–8, 1970 (Asia); **50**: 1–7, 1971 (America); **51**: 1–7, 1972 (Europe); **54**: 963–6, 1975 (Australia & Oceania).

Fungi and fungal diseases. Stevens (1913), Lehmann *et al.* (1937, black rust of wheat), Holton & Heald (1941, wheat bunt), Chester (1946, brown rust of wheat), Ou (1965, rice blast), Byrde & Willets (1977, brown rot of fruit).

Bacteria and bacterial diseases. Smith (1905–14), Smith (1920), Elliott (1930), Dowson (1949), Stapp (1958), Gorlenko (1965), Rangaswami & Rajagopalan (1973).

Viruses and virus diseases. Smith (1937), Bawden (1939), Holmes (1939), Matthews (1957, serology), Corbett & Sisler (1964), Beale (1976, bibliog.), Gibbs & Harrison (1976).

Teratology. Moquin-Tandon (1841), Masters (1869), Penzig (1890), Küster (1903), Worsdell (1915–16), Butler (1930), Yoshi & Kawamura (1947).

Fungicides. Lodeman (1896), Bourcart (1910), Martin (1928), Mason (1928), Cunningham (1935), Horsfall (1945, 1956), Torgeson (1967–8), Marsh (1972).

Legislation and quarantine. McCubbin (1954), Ebbels & King (1979).

Before 1500

BC

8–5th   *Old Testament*: Genesis 41.23; Deuteronomy 28.22; Kings 8.37 (II Chronicles
cent.   6.28); Amos 4.9; Haggai 1.9, 2.16–17.
3rd cent. Theophrastus. *Historia plantarum* (in IX Books). [Eng. trans. by A. Hort,
          Theophrastus. Enquiry into plants, 2 vols., 1916 (Loeb Classical Library).] *De causis
          plantarum* (in VI Books). [Eng. trans. by B. Einarson & G. K. K. Link, 3 vols.
          (vol. 1, Books I and II, 1976; vols. 2, 3 to follow) (Loeb Classical Library); Latin
          trans. by G. Sneider, 1818 (Leipzig); F. Wimmer, 1854 (Leipzig), 1866 (Paris).]

AD

1st cent. Plinius *secundus*, Caius. *Historia naturalis* (in XXXVII Books). [Eng. trans. by
          J. Bostock & H. T. Riley, *The natural history of Pliny*, 6 vols., 1855–7 (and later
          reprints), London (Bohn's Classical Library).]
6–7th   Bassus, Cassianus. *Geoponika. De re rustica selectorum libri xx graeci*, 1539.
cent.   Basle.
9th cent. Raychaudhuri, S. P. & Kaw, R. K. (editors). *Agriculture in ancient India*, 1964. New
          Delhi (Ind. Counc. agric. Res.). [Chap. 7, Protection of crops from diseases and
          pests, pp. 84–100 includes extracts from Surapala's *Vrksyuveda*, AD 800.]
12th cent. Ibn-al-Awam. *Le livre de agriculture d' Ibn-al-Awam (Kitab-al-Felahah)*. Traduit de
          l'Arabe par J.-J. Clément-Mullet, vol. 1, 1864 (Paris). [Diseases of trees and other
          plants, chap. 14:453–97.]

1500

1556   Bock, J. [Tragus, H.]. *Kreuter Buch*. Strasbourg.
1574   Scot, Reynolde [Scott, Reginald]. *A perfite platforme of a hoppe garden, and necessarie
         instructions for the making and mayntenance thereof*. London.
1588   Tabernaemontanus [Dietrich, Jacob]. *Neuw Kreuterbuch*. Frankfurt-on-Main.

1600

1657   Austen, R. A. *A treatise of fruit-trees. Together with the spiritual use of an orchard...*
         Oxford. [Diseases, pp. 67–9.]
1665   Hooke, R. *Micrographia*. London.
1679   Malphigi, M. *Anatome plantarum pars altera*. P. 52, Tab. xxii. London.

1700

1705   de Tournefort, J. P. Observations sur les maladies des plantes. *Mém. Acad. Sci.
         Paris 1705*:332–45.
1714   Cogrossi, C. F. *Nuova idea del male contagioso de' buoi*. Milan. [Facsimile, with Eng.
         trans. by Dorothy M. Schullian, 1953, Rome (6th Internat. Congr. Microbiol.).]
1718   Bradley, R. *New improvements of planting and gardening, both philosophical and practical*,
         ... The third and last part. Of fruit trees. Chap. v. Of blights: 53–100. London.
1720   Cane, H. On the change of colour in grapes and jessamine. *Proc. roy. Soc.* **31**:102.
1723   Eysfarth, C. S. *Dissertatio de morbis plantarum*. 48 pp. Leipzig.
1727   Hales, S. *Vegetable staticks*. London. [Reprint, 1961, London (Oldbourne).]
1728   Duhamel de Monceau, H. L. Explication physique d'une maladie qui fait périr
         plusiurs plantes dans le Gastinois et particulièrement le safran. *Mém. Acad. Sci.
         Paris 1728*:100–12.
1729   Micheli, P. A. *Nova plantarum genera*. Florence.
1733   Tull, J. *The horse-hoing husbandry*. London.

1750

1755   Tillet, M. *Dissertation sur la cause qui corrompt et noircit les grains de bled dans les épis; et sur les moyens de prevenir ces accidens.* Bordeaux.

Suite des expériences et réflexions relatives à la dissertation sur la cause qui corrompt et noircit les grains de bled dans les épis ... Paris. [Eng. trans. of both by H. B. Humphrey, *Phytopath. Classics* 5, 191 pp., 1955.]

1756   Tillet, M. *Précis des expériences qui ont été faites par ordre du Roi à Trianon, sur la cause de la corruption des bleds et sur les moyens de la prévenir; à la suit duquel est une instruction propre à guider les laboureurs dans la manière dont ils doivent préparer le grain avant de le semer.* Troyes; Paris. [Reprinted 1787, Paris.]

1759   Ginanni, Conte Francesco. *Delle malattie del grano in erba;* ... Pesaro.

1760   Tillet, M. Observation sur la maladie du mais ou blé de Turquie. *Mém. Acad. Sci. Paris 1760:*254–61. [Vol. dated 1766.]

1761   Schulthess, H. Vorschlag einiger durch die Erfahrung bewahrter Hilfsmittel gegen den Brand in Korn. *Abehandl. Zürich Naturf. Ges.* 1:498–506.

1763   Adanson, M. *Familles des plantes* 2:4–12. Paris.

1767   Fontana, F. *Osservazioni sopra la ruggine del grano.* Lucca. [Eng. trans. by P. P. Pirone, *Phytopath. Classics* 2, 40 pp., 1932.]

Targioni-Tozzetti, G. *Alimurgia o sia modo di render meno gravi la caresti, proposto per sollievo de' poveri.* Florence. [Reprinted as *Reale Accademia d'Italia. Studi e documenti* 12, 1943; Eng. trans. of Chap. 5 ('True nature, causes and sad effects of the rust, the bunt, the smut, and other maladies of wheat, and of oats in the field') by L. R. Tehon, *Phytopath. Classics* 9, 139 pp., 1952.]

1773   Zallinger, J. B. *De morbis plantarum.* Innsbruck. [Germ. trans. 1779, 1809, fide Whetzel (1918):28.]

1774   Fabricius, J. C. Forsøg til en Afhandling om Planternes Sygdomme. *Det. Kongeelige Norske Videnskabers Selskabs Skrifter* 5:431–92. [Eng. trans. by M. K. Ravn, *Phytopath. Classics* 1, 66 pp., 1926.]

1783   Tessier, A. H. *Traité des maladies des grains.* Paris.

1784   Bryant, H. *A particular enquiry into the causes of that disease in the wheat commonly called brand,* ... Norwich.

1791   Forsyth, W. *Observations on the diseases, defects, and injuries, in all kinds of fruit and forest trees. With an account of a particular method of cure invented and practised by William Forsyth.* 71 pp. London. [Reprinted with additions in Forsyth (1802): 285–344; Germ. trans. (by G. Foister), 1791 (Mainz).]

1794   Plenck, J. J. von. *Physiologia et pathologia plantarum.* [Trans. into Germ., 1795, 1801, 1818; French, 1802; Ital., 1799.]

1795   Ehrenfels, J. M. Ritter von. *Ueber die Krankheiten und Verlezzungen der Frucht- oder Gartenbäume. Ein Buch für Landburger und Gartenfreund.* Breslau, etc.

Schreger, B. N. G. *Erfahrungsmässige Anweisung zur richtigen Kentniss der Krankheiten der Wald- und Gartenbäume* ... Leipzig.

1799   Knight, T. A. An account of some experiments on the fecundation of vegetables. *Phil. Trans.* 89:195–204.

1800

1802   Forsyth, W. *A treatise on the culture and management of fruit-trees* ... London.

1805   Banks, J. *A short account of the cause of the disease in corn, called by farmers the blight, the mildew, and the rust.* (With 2 pl. by F. Bauer.) London.

1806   Windt, L. G. *Der Beberitzenstrauch, ein Feind des Wintergetreides. Aus Erfahrungen, Versuchen und Zeugnissen.* Bückeburg ü. Hannover. [Ms. Eng. trans. in Brit. Mus. (Nat. Hist.).]

1807   Prévost, B. *Mémoire sur la cause immédiate de la carie ou charbon des blés, et de plusieurs*

autres maladies des plantes, et sur les préservatifs de la carie. Paris. [Eng. trans. by
G. W. Keitt, *Phytopath. Classics* **6**, 1939.]

Ré, F. *Saggio teorica practico sulle mallatie delle piante.* Venice. [Edn 2, 1817 (Milan);
Eng. trans. by M. J. Berkeley, *Gdnrs' Chron.* 1849:228–1850:469.]

1814   Dickson, T. Observations on the disease in the potato, generally called the curl.
Mem. Caledonian hort. Soc. **1**:49–59.

Weighton, D. On destroying insects, and removing mildew, and canker in
fruit-trees. *Mem. Caledonian hort. Soc.* **1**:131–42.

1818   Knight, T. A. On the mode of propagation of the *Lycoperdon cancellatum*, a species
of fungus, which destroys the leaves and branches of the pear tree. *Trans. hort.
Soc. Lond.* **2**:178.

1824   Robertson, J. On the mildew and some other diseases incident to fruit trees.
Trans. hort. Soc. Lond. **1**:175–85.

1825   Loudon, J. C. *An encyclopaedia of agriculture.* London.

Young, J. On the cause of curl in the potato, and means of prevention. *Mem.
Caledonian hort. Soc.* **3**:278–86.

1833   Unger, F. *Die Exantheme der Pflanzen und einige mit diesen verwandte Krankheiten der
Gewächse, pathogenetisch und nosographisch dargstellt.* Vienna.

1839   Wiegmann, A. F. *Die Krankheiten und Krankhaften Missbildungen der Gewächse.*
Brunswick.

1841   Meyen, F. J. F. *Pflanzenpathologie.* Berlin.

Moquin-Tandon, A. *Elémens de tératologie végétale,* ... Paris. [Germ. trans. by J. C.
Schauer, 1842 (Berlin).]

1842   Knight, T. A. Upon the causes of the diseases and deformities of the peach-tree.
Trans. hort. Soc. Lond. Ser. 2, **2**:27–9.

1845   Morren, C. F. A. Sur le maladie des pommes-de-terre. *Bull. Acad. Sci. Bruxelles*
12:372–6.

1846   Berkeley, M. J. Observations, botanical and physiological, on the potato
murrain. *J. hort. Soc. Lond.* **1**:9–34. [Reprinted in *Phytopath. Classics* **8**, 1948.]

Taylor, S. Smut in wheat, *Gdnrs' Chron. 1846*:242–3.

1847   Berkeley, M. J. Observations on the propagation of bunt (*Uredo caries*, D.C.)
made with especial reference to the potato disease. *J. hort. Soc. Lond.* **2**:107–14.

1850

1850   Anon. American mode of preventing potato rot. *Gdnrs' Chron. 1850*:503.

Mitscherlich, E. 18 Marz. Sitzung der physikalischmathematischen Klasse.
Monatsschr. Kais. Akad. Wiss. Berlin 1850:102–10.

1852   Bergman. *Gdnrs' Chron. 1852*:419–20.

Schleiden, [M. J.] Observations on the diseases of plants, as illustrated by the
potato murrain. *J. hort. Soc. Lond.* **7**:178–93.

1853   de Bary, A. *Untersuchungen über die Brandpilze und die durch sie verursachten Krankheiten
der Pflanzen* ... Berlin. [Eng. trans. by R. M. S. Heffner, D. C. Arny & J. D.
Moore, *Phytopath. Classics* **11**, 1969.]

1854   Regel, C. *Die Schmarotzergewächse und die mit denselben in verbundung stehenden
Pflanzen-Krankheiten.* Zurich.

1854–7 B[erkeley], M. J. Vegetable pathology [173 articles]. *Gdnrs' Chron.* 7 Jan. 1854:20
– 3 Oct. 1857:549. [Selection reprinted in *Phytopath. Classics* **8**, 1948.]

1857   Speerschneider, J. Dass das Faulen der Kartoffelknollen bei der sogenannten
Kartoffelkrankheit durch die ausgestreuten und keimenden Sporen des
Blattpilzes (*Peronospora devastatrix*) verursacht wird, durch Experimente beweisen.
Flora **40** (NS **15**): 81–7.

1858   Kühn, J. *Die Krankheiten der Kulturgewächse, ihre Ursachen und ihre Verhütung.* Berlin.

1861   de Bary, A. *Die gegenwärtig herrschende Kartoffelkrankheit, ihre Ursache und ihre Verütung*. 75 pp. Leipzig.

de Bary, A. Recherches sur le développement de quelques champignons parasitaires. *Ann. sci. nat.* Sér. 4 Bot. **20**:5–148.

1863   Ørsted, A. S. *Om Sygdomme hos Planterne, som foraarsages af Snyltesvampe, navnlig om Rust og Brand* ... Copenhagen.

1864   * Jessen, K. F. W. *Botanik der Gegenwart und Vorseit in culturhistorischer Entwicklung.* Leipzig. [Chronica Botanica reprint, 1948.]

1865   Cooke, M. C. *Rust, smut, mildew and mould.* London. [Edn 6, 1897.]

Ørsted, A. S. *Om Växtsjukdomar.* Örebro.

1865–6   de Bary, A. Neue Untersuchungen über Uredineen, inbesondere die Entwicklung der *Puccinia graminis. Monatsber. Kon. Akad. Wiss. Berlin* 1865:15–49; 1866:205–15.

1866   Davaine, C. Recherches sur la pourriture des fruits. *C. R. Acad. Sci. Paris* **63**:276–9; ... et des autres parties des végétaux vivants. *Ibid.* **63**:344–6.

1868   Davaine, C. (a) Sur la nature des maladies charbonneuses. *Archiv. gén. Med.* **1**:147–8.

Davaine, C. (b) Recherches physiologiques et pathologiques sur les bactéries. *C. R. Acad. Sci. Paris* **66**:499–503.

Hallier, E. *Phytopathologie. Die Krankheiten der Culturgewächse für Land-und Forstwirthe, Gärtner und Botaniker.* Leipzig.

1869   Masters, M. T. *Vegetable teratology.* London (Ray Soc.).

1871   * Pritzel, G. A. *Thesaurus literaturae botanicae.* Leipzig. [Reprint, 1950 (Milan).] [Valetudo et morbi plantarum, pp. 526–7.]

1872–   Brefeld, O. *Untersuchungen aus dem Gesammtgebeit der Mykologie.* Heft I, 1872; II,
1912   1874; III, 1877; IV, 1881; V, 1883; VI, 1884; VII, 1888; VIII, 1889; IX, X, 1891; XI, XII, 1895; XIII, 1905; XIV, 1908; XV, 1912. Leipzig; Munster; Berlin.

1873   Kühn, J. G. Die Anwendung des Kupfervitrioles also Schutzmittel gegen den Steinbrand des Weizens. *Bot. Z.* **31**:502–5.

1874   Hartig, R. *Wichtige Krankheiten der Waldbäume.* Berlin [Eng. trans. by W. Merrill, D. H. Lambert & W. Liese, *Phytopath. Classics* **12**, 1975.]

Sorauer, P. *Handbuch der Pflanzenkrankheiten.* 1 vol. Berlin. [Aufl. 2, 2 vols., 1886; Aufl. 3, 3 vols., 1905–13; currently Aufl. 7, a multi-author multi-volume work, 1962– .] See also Sorauer (1914).

1875

1875   Smith, W. G. Resting-spores of the potato fungus. *Nature* **12:234**.

1876   de Bary, A. Researches into the nature of the potato-fungus, *Phytophthora infestans. J. roy. agric. Soc.* Ser. 2, **12**:239–69. [Also *J. Bot. Lond.* **14**:105–26, 149–54, 1876.]

1877   Farlow, W. G. Onion smut: an essay presented to the Massachusetts Society for Promoting Agriculture. *24th Ann. Rep. Sec. Mass. Sta. Bd Agric.*: 164–76.

1878   d'Arbois de Jubanville & Vesque, J. *Les maladies des plantes cultivées, des arbres frutiers et forestiers.* Paris.

Burrill, J. T. Pear-blight. *Trans Ill. Sta hort. Soc.* NS **11**(1877):114–16.

Winter, G. *Die durch Pilze verursachten Krankheiten der Kulturgewächse.* Leipzig.

Woronin, M. *Plasmodiophora brassicae*, Urheber der Kohlpflanzen-Hernie. *Jahrb. wiss. Bot.* **11**:548–74. [Eng. trans. by C. Chupp, *Phytopath. Classics* **4**, 1934.]

1879   Burrill, T. J. (On fire blight.) *Trans. Ill. Sta. hort. Soc.* NS **12** (1878): 79–80.

Thümen, F. K. A. *Fungi pomicola.* Vienna.

1880   Frank, A. B. *Die Krankheiten der Pflanzen. Ein Handbuch* ... Breslau. [Edn 2, 3 vols., 1895–6.]

1882   Burrill, T. J. The bacteria: an account of their nature and effects together with a
        systematic description of the species. *Rep. Ill. Industr. Univ.* 11:126, 134.
        Comes, O. *Le crittogame parassite delle piante agrarie*. Naples.
        Hartig, R. *Lehrbuch der Baumkrankheiten*. Berlin. [Edn 2, 1889 (Eng. trans. by
        W. Somerville & H. M. Ward as 'Textbook of diseases of trees', 1894
        (London)); edn. 3 (as *Lehrbuch der Pflanzenkrankheiten*), 1900.]
        Sorauer, P. *Die Obstbaumkrankheiten*. Berlin.
        Ward, H. M. Researches on the life-history of *Hemileia vastatrix*, the fungus of
        'coffee-leaf disease'. *J. Linn. Soc. (Bot.)* 19:299–335. [See also *Quart. J. microscop.
        Soc.* NS 22:1–11, 1882 (morphology of *H. vastatrix*).]
1882–   Saccardo, P. A. *Sylloge fungorum hucusque cognitorum*. 26 vols. Pavia. [Vol. 13 (host
1972   index), 1898.]
1883   Frank, A. B. Die Fleckenkrankheit der Bohnen, veranlasst durch *Gloeosporium
        lindemuthianum* Sacc. et Magnus. *Landw. Jahrb.* 12:511–23.
        Wakker, J. H. Vorlaufige Mittheilungen über Hyacinthenkrankheiten. *Bot.
        Centralb.* 14:315–17.
1884   Smith, W. G. *Diseases of field and garden crops, chiefly such as are caused by fungi*.
        London.
1885   Arthur, J. C. Proof that bacteria are the direct cause of the disease in trees
        known as pear blight. *Proc. Am. Ass. Adv. Sci.* 23:295–8 (also *Bot. Gaz.* 10:343–5,
        1885; *Gdnrs' Chron.* 24:586, 1885).
        Millardet, P. M. A. Traitement du mildiou et du rot. *J. Agric. pract. Paris*
        2:513–16. [For Eng. trans. of this and other papers by Millardet relating to
        Bordeaux mixture see F. J. Schneiderhan, *Phytopath. Classics* 3 (1933).]
        Thümen, F. von. *Die Bekämpfung der Pilzkrankheiten unserer Culturgewächse*. Vienna.
        Vialla, P. *Les maladies de la vigne*. Montpellier. [Edn 3, 1893 (Paris).]
1886   Arthur, J. C. History and biology of pear blight. *Proc. Philadelphia Acad. nat. Sci.
        1886*:322–41.
        de Bary, A. Ueber einige Sclerotinien und Sclerotienkrankheiten. *Bot. Zeit.*
        44:377–87, 392–403, 408–26, 432–41, 448–61, 464–74.
        Lamson-Scribner, F. Report on the fungus diseases of the grape vine. *Bull. Dep.
        Agric. Bot. Div. Sect. Pl. Path.* 11, 136 pp.
        Mayer, A. Ueber die Mosaikkrankheit des Tabaks. *Landw. Vers.-Stn* 32:451–67.
        [Eng. trans. by J. Johnson, *Phytopath. Classics* 7:11–24, 1942.]
1887   Brunchorst, J. *De vigtste plantesydomme*. Bergen.
        Masson, E. Nouveau procédé bourguignon contre mildou. *J. Agric. pract. Paris*
        51:814–16.
        Patrigeon, G. Nouveaux procédés de traitement du mildou à l'aide de
        hydrocarbonate de cuivre. *J. Agric. pract. Paris* 51:879–84.
        Savastano, L. Tubercolosi ipeplasie et tumori dell'olivo. *Ann. R. Soc. sup. Agric.
        Portici* 5(4), 127 pp.
        Shipley, A. E. Onion disease in Bermuda, caused by *Peronospora schleideniana*. *Kew
        Bull. 1887*(10):1–22.
        Thümen, F. K. A. von. *Die Pilze der Obstgewächse*. Vienna.
        Wolf, R. *Krankheiten der landwirtschaftlichen Nutzpflanzen durch Schmarotzerpilze*.
        Berlin.
1887–93 Sorauer, P. *Atlas der Pflanzenkrankheiten*. Heft 1–6, 48 col. Taf. Berlin.
1888   Brefeld, O. Neue Untersuchungen über die Brandpilze und die
        Brandkrankheiten. *Nachr. Klub. Landw. Berlin*; 220:1577–84; 221:1588–94;
        222:1597–1601. [Eng. trans. by E. F. Smith, *J. Mycol.* 6:1–8, 59–71, 153–64,
        1890. See also Brefeld (1872–1912).]
        Frank, B. Ueber den Einfluss welchen das Sterilisieren des Erdbodens auf die

Pflanzenentwicklung ausübt. *Ber. dtsch. bot. Ges.* **6**:lxxxvii–xcvii.

Ward, H. M. A lily disease. *Ann. Bot. Lond.* **2**:319–82.

1888–94   Smith, E. F. Peach yellows: a preliminary report. *Bull. Dep. Agric. Bot. Div. Sect. Pl. Path.* **9**, 254 pp., 1888; Additional evidence on the communicability of peach yellows and peach rosette, *Bull. USDA Div. veg. Path.* **1**:65 pp., 1891; Experiments with fertilizers for the prevention and cure of peach yellows, *ibid.* **4**: 197 pp., 1893; Peach yellows and peach rosette, *USDA Farmers' Bull.* **17**: 20 pp., 1894.

1889   Galloway, B. T. A new modification of the Vermorel nozzle. *J. Mycol.* **5**:96–7.

Jensen, J. L. The propagation and prevention of smut in oats and barley. *J. roy. agric. Soc. Engl. Ser.* 2, **24**:397–415.

Plowright, C. B. *A monograph of the British Uredineae and Ustilagineae.* London.

Ward, H. M. *Diseases of plants.* London.

Weed, C. M. On the combination of insecticides and fungicides. *Agric. Sci.* **3**:263–4.

1890   Bolley, H. L. Potato scab a bacterial disease. *Agric. Sci.* **4**:243–56, 277–87.

Kellerman, W. A. & Swingle, W. T. Prevention of smut in oats and other cereals. *J. Mycol.* **6**:26–9.

Kirchner, O. *Die Krankheiten und Beschädigungen unserer landwirtschaftlichen Kulturpflanzen.* Stuttgart.

Lamson-Scribner, F. *Fungus diseases of the grape and other plants and their treatment.* Little Silver, N.J.

Penzig, O. *Pflanzen-Teratologie.* 2 vols. Genoa. [Edn 2, 3 vols., 1921–2 (Berlin).]

*Sachs, J. von. *History of botany (1530–1860),* trans. by H. E. F. Garnsey and revised by I. B. Balfour. Oxford, J. R. Green, *A history of botany, 1860–1900, being a continuation of Sachs' History of botany,* 1909. Oxford.

*Savastano, L. La patologia vegetale dei Greci, Latini ed Arabi. *Memoria:* 1–75. Portici. (Scula sup. Agric. Portici Ann. **6**:1890–1.)

Thaxter, R. On certain diseases of the onion (*Allium cepa*). *Ann. Rep. Conn. agric. Exp. Stn 1889:* 129–67.

Ward, H. M. Croonian lecture. On some relations between host and parasite in certain epidemic diseases of plants. *Proc. roy. Soc.* **47**:393–443.

1891   Chester, F. D. *Rep. Delaware agric. Res. Stn 1891:*71.

Comes, O. *Crittogamia agraria.* Naples.

Galloway, B. T. Description of a new knapsack sprayer. *J. Mycol.* **6**:52–9.

*Sorauer, P. Errichtung einer phytopathologischen Versuchsstation in Rumanien. *Z. PflKrankh.* **1**:257.

Thaxter, R. Further experiments on the 'smut' of onions. *Ann. Rep. Conn. agric. Exp. Stn. 1890:*103–4.

Waite, M. B. Results from recent investigations in pear blight. *Bot. Gaz.* **16**:259.

1892   Beach, S. A. Blight of common beans. Blight of lima beans. *Bull. N. Y. agric. Exp. Stn* **48**:329–31.

Frank, A. B. & Sorauer, P. *Pflanzenschutz. Leitfaden für den praktischen Landwirth zur Erkennung und Bekämpfung der Beschädigungen der Culturpflanzen.* Berlin. [Aufl. 9, by E. Riehm & M. Schwartz, 1935.]

Ivanovski, D. Ueber die Mosaikkrankheit der Tabakspflanze. *St Petersb. Acad. Imp. Sci. Bull.* **35** (ser. 4, vol. 3): 67–70. [Eng. trans. by J. Johnson, *Phytopath. Classics* **7**:27–30, 1942.]

Wortmann, J. Ueber die sogen[annte] 'Stippen' der Aepfel. *Landw. Jahrb.* **21**:663–75.

1893   Bolley, H. L. Prevention of potato scab. *Bull N. Dakota agric. Exp. Stn* **9**: 27–41.

    * Sorauer, P. Die Bewegung auf Phytopathologischen Gebiete in der Schweiz. *Z. PflKrankh.* **3**:321–2.

1893–4 Shirai, M. *Shokubutsu Byòri Gaku.* [1st Japanese book on plant disease fide Akai (1974):15.]

1894 Aderhold, R. Die Perithecienform von *Fusicladium dendriticum. Ber. dtsch. bot. Ges.* **12**:338–42.

Eriksson, J. Ueber die Specialisirung des Parasitismus bei den Getreiderostpilzen. *Ber. dtsch. bot. Ges.* **12**:292–331.

Weed, C. M. *Fungi and fungicides. A practical manual* ... New York.

1895 Cobb, N. A. The cause of gumming in sugar cane. *Agric. Gaz. N.S.W.* **6**:683–9.

Pammel, L. H. Bacteriosis of rutabaga (*Bacillus campestris* n.sp.). *Bull. Iowa agric. Exp. Stn.* **27**:130–4.

Tubeuf, K. F. von. *Pflanzenkrankheiten durch kryptogame Parasiten verursacht.* Berlin. [Eng. trans. by William G. Smith, 'Diseases of plants induced by cryptogamic parasites', 1896 (London).]

1895–7 Prillieux, E. E. *Maladies des plantes agricoles et des arbres fruitiers et forestiers causées par des parasites végétaux.* 2 vols. Paris.

Smith, E. F. *Bacillus tracheiphilus* sp. nov., die Ursache des Welkens verschiedener Cucurbitaceen. *Zbl. Bakt.* Abt. II **1**: 364–73.

1896 Eriksson, J. & Henning, E. *Die Getreiderost, ihre Geschichte und Natur, sowie Massregeln gegen dieselben.* Stockholm.

Lodeman, E. G. *The spraying of plants.* New York & London.

Maddox, F. Smut and bunt. *Agric. Gaz. Tasmania* **4**:92–5.

Smith, E. F. A bacterial disease of the tomato, eggplant and Irish potato. *Bull. USDA Div. veg. Physiol. Path.* **12**, 28 pp.

1896– Kirchner, O. & Boltshauser. *Atlas der Krankheiten und Beschädigungen unserer*
1902 *landwirtschaftlichen Kulturpflanzen.* 126 col. pl. (in 6 series). Stuttgart.

1897 Bolley, H. L. New studies upon the smut of wheat, oats, and barley, with a résumé of treatment experiments for the past three years. *Bull. N. Dakota agric. Exp. Stn* **27**:109–62.

1897– Eriksson, J. Vie latente et plasmatique de certain Uredinées. *C.R. Acad. Sci.*
1921 *Paris.* **124**:475–7, 1897; The mycoplasm theory – is it dispensable or not? *Phytopathology* **11**:385–8, 1921.

1898 Beijerinck, M. W. Ueber ein contagium vivum fluidum als Ursache der Fleckenkrankheiten der Tabaksblatter. *Verh. Kon. Akad. Wettenshappen Amsterdam* **65**(2):3–21. [Eng. trans. by J. Johnson, *Phytopath. Classics* **7**:33–52, 1942.]

1899 Arthur, J. C. Formalin for grain and potatoes. *Bull. Indian agric. Exp. Stn* **77**:38–44.

McAlpine, D. *Fungus diseases of citrus trees in Australia and their treatment.* Melbourne.

Massee, G. *A text-book of plant diseases caused by cryptogamic parasites.* London. [Edn 3, 1907.]

Peglion, V. *Le malattie crittogamiche delle piante coltivate.* Casale. [dn 5, 1928.]

Smith, E. F. Wilt disease of cotton, watermelon, and cowpea (*Neocomospora nov. gen.*). *Bull. USDA Div. veg. Physiol. Path.* **17**, 72 pp.

1900

1900 Orton, W. A. The wilt disease of cotton and its control. *Bull. USDA Div. veg. Physiol. Path.* **27**, 15 pp.

Salmon, E. S. A monograph of the Erysiphaceae. *Mem. Torrey bot. Club* **9**, 292 pp. (Suppl. *Bull. Torrey bot. Club* **29**:1–22, etc., 1902.)

Selby, A. D. Onion smut. Preliminary experiments. *Bull. Ohio agric. Exp. Stn* **122**:71–84.

1901   Bolley, Flax wilt and flax sick soil. *Bull. N. Dakota agric. Exp. Stn* **50**:27–60.

Delacroix, G. *Atlas des conférences de pathologie végétale.* 56 pl. Paris.

Henry, E. Action du sulfure de carbone sur la végétation de quelques plantes forestiers. *Bull. Séanc. Soc. Sci. Nancy* Sér. 3, **2**:27–33.

Ideta, A. Jitsuyō Shokubutsu Byōrigaku [Practical handbook of vegetable pathology. Jap.]. Tokyo. [Germ. trans. 'Lehrbuch der Pflanzenkrankheiten in Japan', 1903 (Tokyo).]

Jones, L. R. *Bacillus carotovorus* n.sp., die Ursache einer weichen Fäulnis der Möhre. *Zbl. Bakt.* Abt. II **7**:12–21, 61–8.

Smith, E. F. The cultural characters of *Pseudomonas hyacinthi, Ps. campestris, Ps.phaseoli,* and *Ps.stewarti* – four one-flagellate yellow bacteria parasitic on plants. *Bull. USDA Div. veg. Physiol Path.* **28**:7–153.

Ward, H. M. *Disease in plants.* London.

1902   Delacroix, G. *Maladies des plantes cultivées.* Paris. [Edn 3, 2 vols., 1926.]

McAlpine, D. *Fungus diseases of stone-fruit trees in Australia and their treatment.* Melbourne.

Orton, W. A. Some diseases of cowpea. *Bull. USDA Bur. Pl. Industr.* **17**, 36 pp.

1902   Rostrup, E. *Plantepatologi. Haandbogi i Laeren om Plantesygdomme for Landbrugere, Havebrugere og Skorbrugere.* Copenhagen.

Van Hall, C. J. J. Bijdragen tot de kennis der bacterieelle plantenziekten. Thesis, Univ. Amsterdam.

Woods, A. F. Observations on the mosaic disease of tobacco. *Bull. USDA Bur. Pl. Industr.* **18**, 24 pp.

1902–3   Ward, H. M. The bromes and their rust-fungus (*Puccinia dispersa*). *Ann. Bot. Lond.* **15**:560–2, 1902; Further observations on the brown rust of the bromes, ... *Ann. mycol. Berl.* **1**:132–51.

1903   *Jaczewski, A. de. Le laboratoire central de pathologie végétale du Ministère de l'Agriculture à St-Petersburg. *Bull. Soc. mycol. Fr.* **19**:324–9.

Küster, E. *Pathologische Pflanzenanatomie.* Jena. [Edn 3, 1925].

Massee, G. On a method of rendering cucumber and tomato plants immune against fungus parasites. *J. roy. hort. Soc.* **28**:142–5.

Smith, E. F. Observations on a hitherto unreported bacterial disease, the cause of which enters the plant through ordinary stomata. *Science* **17**:456–7.

Ward, H. M. On the histology of *Uredo dispersa* Erikss., and the 'mycoplasm' hypothesis. *Phil. Trans.* B**196**:29–46. [See also *Ann. Bot. Lond.* **19**:21–35, 1905.]

1904   Bauer, E. Zur Aetiologie der infectiösen Panachierung. *Ber. dtsch. bot. Ges.* **22**:453–60. [Eng. trans. by J. Johnson, *Phytopath. Classics* **7**:55–62, 1942.]

Biffen, R. H. Experiments with wheat and barley hybrids illustrating Mendel's Laws of heredity. *J. roy. agric. Soc.* **65**:337–45.

Klebahn, H. *Die wirtswechselnden Rostpilze.* Berlin.

McAlpine, D. Take-all and whiteheads in wheat. *J. Dep. Agric. Victoria* **2**:410–26.

Ridley, H. N. Parasitic fungi on *Hevea brasiliensis. Agric. Bull. Straits Settlements* **3**:173–5.

Salmon, E. S. Recent researches on specialization of parasitism in the Erysiphaceae. *New Phytol.* **3**:55–60.

Smith, E. F. & Swingle, D. B. The dry rot of potato due to *Fusarium oxysporum. Bull. USDA Bur. Pl. Industr.* **55**, 64 pp.

1905   Aderhold, R. & Ruhland, W. Zur Kentniss der Obstbaum-Sklerotinien. *Arb. biol. Abt. (Anst. Reichanst.) Berlin* **4**:427–42.

Butler, E. J. The bearing of Mendelism on the susceptibility of wheat to rust. *J. agric. Sci.* **1**:361–3.

Freeman, E. M. *Minnesota plant diseases.* St Paul, Minn.

Jones, L. R. The cytolic enzyme produced by *Bacillus carotovorus* and certain other soft-rot bacteria. *Zbl. Bakt.* Abt. II **14**:257–72.

1905–14 Smith, E. F. *Bacteria in relation to plant diseases.* **1**, 1905; **2**, 1911 [History, pp. 7–22]; **3**, 1914. Washington (Carnegie Inst.).

1906 Bolley, H. L. Tree feeding. *Rep. N. Dakota agric. Exp. Stn* **17**:104.

Cooke, M. C. *Fungoid pests of cultivated plants.* London. [Reprinted from *J. roy. hort. Soc.* **27**–9, 1902–4.]

McAlpine, D. *The rusts of Australia. Their structure, nature and classification.* Melbourne.

1907 Smith, E. F. & Townsend, C. O. A plant tumour of bacterial origin. *Science* **25**:671–3.

1908 Biffen, R. H. Rust in wheat. *J.Bd. Agric.* **15**:241–53.

Cordley, A. B. Lime-sulfur spray as a preventive for apple scab. *Rural New Yorker,* March 1: 202.

Ducomet, V. *Pathologie végétale. Maladies parasitaires, champignons-bactéries.* Paris.

Potter, M. C. On a method for checking parasitic disease in plants. *J. agric. Sci.* **3**: 102–7.

Scott, W. M. Self-boiled lime-sulphur mixture as a promising fungicide. *Cir. USDA Bur. Pl. Industry* **1**; 18 pp.

Smith, E. F. Recent studies of the olive-tubercle organism. *Bull. USDA Bur. Pl. Industr.* **131** (part iv):25–43.

1908–9 Delacroix, G. & Maublanc, A. *Maladies des plantes cultivées.* I, *Maladies non-parasitaires,* 1908; II, *Maladies parasitaires,* 1909. Paris.

1908–17 *Lindau, G. & Sydow, P. Thesaurus litteraturae mycologicae et lichenologicae,* 5 vols. Leipzig. [Krankheiten der Pflanzen durch Pilze, **4**:244–609, 1915.] *Supplementum,* 1911–30 by R. Ciferri, 4 vols., 1957–60 (Cortina).

1909 Brown, Nellie A. A new bacterial disease of the sugar-beet leaf. *Science* **29**:915.

Duggar, B. J. *Fungous diseases of plants, with chapters on physiology, culture methods and techniques.* Boston, etc.

Faes, H. *Les maladies des plantes cultivées et leur traitement.* Paris: Lausanne.

Freeman, E. M. & Johnson, E. C. The loose smut of barley and wheat. *Bull. USDA Bur. Pl. Industr.* **152**, 48 pp.

Salmon, E. S. On the making and application of Bordeaux mixture; with notes on 'Bordeaux injury'. *J.S.E. agric. Coll. Wye* **18**:240–66. [Also *J.Bd. Agric.* **16**:793–810, 1910.]

1910 Agulhon, H. Emploi du boron comme engrais catalytique. *C. R. Acad. Sci. Paris* **150**:288–91; Accoutumance du maïs au bore, *ibid.* **151**:1382–3.

Arthur, J. C. & Johnson, A. G. The loose smut of oats and stinking smut of wheat and their prevention. *Circ. Purdue Univ. agric. Exp. Stn* **22**, 15 pp.

Bancroft, K. *A handbook of the fungus diseases of West Indian plants.* London.

Bourcart, E. *Les maladies des plantes. Leur traitement raisonné et efficace en agriculture et en horticulture.* Paris. [Eng. trans. by D. Grant, 1913; of edn 2, by T. R. Burton, 1925 (London).]

Eriksson, J. *Landtbruksväxternas svampsjukdomar.* Stockholm. [Eng. trans. by Anna Molander, 1912; of edn 2 by W. Goodwin, 1930 (London); Germ. trans. 1913.]

Güssow, H. T. Outbreak of potato canker (*Chrysophlyctis endobiotica* Schilb.) in Newfoundland, and the dangers of its introduction into the United States. *Science* **31**:796.

Harding, H. A., Morse, W. J. & Jones, L. R. The bacterial soft rots of certain vegetables. *Bull. Vt agric. Exp Stn* **147**:241–360 [Part 1 (pp. 243–79). The mutual relationships of the causal organisms by Harding & Morse; Part 2 (pp. 281–360) Pectinase, the cytolytic enzyme produced by *Bacillus carotovorus* by Jones (see also L. R. Jones, 1905).]

McAlpine, D. *The smuts of Australia. Their structure, life history, treatment, and classification.* Melbourne.

Massee, G. *Diseases of cultivated plants and trees.* London. [Edn. 2, 1915.]

Reddick, D. & Wallace, E. On a laboratory method of determining the fungicidal value of a spray mixture or solution. *Science* **31**:798.

*Shear, C. L. The American Phytopathological Society. *Science* **31**:746–57, 790–8. [See also *Phytopathology* **9**:165–70, 1919.]

Smith, E. F. (a) A Cuban banana disease. *Science* **31**:754–5.

Smith, E. F. (b) A new tomato disease of economic importance. *Science* **31**:794–6.

Stevens, F. L. & Hall, J. G. *Diseases of economic plants.* New York. [Edn 2, 1921.]

1911   Appel, O. & Schlumberger, O. Die Blattrollkrankheit und unsere Kartoffellernten. *Arbeit. dtsch. Landw. Ges.* Heft **190**:1–102.

Clinton, G. P. Oospores of potato blight. *Science* **33**:744–7.

Delacroix, G. *Maladies des plantes cultivées dans les pays chaud. Ouvrage terminé et publié par A. Maublanc.* Paris.

Freeman, E. M. & Johnson, E. C. The rusts of grains in the United States. *Bull. USDA Bur. Pl. Industr.* **216**, 87 pp.

Gifford, G. H. The damping off of conifer seedlings. *Bull. Vt agric. Exp. Stn* **157**:143–71.

McAlpine, D. *Handbook of fungus diseases of the potato in Australia and their treatment.* Melbourne.

Rorer, J. R. A bacterial disease of bananas and plantains. *Phytopathology* **1**:45–9.

1911–16 McAlpine, D. *Bitter pit investigations. First progress report,* 197 pp., 1911–12; *Second,* 224 pp., 1912–13; *Third,* 176 pp., 1913–14; *Fourth,* 178 pp., 1914–15; *Fifth,* 144 pp., 1915–16. Melbourne.

1912   Bancroft, K. A root disease of the Para rubber tree (*Fomes semitostus*). *Bull. Dep. Agric. Fed. Malay States* **13**, 30 pp.

Cunningham, G. C. The relationship of *Oospora scabies* to the higher bacteria. *Phytopathology* **2**:97 (abst.).

Klebahn, H. *Grundzüge der allgemeinen Phytopathologie.* Berlin.

Lang, W. Zum Parasitismus der Brandpilze. *Jahresb. Ver. angew. Bot.* **10**:172–80.

Spaulding, P. & Field, Ethel C. Two dangerous imported plant diseases. *USDA Farmers' Bull.* **489**, 29 pp.

1913   Blogett, F. M. Hop mildew. *Bull. Cornell agric. Exp. Stn* **328**:277–312.

Cook, M. T. *The diseases of tropical plants.* London.

Quanjer, H. M. Die Nekrose des Phloëms der Kartoffelpflanze, die Ursache der Blattrollkrankheit. *Meded. Rijks Hoogere Land-, Tuin- en Boschbouschool* **6**(2):41–80.

Ferraris, T. *I parasiti vegetale delle piante coltivate od utile.* Alba. [Mimeographed issue, 1910–11; edn 5 (revised by Ciferri & Baldacci, 1948.)]

*Lind, J. *Danish fungi as represented in the herbarium of E. Rostrup.* Copenhagen. [History of phytopath. pp. 19–25.]

Riehm, E. Prüfung einiger Mittel zur Bekämpfung des Steinbrandes. *Mitt. Kaiserl. Biol. Anstalt. Land-Forstw.* **14**:8–9. [*Zbl. Bakt.* Abt. II **40**:424, 1914 (abst.).]

Smith, E. F. A new type of bacterial disease. *Science* **38**:926.

Spinks, G. T. Factors affecting susceptibility to disease in plants. *J. agric. Sci. Cambr.* **5**:231–47.

Stakman, E. C. A study of cereal rusts: physiological races. Thesis, Univ. Minnesota (*Bull. Minn. agric. Exp. Stn* **138**), 56 pp.

Stevens, F. L. *The fungi which cause plant disease.* New York. [Replaced by *Plant disease fungi,* 1925.]

1913–15 Bartholomew, E. T. Black heart of potatoes. *Phytopathology* **3**:180–2, 1913; A pathological and physiological study of the black heart of potato tubers, *Zbl. Bakt.* Abt. II **43**:609–39, 1915.

1914   Allard, H. A. The mosaic disease of tobacco. *Bull. USDA* **40**, 33 pp.

Anderson, P. J. & Rankin, W. H. Endothia canker of chestnut. *Bull. Cornell agric. Exp. Stn* **347**:529–618.

Brooks, C. Blossom-end rot of tomatoes. *Phytopathology* **4**:345–74.

Güssow, H. T. The systematic position of the organism of the common potato scab. *Science* **39**:431–2.

Lutman, B. F. & Cunningham, G. C. Potato scab. *Bull. Vt agric. Exp. Stn* **184**, 64 pp.

Orton, W. A. Potato wilt, leaf-roll, and related diseases. *Bull. USDA* **64**, 45 pp.

Ravn, F. K. *Smitsomme sygdomme hos landbrugsplanterne.* Copenhagen.

*Sorauer, P. Historical survey. *Manual of plant diseases*, edn 3 (trans. by Frances Dorrance) **1**:37–68.

1914–21   Mangin, L. *Parasites végétaux des plantes cultivées.* I, *Céréales...*,1914; II, *Vigne...*, 1921. Paris.

1915   Brown, W. Studies in the physiology of parasitism. I. The action of *Botrytis cinerea. Ann. Bot. Lond.* **29**:313–48.

Hasse, Clara. *Pseudomonas citri*, the cause of citrus canker. (A preliminary report.) *J. agric. Res.* **4**:97–100.

*Hecke, L. *Die wissenschaftliche Entwicklung der Phytopathologie.* Vienna.

Lind, J. Berberisbusken og berberisloven. *Tidsskr. Planteavl* **22**:729–80.

Smith, E. F. & Bryan, Mary K. Angular leaf-spot of cucumbers. *J. agric. Res.* **5**:465–76.

Wiltshire, S. P. Infection and immunity studies on the apple and pear scab fungi (*Venturia inaequalis* and *V. pirini*). *Ann. appl. Biol.* **1**:335–50.

1915–16   Worsdell, W. C. *The principles of plant-teratology.* 2 vols. London (Ray Soc.).

1915–19   Mazé, P. Détermination des éléments minéraux rares nécessaire au développement du maïs. *C. R. Acad. Sci. Paris* **160**:211–14, 1915; Recherche d'une solution purement minérale capable d'assurer l'évolution complète du maïs cultivé a l'abri des microbes, *Ann. Inst. Pasteur Paris* **33**:139–73, 1919.

1915–25   Jones, L. R. *et al.* The control of cabbage yellows through disease resistance. *Res. Bull. Wisc. agric. Exp. Stn* **38**, 69 pp., 1915; Fusarium resistant cabbage, *ibid.* **48**, 34 pp., 1920; Fusarium resistant cabbage: progress with second early varieties. *J. agric. Res.* **30**:1027–34, 1925.

1916   Allard, H. A. Some properties of the virus of the mosaic disease of tobacco. *J. agric. Res.* **6**:649–74.

Eyre, J. V. & Salmon, E. S. A new fungicide for use against American gooseberry mildew. *J. Bd Agric.* **22**:118–25. [See also *ibid.* **25**: 1494–7, 1919.]

Mix, A. J. Sun-scald of fruit trees, a type of winter injury. *Bull. Cornell agric. Exp. Stn* **382**:235–84.

Whetzel, H. H., Hesler, L. R., Gregory, C. T. & Rankin, W. H. *Laboratory outlines of plant pathology.* Philadelphia. [Edn 2,1925 (revised by Whetzel).]

1916–17   Waksman, S. A. Do fungi live and produce mycelium in the soil?, *Science* **44**:320–2, 1916; Is there any fungus flora of the soil?, *Soil Sci.* **3**:565–89, 1917.

1917   Allard, H. A. Further studies of the mosaic disease of tobacco. *J. agric. Res.* **10**:615–32.

Butler, E. J. The dissemination of parasitic fungi and international legislation. *Mem. Indian Dep. Agric. Bot. Ser.* **9**:1–73.

Chupp, C. Studies on club root of cruciferous plants. *Bull. Cornell agric. Exp. Stn* **387**:419–52.

Conner, S. D. The injurious effect of borax in fertilizers on corn. *Proc. Ind. Acad. Sci. 1917*:195–9.

Darnell-Smith, G. P. The prevention of bunt. Experiments with various fungicides. *Agric. Gaz. N.S.W.* **28**:185–9.

Doidge, Ethel M. A bacterial rot of citrus. *Ann. appl. Biol.* **3**:53–81.

Munn, M. T. Neck-rot disease of onions. *Bull. N. Y. Sta. agric. Exp. Stn* **437**:365–455.

Stahel, G. De Zuid-Amerikaansche Hevea-bladziekte veroonzaakt door *Melanopsammopsis ulei* nov. gen. (=*Dothidella ulei* P. Hennings). *Bull. Dep. Landb. Suriname* **34**, 111 pp., 29 pl.

Stakman, E. C. & Piemeisel, F. J. Biologic forms of *Puccinia graminis* on cereals and grasses. *J. agric. Res.* **10**:429–95.

1918   Beinhart, E. G. Steam sterilization of seed beds for tobacco and other crops. *USDA Farmers' Bull.* **996**, 15 pp. [Replaced by J. Johnson, *ibid.* **1629**, 1930.]

Brown, Nellie A. Some bacterial diseases of lettuce. *J. agric. Res.* **13**:367–88.

Butler, E. J. *Fungi and disease in plants.* Calcutta & Simla.

Gardner, M. W. The dissemination of fungous and bacterial diseases of plants. *Rep. Mich. Acad. Sci.* **20**:357–423. [220 refs.]

Gäumann, E. *Über die Formen der* Peronospora parasitica (*Pers.*) Fries. 132 pp. Dresden (Thesis, Univ. Bern).

Jensen, C. O. Undersøgelser vedrørende nogle svulstlignende dannelser hos planter. *Kgl. Veterin. Landb. Aarssk. Serumlab.* **54**:91–143.

Lyman, G. R. The relation of phytopathologists to plant disease survey work. *Phytopathology* **8**:219–28.

Nishimura, M. A carrier of the mosaic disease. *Bull. Torrey bot. Club.* **45**:219–33.

Sanders, G. E. & Kelsall, A. A copper dust. *Proc. Entomol. Soc. Nova Scotia* **4**:32–7.

Shear, C. L. Pathological aspects of the Federal Fruit and Vegetable Inspection Service. *Phytopathology* **8**:155–60.

Stakman, E. C., Piemeisel, F. J. & Levine, M. N. Plasticity of biologic forms of *Puccinia graminis*. *J. agric. Res.* **15**:221–50.

*Whetzel, H. H. *An outline history of plant pathology.* Philadelphia

1919   Brooks, F. T. & Bailey, M.A. Silver-leaf disease, III (including observations on the injection of trees with antiseptics). *J. agric. Sci.* **9**:189–215.

Darnell-Smith, G. P. & Ross, H. A dry method of treating seed wheat for bunt. *Agric. Gaz. N.S.W.* **30**:685–92.

Johnson, J. The influence of heated soils on seed germination and plant growth. *Soil Sci.* **7**:1–103.

Lutman, B. F. Tip burn of the potato and other plants. *Bull. Vt agric. Exp. Stn* **214**, 28 pp.

Reddick, D. & Stewart, V. B. Transmission of the virus of bean mosaic in seed and observations on the thermal death-point of seed and virus. *Phytopathology* **9**: 443–50.

Stahel, G. Bijdrage tot kennis der Krullotenziekte. *Bull. Dep. Landb. Suriname* **39**, 34 pp., 5 pl. [Eng. trans. (without pl.) *Trop. Agric.* **9**:167–76, 1932.]

1919–24   Oudemans, C. A. J. A. *Enumeratio systematica fungorum.* 5 vols. The Hague.

1920   Doolittle, S. P. The mosaic disease of cucurbits. *Bull. USDA* **879**, 69 pp.

McCulloch, Lucia. Basal glumerot of wheat. *J. agric. Res.* **18**:543–53.

Rumbold, Caroline. The injection of chemicals into chestnut trees. *Am. J. Bot.* **7**: 1–20.

Smith, E. F. *Introduction to bacterial diseases of plants.* Philadelphia.

Taubenhaus, J. J. *Diseases of greenhouse crops and their control.* New York.

1921   Bewley, W. F. Control of 'damping-off' and 'foot-rot' of tomatoes. *J. Ministr. Agric. Lond.* **28**:653–4.

*Butler, E. J. The Imperial Bureau of Mycology. *Trans. Br. mycol. Soc.* **7**:168–72. [See also Wiltshire in *Herb. I.M.I. Handbook*, 1960: 1–23.]

Elvedon, Viscount. A contribution to the investigation into the results of partial soil sterilisation of the soil by heat. *J. agric. Sci.* 11:197–210.

Krout, W. S. Treatment of celery seed for the control of septoria blight. *J. agric. Res.* 21:369–72.

McCulloch, Lucia. A bacterial disease of gladiolus. *Science* 54:115–16.

*Morstatt, H. Zur Ausbildung für den Pflanzenschutzdienst. *Z. PflKrankh.* 31:89–94.

Petch, T. *Diseases and pests of the rubber tree.* London. [A replacement for *The physiology and diseases of* Hevea brasiliensis, 1911 (London).]

1921–4   Walker, J. C. Onion smudge. *J. agric. Res.* 20:685–722, 1921; Disease resistance to onion smudge, *ibid.* 24:1019–40, 1923; Further studies on the relation of onion scale pigmentation to disease resistance, *ibid.* 29:507–14.

1922   Atanasoff, D. A study into the literature on stipple-streak and related diseases of the potato. *Meded. Landbouwhoogeschoole Wageningen,* 26:52 pp.

Hedges, Florence. A bacterial wilt of the bean caused by *Bacterium flaccumfaciens* nov. sp. *Science* 55:433–4.

Jones, L. R. Experimental work on the relation of soil temperature to disease in plants. *Trans. Wisc. Acad. Sci. Arts Lett.* 20:433–59.

Nelson, R. The occurrence of protozoa in plants affected with mosaic and related diseases. *Tech. Bull. Mich. agric. Exp. Stn* 58:28 pp.

Nicolle, M. & Magrou, J. *Les maladies parasitaires des plantes, infestation – infection.* Paris.

Reddy, C. S. & Brentzel, W. E. Investigations of heat canker of flax. *Bull. USDA* 1120: 18 pp.

Stakman, E. C. and Levine, M. N. The determination of biologic forms of *Puccinia graminis* on *Triticum* spp. *Tech. Bull. Minn. agric. Exp. Stn* 8: 10 pp.

Voglino, P. Servizio di segnalazione degli attacchi di *Plasmopara viticola* nel 1921 nelle Province di Totino, Cuneo, Novara. *Nuovi Ann. Min. Agric.* 2:72–80. [*Rev. appl. Mycol.* 2:6.]

1923   Bewley, W. F. *Diseases of glasshouse plants.* London.

Chaptal, J. Les avertissements agricoles et la phytopathologie. *Congrès Path. vég. (Centenaire de Pasteur) Strasbourg, 1923:* 71–4. [*Rev. appl. Mycol.* 3: 48.]

Farley, A. J. Dry-mix sulfur lime. A substitute for self-boiled lime-sulfur, and summer-strength concentrated lime-sulfur. *Bull. N. J. agric. Exp. Stn* 379: 16 pp.

Garner, W. W., McMurtrey, J. E., Bacon, C. W. & Moss, E. G. Sand drown, a chlorosis of tobacco due to magnesium deficiency, and the relation of sulphates and chlorides of potassium to the disease. *J. agric. Res.* 23:27–40.

Gäumann, E. Beiträge zu einer Monographie der Gattung *Peronospora* Corda. *Beitr. Kryptogamedfl. Schweiz* 5(4), 360 pp.

Honing, J. A. *Nicotiana deformis* n. sp. und die Enzymtheorie der Erblichkeit. *Genetica* 5:455–76.

McKinney, H. H. Influence of soil temperature and moisture on infection of wheat seedlings by *Helminthosporium sativum. J. agric. Res.* 26:195–217.

MacMillan, H. G. The cause of sunscald of bean. *Phytopathology* 13:376–80.

Newhall, A. G. Seed transmission of lettuce mosaic. *Phytopathology* 13:104–6.

Nowell, W. *Diseases of crop plants in the Lesser Antilles.* London.

Paine, S. G. & Lacey, Margaret S. Studies in bacteriosis. X. The use of serum-agglutination in the diagnosis of plant parasites. *Ann. appl. Biol.* 10:204–9.

Petch, T. *The diseases of the tea bush.* London.

Stakman, E. C., Henry, A. W., Curran, G. C. & Christopher, W. N. Spores in the upper air. *J. agric. Res.* 24:599–606.

Warrington, Kathleen. The effect of boric acid and borax on the broad bean and certain other plants. *Ann. Bot. Lond.* **37**:629–72.

Wilbrink, Gerarda. Warmwaterbehandeling van stekken als geneesmiddel tergen de serehziekte van het suikerriet. *Arch. Suikerind. Ned. Ind. Surabaya* **31**:1–15.

1924   Carsner, E. & Stahl, C. F. Studies on curly-top disease of the sugar beet. *J. agric. Res.* **28**:297–320.

*Fryer, J. C. F. & Pethybridge, G. H. The phytopathological service of England and Wales. *J. Ministr. Agric.* **34**:331–40.

Güssow, H. T. International plant disease legislation as viewed by a scientific officer of an importing country. *Rep. internat. Congr. Phytopath. Econ. Entom. Holland 1923*:96–107.

*Hartley, C. & Rands, R. D. Plant pathology in the Dutch East Indies. *Phytopathology* **14**:8–23.

McCubbin, W. A. Peach yellows and little peach. *Bull. Pa Dep. Agric.* **382**, 16 pp.

Pratt, Clara A. The staling of fungus cultures. I. General and chemical investigation of staling by *Fusarium*. II. The alkaline metabolic products and their effects on the growth of fungus spores. *Ann. Bot. Lond.* **38**:563–95, 599–615.

Roussakov, L. F. [Peculiarities of the microclimate in the midst of plants in connection with the development of cereal rusts.] [Russ.] *Trans. 4th All-Russian Entomo-Phytopath. Congr. Moscow Dec. 8–14, 1922*:201–16. [*Rev. appl. Mycol.* **4**: 471–3.]

Voglino, P. *Patologia vegetale.* Turin.

Woolman, H. M. & Humphrey, H. B. Summary of literature on bunt, or stinking smut, of wheat. *Bull. USDA* **1210**, 44 pp.

1924–33   Dressel, A. *Atlas der Krankheiten der landwirtschaften Kulturpflanzen.* (Text by O. Appel & E. Riehm.) Berlin.

1925   Bunting, R. H. & Dade, H. A. *Gold Coast plant diseases.* London.

Cunningham, G. H. *Fungous diseases of fruit-trees in New Zealand and their remedial treatment.* Auckland.

Dickson, B. T. Tobacco and tomato mosaic. (2). Streak of tomato in Quebec a 'double-virus' disease. *Science* **62**:398.

Duffield, C. A. Nettlehead in hops. *Ann. appl. Biol.* **12**:526–43.

Johnson, J. Transmission of viruses from apparently healthy potatoes. *Res. Bull. Wisc. agric. Exp. Stn* **63**, 12 pp.

McCulloch, Lucia. *Aplanobacter insidiosum* n. sp., the cause of an alfalfa disease. *Phytopathology* **15**:496–7.

Marchal, E. *Elements de pathologie végétale appliquée à l'agronomie et à la sylviculture.* Paris & Gembloux.

St John-Brooks, R., Nain, K. & Rhodes, Mabel. The investigation of phytopathogenic bacteria by serological and biochemical methods. *J. Path. Bact.* **28**:203–9.

Salmon, E. S. & Ware, W. M. The downy mildew of the hop and its epidemic occurrence in 1924. *Ann. appl. Biol.* **12**:121–51.

Schultz, E. S. A potato necrosis resulting from cross-inoculation between apparently healthy potato plants. *Science* **62**:571–2.

1926   Heald, F. D. *Manual of plant diseases.* New York. [Edn 2, 1933.]

Jones, L. R., Johnson, J. & Dickson, J. G. Wisconsin studies upon the relation of soil temperature to plant disease. *Res. Bull. Wisc. agric. Exp. Stn* **71**, 144 pp.

Keitt, G. W. & Jones, L. K. Studies on the epidemiology and control of apple scab. *Res. Bull. Wisc. agric. Exp. Stn* **73**:104 pp.

Müller, A. *Die innere Therapie der Pflanzen.* Berlin.

Noumov, N. A. *Obschiĭ kurs fitopatologii* [Textbook of phytopathology. Russ.] Moscow.

Sanford, G. B. Some factors affecting the pathogenicity of *Actinomyces scabies*. *Phytopathology* **16**:525–47.

Stevenson, J. A. *Foreign plant diseases*. 198 pp. Washington, D.C. (USDA).

1926–35 Van Everdingen, E. Het verband tusschen de weergesteldheid en de aardappelziekte (*Phytophthora infectans*). *Tijdschr. PlZiekt.* **32**:129–40, 1926; **41**: 125–33, 1935.

1927 Brown, W. & Harvey, C. C. Studies in the physiology of parasitism. X. On the entrance of parasitic fungi into the host plant. *Ann. Bot. Lond.* **41**:643–62.

Craigie, J. H. Experiments in sex on rust fungi. *Nature* **120**:116–17; Discovery of the function of the pycnia of the rust fungi, *ibid.* **120**:765–7.

Dvorak, Mayme. The effect of mosaic on the globulin of potato. *J. infect. Dis.* **41**: 215–21.

Johnson, J. The classification of plant viruses. *Bull. Wisc. agric. Exp. Stn* **76**, 16 pp.

Millard, W. A. & Taylor, C. B. Antagonism of micro-organisms as the controlling factor in the inhibition of scab by green-manuring. *Ann. appl. Biol.* **14**: 202–16.

Murphy, P. A. The production of the resting spores of *Phytophthora infestans* on potato tubers. *Sci. Proc. roy. Dublin Soc.* **18**:407–12.

Tucker, J. Canadian certified seed potatoes. Rules and regulations governing their production. *Canada Dep. Agric. Pamphl.* **84** (NS), 11 pp.

Vinson, C. G. Precipitation of the virus of tobacco mosaic. *Science* **66**:357–8.

1928 Brooks, F. T. *Plant diseases*. London. [Edn 2 (by W. C. Moore), 1953.]

Lesley, J. W. & Lesley, M. M. The wiry tomato. A recessive mutant form resembling a plant affected with mosaic disease. *J. Heredity* **19**:337–44.

Martin, H. *The scientific principles of plant protection*. London. [Edn 4 (1959) – as *The scientific principles of crop protection*; Edn 6, 1973.]

Mason, A. F. *Spraying, dusting and fumigating of plants. A popular handbook on crop protection*. New York.

Owens, C. E. *Principles of plant pathology*. New York.

Purdy, Helen A. Immunologic reactions with tobacco mosaic virus. *Proc. Soc. exp. Biol. Med.* **25**:702–3.

1928–33 Stoughton, R. H. The influence of environmental conditions on the development of angular leaf-spot of cotton. *Ann. appl. Biol.* **15**:333–41, 1928; **17**:493–503, 1930; **18**:524–34, 1931; **19**:370–7, 1932; **20**:590–611, 1933.

1929 Arthur, J. C. *et al. The plant rusts (Uredinales)*. New York. [Historical review, pp. 30–72.]

*Butler, E. J. Presidential address. The development of economic mycology in the Empire overseas. *Trans. Br. mycol. Soc.* **14**:1–18.

Gäumann, E. A. & Fischer, E. *Biologie der pflanzenbewohnenden parasitischer Pilze*. Jena.

Güssow, H. T. International plant disease legislation – is it practical? *Proc. internat. Congr. Pl. Sci. Ithaca 1926*: 1334–42.

Holmes, F. O. (a) Local lesions in tobacco mosaic. *Bot. Gaz.* **87**:39–55.

Holmes, F. O. (b) Inoculating methods in tobacco mosaic studies. *Bot. Gaz.* **87**:56–63.

Jacob, H. E. The use of sulfur dioxide in shipping grapes. *Bull. Calif. agric. Exp. Stn* **471**:24 pp.

Purdy, Helen A. Immunologic reactions with tobacco mosaic virus. *J. exp. Med.* **49**:919–35.

Samuel, G. & Piper, C. S. Manganese as an essential element for plant growth. *Ann. appl. Biol.* **16**:493–524.

Seymour, A. B. *Host index of North American fungi.* Cambridge, Mass.

*Smith, E. F. Fifty years of plant pathology. *Proc. internat. Congr. Pl. Sci. Ithaca 1926*: 13–46, 32 pl.

Stoughton, R. H. Apparatus for growing plants in a controlled environment. *Ann. appl. Biol.* **17**:90–106.

Whetzel, H. H. The terminology of plant pathology. *Proc. internat. Congr. Pl. Sci. Ithaca 1926*: 1204–15.

1930   Angell, H. R., Walker, J. C. & Link, K. P. The relation of pyrocatechuic acid to disease resistance in onion. *Phytopathology* **20**:431–8.

Butler, E. J. Some aspects of morbid anatomy of plants. *Ann. appl. Biol.* **17**:175–212.

Elliott, Charlotte. *Manual of bacterial plant pathogens.* London. [Edn 2, 1951 (Waltham, Mass.).]

Gravatt, G. F. & Gill, L. S. Chestnut blight. *Bull. USDA* 1641.

Hara, K. *Pathologia agriculturalis plantarum.* [Jap.] Tokyo.

Hughes, A. W. McKenny. Aphis as a possible vector of 'breaking' in tulip species. *Ann. appl. Biol.* **17**:36–42.

McCallan, S. E. A. Studies on fungicides. II. Testing protective fungicides in the laboratory. *Mem. Cornell agric. Exp. Stn* **128**:8–24.

Mes, Margaretha. Physiological disease symptoms in tobacco. *Phytopath. Z.* **2**:593–614.

Salaman, R. N. & Le Pelley, R. H. Para-crinkle: a potato disease of the virus group. *Proc. roy. Soc.* **B106**:50–83.

Tehon, L. R. & Stout, G. L. Epidemic diseases of fruit trees in Illinois. *Ill. nat. Hist. Survey Bull.* **18**:415–502.

Wilcoxon, F. & McCallan, S. E. A. The fungicidal action of sulphur. I. The alleged role of pentathionic acid. *Phytopathology* **20**:391–417.

1930–7   Samuel, G., Bald, J. G. & Pittman, H. A. Investigations on 'spotted wilt' of tomatoes. *Bull. C.S.I.R. Australia*, **44**:64 pp. [II, Bald & Samuel, *ibid.* **54**:24 pp., 1931, III, Bald, *ibid.* **106**, 32 pp., 1937.]

1931   Anon. Practical soil sterilization with special reference to glasshouse crops. *Bull. Ministr. Agric.* **22**, 23 pp. [See also, W. F. Bewley, *J. ministr. Agric.* **33**:297–311, 1926 and later editions of the bulletin.]

1931   Butler, E. J. & Bisby, G. R. *The fungi of India.* Calcutta (Sci. Monogr. Counc. agric. Res.). [Revised by R. S. Vasudeva, 1960.]

Barger, G. *Ergot and ergotism.* London.

Beale, Helen Purdy. Specificity of the precipitin reaction in tobacco mosaic virus disease. *Contrib. Boyce Thompson Inst.* **3**:529–39.

Brandenburg, E. Die Herz- und Trochenfäule der Rüben als Bornmangel-Erscheinung. *Phytopath. Z.* **3**:499–517.

Hamilton, J. M. Studies of the fungicidal action of certain dusts and sprays in the control of scab. *Phytopathology* **21**:445–523.

Johnstone, K. H. Observations on the varietal resistance of apple to scab (*Venturia inaequalis*, Aderh.) with special reference to physiological aspects. *J. Pomol. hort. Sci.* **9**:30–52, 195–227.

Orton, C. R. Seed-borne parasites. A bibliography. *Bull. W. Virg. agric. Exp. Stn* **245**, 47 pp.

Preston, N. C. The prevention of finger-and-toe (clubroot) in gardens and allotments. *J. Ministr. Agric.* **38**:272–84.

Smith, K. M. On the composite nature of certain potato virus diseases of the mosaic group as revealed by the use of plant indicators and selective methods of transmission. *Proc. roy. Soc.* **B109**:251–67.

1932   Alben, A. O., Cole, J. R. & Lewis, R. D. New developments in treating pecan rosette with chemicals. *Phytopathology* **22**:979–81.

Henry, A. W. Influence of soil temperature and soil sterilization on the reaction of wheat seedlings to *Ophiobolus graminis* Sacc. *Canad. J. Res.* **7**:198–203.

Horsfall, J. G. (a) Red oxide of copper as a dust fungicide for combating damping-off of seedlings. *Bull. N.Y. Sta. agric. Exp. Stn* **615**, 26 pp.

Horsfall, J. G. (b) Dusting tomato seed with copper sulphate monohydrate for combating damping-off. *Tech. Bull. N.Y. Sta. agric. Exp. Stn* **198**, 34 pp.

Murphy, P. A. & McKay, R. The compound nature of crinkle, and its production by a mixture of viruses. *Proc. roy. Dublin Soc.* **20**(NS):227–47. [See also Phyllis Clinch & J. B. Loughnane, *ibid.* **20**:567–96, 1933.]

Napper, R. P. N. Observations on the root disease of rubber trees caused by *Fomes lignosus*; A scheme of treatment for the control of *Fomes lignosus* in young rubber areas. *J. Rubb. Res. Inst. Malaya* **4**:5–33, 34–38.

Price, W. C. Acquired immunity to ring-spot virus in Nicotiana. *Contrib. Boyce Thompson Inst.* **4**:359–403.

Takahashi, W. N. & Rawlins, T. E. Method for determining shape of colloidal particles: application in study of tobacco mosaic virus. *Proc. Soc. exp. Biol. Med.* **30**:155–7.

Weindling, R. *Trichoderma lignorum* as a parasite of other soil fungi. *Phytopathology* **22**:837–45.

1932–6   Davies, W. Maldwyn. Ecological studies on aphides infesting the potato crop. *Bull. entomol. Res.* **23**:525–48, 1932; II, *Ann. appl. Biol.* **21**:283–99, 1934; III, IV (with T. Whitehead), *ibid.* **22**:106–15, 549–56, 1935; V, *ibid.* **23**:401–8, 1935.

1933   Beaumont, A. *et al.* Symposium and discussion on the measurement of disease intensity. *Trans. Br. mycol. Soc.* **18**:174–86.

*Braun, K. Überblick über die Geschichte der Pflanzenkrankheiten und Pflanzenschädlinge (bis 1880). In Sorauer, *Handb. der PflKrankh.* Aufl. 6, **1**(1): 1–79. [See also Braun, 1965.]

Gratia, A. Pluralité antigéniques et identification sérologique des virus de plantes; Pluralité, hétérogénéité, autonomie antigéniques des virus de plantes et des bactériophages. *C. R. Soc. Biol. Paris* **114**:923–4, 1382–3.

Link, K. P. & Walker, J. C. The isolation of catechol from pigmented onion scales and its significance in relation to disease resistance in onion. *J. biol. Chem.* **100**:379–83.

*McKay, M. B. & Warner, M. F. Historical sketch of tulip mosaic or breaking. The oldest plant virus disease. *Nat. hort. Mag.* **12**:179–216. [See also *Science* **79**:385, 1934.]

Rawlins, T. E. *Phytopathological and botanic research methods.* New York.

Reinking, O. A. & Manns, M. M. Parasitic and other *Fusaria* counted in tropical soils. *Z. Parasitenk.* **6**:23–75.

Salaman, R. N. Protective inoculation against a plant virus. *Nature* **131**:468.

*Stevens, N. E. Phytopathology. The dark ages in plant pathology in America: 1830–1870. *J. Wash. Acad. Sci.* **23**:435–46.

Tanaka, S. Studies in the black spot disease of the Japanese pears (*Pyrus serotina*). *Mem. Coll. Agric. Kyto Imp. Univ.* **28**:1–31.

1934   Arthur, J. C. *Manual of the rusts of the United States and Canada.* New York. [Re-issued with suppl. by G. Cummins, 1962.]

Atanasoff, D. *Bolesti na kulturnit' rastenii* [Diseases of cultivated plants]. [Rumanian.] Sofia.

Birkland, J. M. Serological studies of plant viruses. *Bot. Gaz.* **95**:419–36.

Garrett, S. D. Factors affecting the severity of take-all. *J. Agric. S. Australia* **37**: 664–74.

Kunkel, L. O. Studies on acquired immunity with tobacco and aucuba mosaic. *Phytopathology* **24**:437–66.

Luthra, J. C. & Sattar, A. Some experiments on the control of loose smut, *Ustilago tritici* (Pers.) Jens., of wheat. *Indian J. agric. Sci.* **4**:177–99.

Newhall, A. G., Chupp, C. & Guterman, C. E. F. Soil treatment for the control of diseases in the greenhouse and the seedbed. *Ext. Bull. Cornell agric. Exp. Stn* **217**.

Rawlins, T. E. & Tomkins, C. M. The use of carborundum as an abrasive in plant-virus inoculations. *Phytopathology* **24**:1147 (abst).

Roark, R. C. A bibliography of chloropicrin 1848–1932. *Misc. Publ. USDA* **176**, 88 pp., 533 refs.

*Stevens, N. E. Plant pathology in the penultimate century. *Isis* **21**:98–112.

1935   Angell, H. R., Hill, A. V. & Allen, J. M. Downy mildew (blue mould) of tobacco: its control by benzol and toluol vapours in covered seed-beds. *J. Austral. C.S.I.R.* **8**:203–13.

Cunningham, G. H. *Plant protection by the aid of therapeutants.* Dunedin.

*Johnson, G. F. The early history of copper fungicides. *Agric. Hist.* **9**:67–79.

Newhall, A. G. & Nixon, M. W. Disinfecting soils by electric pasteurization. *Bull. Cornell agric. Exp. Stn* **636**, 20 pp.

Ogilvie, L. & Brian, P. W. Hot-water treatment for mint rust. *Gdnrs' Chron.* **98**:65.

de Ong, E. R. The use of oil-soluble copper as a fungicide. *Phytopathology* **25**: 368–70.

*Sirks, M. J. (Editor). *Botany in the Netherlands.* Leiden (6th Internat. bot. Congr.). [Plant pathology, pp. 41–3, 56–8, 74–7, 89–90.]

Stanley, W. M. Isolation of a crystalline protein possessing the properties of tobacco-mosaic virus. *Science* **81**:644–5. [See also *Phytopathology* **26**:305–20, 1936.]

Stephanov, K. M. Dissemination of infective diseases of plants by air currents. *Bull. Pl. Prot. Leningrad*, Ser. 2, Phytopathology, No. **8**, 68 pp. [Russ.; Eng. title.]

Vanterpool, T. C. Studies on browning foot rot of cereals. III. Phosphorus-nitrogen relations of infested fields. IV. Effects of fertilizer amendments. V. Preliminary plant analyses. *Canad. J. Res.* **13**:220–50.

Wormald, H. The brown rot diseases of fruit. *Bull. Ministr. Agric.* **88**:50 pp. [Edn 2, *Tech. Bull. Ministr. Agric.* **3**, 112 pp., 1954.]

1935–61 Wardlaw, C. W. *Diseases of the banana and the manila hemp plant*, 1935; *Banana diseases including plantains and abaca*, 1961. London.

1936   Bawden, F. C., Pirie, N. W., Bernal, J. D. & Fankuchen, I. Liquid crystalline substances from virus-infected plants. *Nature* **138**:1051.

Chester, K. S. Separation and analysis of virus strains by means of precipitin tests. *Phytopathology* **26**:778–85.

Clinton, G. P. & McCormick, Florence A. Dutch elm disease, *Graphium ulmi. Bull. Conn. agric. Exp. Stn* **389**:701–52.

Güssow, H. T. Plant quarantine legislation – a review and a reform. *Phytopathology* **26**:465–82.

Katsura, S. The stunt disease of Japanese rice, the first plant virosis shown to be transmitted by an insect vector. *Phytopathology* **26**:887–95.

Kunkel, L. O. Heat treatment for the cure of yellows and other virus diseases of peach. *Phytopathology* **26**:809–30.

Parker, K. G. Fire blight: overwintering, dissemination and control of the pathogen. *Mem. Cornell agric. Exp. Stn* **193**, 42 pp.

Riker, A. J. & Riker, Regina S. *Introduction to research on plant disease.* St Louis, Mo.

Sheffield, Frances M. L. The role of plasmodesms in the translocation of virus. *Ann. appl. Biol.* **23**:506–8.

Sigrianskiĭ, A. M. *The agronomist's manual on the control of agricultural plants.* Moscow. [Russ.]

Weindling, R. & Fawcett, H. S. Experiments in the control of Rhizoctonia damping-off of citrus seedlings. *Hilgardia* **10**:1–16.

1937 Bawden, F. C. & Pirie, N. W. The isolation and some properties of liquid crystalline substances from solanaceous plants infected with three strains of tobacco mosaic virus. *Proc. roy. Soc.* **B123**:274–320.

Chapman, H. D., Vanselow, A. P. & Liebig, G. F. Jr. The production of mottle-leaf in controlled nutrient cultures. *J. agric. Res.* **55**:365–79.

Chester, K. S. Serological studies on plant viruses. *Phytopathology* **27**:903–12.

Heald, F. D. *Introduction to plant pathology.* New York. [Edn 2, 1943.]

Leach, L. D. Observations on the parasitism and control of *Armillaria mellea. Proc. roy. Soc.* **B121**:561–73.

Lehmann, E., Kummer, H. & Dannemann, H. *Der Schwarzrost, seine Geschichte und seine Bekämpfung in Verbindung mit der Berberitzenfrage.* Munich.

Neal, D. C. Crinkle leaf, a new disease of cotton in Louisiana. *Phytopathology* **27**:1171–5.

Smith, K. M. *A textbook of plant virus diseases.* London. [Edn 6, 1977.]

*Wehnelt, B. Mathieu Tillet. Tilletia. Die Geschichte einer Endeckung. *Nachr. Schädlingsbekämpfung* **12**(2):45–146.

Wilson, A. R. Apparatus for growing plants under controlled environmental conditions; The chocolate spot disease of beans (*Vicia faba* L.) caused by *Botrytis cinerea* Pers. *Ann. appl. Biol.* **24**:911–31, 258–88.

1938 Ainsworth, G. C., Oyler, Enid & Read, W. H. Observations on the spotting of tomato fruits by *Botrytis cinerea* Pers. *Ann. appl. Biol.* **25**:308–21.

*Bulloch, W. *The history of bacteriology.* London. [Reprinted 1960.]

Clayton, E. E. Paradichlorobenzene as a control for blue mould of tobacco. *Science* **88**:56.

*Cooke, M. T. Pioneers in the study of virus diseases of plants. *Scientific Monthly* **46**:41–6.

Doyer, Lucie C. *Manual for the determination of seed-borne diseases.* Wageningen.

*Hino, I. [Nakata & Hino's *New system of phytopathology*] **1**. Tokyo. [Jap.; history of phytopathology pp. 63–184.]

Roach, W. A. Plant injection for diagnosis and curative purposes. *Tech. Commun. Imp. Bur. Plantation Crops* **10**, 78 pp.

Van Luijk, A. Antagonisms between various micro-organisms and different species of the genus *Pythium* parasitizing upon grasses and lucerne. *Meded. phytopath. Lab. Scholten* **14**:45–83.

1939 Bawden, F. C. *Plant viruses and virus diseases.* Leiden. [Edn 4, 1964 (New York).]

Holmes, F. O. Proposal for extension of the binomial system of nomenclature to include viruses. *Phytopathology* **29**:431–6; *Handbook of phytopathogenic viruses.* Minneapolis, Minn.

Melhus, I. E. & Kent, G. C. *Elements of plant pathology.* New York.

Napper, R. P. N. Root disease investigations. *Rep. Rubb. Res. Inst. Malaya 1938*:116–23.

*Pethybridge, G. H. History and connotation of the term 'Blattrollkrankheit' (leaf roll disease) as applied to certain potato diseases. *Phytopath. Z.* **12**:283–91.

Roach, W. A. Plant injection as a physiological method. *Ann. Bot Lond.* NS **3**:155–226.

Watson, Marian A. & Roberts, F. M. A comparative study of the transmission of *Hyoscyamus* virus 3, potato virus Y and cucumber virus 1 by the vectors *Myzus persicae* (Sulz), *M. circumflexus* (Buckton), and *Microsiphon gei* (Koch). *Proc. roy. Soc.* **B127**:543–76.

1940   Bennett, C. W. Acquisition and transmission of viruses by dodder (*Cuscuta subinclusa*). *Phytopathology* **30**:2 (abst.). [See also *ibid.* **34**:905–32, 1944.]

Craigie, J. H. Aerial dissemination of plant pathogens. *Proc. 6th Pacific Sci. Congr.* **4**:753–67.

Horsfall, J. G., Heuberger, J. W., Sharvelle, E. G. & Hamilton, J. M. A design for laboratory assay of fungicides. *Phytopathology* **30**:545–63.

Large, E. C. *The advance of the fungi.* London.

Leach, J. G. *Insect transmission of plant diseases.* New York.

Muskett, A. E. & Colhoun, J. Prevention of seedling blight in the flax crop. *Nature* **146**:32.

Naumov, N. A. Bolezni sel'skokhozaĭstvennȳkh rasteniĭ (fitopathologiya) [Diseases of economic crops (phytopathology). Russ.]. Leningrad. [Edn 2, 1952.]

Person, L. H. & Martin, W. J. Soil rot of sweet potato in Louisiana. *Phytopathology* **30**:913–26.

*Puttemans, A. History of phytopathology in Brazil. *J. Agric. Univ. Puerto Rico* **24**:77–107.

Steenbjerg, F. Das Kupfer im Boden mit besonderer Bezugnahme an die Heidenmoorkrankheit. *Tidsskr. Planteavl* **45**:259–368.

1941   Ciferri, R. *Manuale di patologia vegetale.* Genoa. [Edn 2, 2 vols. & atlas, 1952–5 (Rome).]

Dimond, A. E., Heuberger, J. W. & Stoddard, E. M. Role of the dosage–response curve in the evaluation of fungicides. *Bull. Conn. agric. Exp. Stn* **451**:635–67.

Holton, C. S. & Heald, F. D. *Bunt or stinking smut of wheat (a world problem).* Minneapolis, Minn.

*Klemm, M. Pflanzenschutz in der UdSSR. *Angew. Bot.* **23**:41–62.

Kunkel, L. O. Heat cure of aster yellows in periwinkles. *Am. J. Bot.* **28**:761–9.

Müller, K. O. & Börger, H. Experimentelle Untersuchungen über die *Phytophthora*-resistenz der Kartoffel. *Arb. biol. Reichsanstal. Land.–u. Forstwirtsh. Berlin* **23**:189–231.

Piper, C. S. Marsh spot of peas: a manganese deficiency disease. *J. agric. Sci.* **31**:448–53.

1942   Chester, K. S. *The nature and prevention of plant diseases.* Philadelphia. [Edn 2, 1947.]

Clayton, E. E., Gaines, J. G., Shaw, K. T., Smith, T. E., Foster, H. H. & Lunn, W. M. Gas treatment for the control of blue mold disease of tobacco. *Tech. Bull. USDA* **799**, 38 pp.

Fernandez Valiela, M. V. *Introducción a la fitopatología.* Buenos Aires. [Edn 3, 2 vols., 1969–75.]

Garrett, S. D. The take-all disease of cereals. *Tech. Commun. Imp. Bur. Soil Sci.* **41**, 40 pp.

Horsfall, J. G. & Heuberger, J. W. Measuring magnitude of a defoliation disease of tomatoes. *Phytopathology* **32**:226–32.

Johnson, J. Studies on the viroplasm hypothesis. *J. agric. Res.* **64**:443–54.

*Moore, W. C. Presidential address. Organization for plant pathology in England and Wales – retrospect and prospect. *Trans. Br. mycol. Soc.* **25**:229–45.

*Reed, H. S. *A short history of the plant sciences.* Waltham, Mass. [Chap. 19, Plant pathology.]

Sharvelle, E. G., Young, H. C. Jr & Shema, B. F. The value of spergon as a seed protectant for canning peas. *Phytopathology* **32**:944–52. [See also E. L. Felix, *ibid.* **32**:4, 1942 (abst.).]

1943   Anon. Proprietary products for the control of plant pests and diseases. Scheme for official approval. *Agriculture* **50**:331–4.

Anslow, W. K., Raistrick, H. & Smith, G. Anti-fungal substances from moulds. Part I. Patulin (anhydro-3-hydroxymethylene-tetrahydro-1:4-pyrone 2-carboxylic acid), a metabolic product of *Penicillium patulum* Bainier and *Penicillium expansum* (Link) Thom. *J. Soc. chem. Indust. Lond.* **62**:236–8.

Dimond, A. E., Heuberger, J. W. & Horsfall, J. G. A water soluble protective fungicide with tenacity. *Phytopathology* **33**:1095–7.

Dowson, W. J. On the generic names Pseudomonas, Xanthomonas and Bacterium for certain bacterial plant pathogens. *Trans. Br. mycol. Soc.* **26**:4–14.

Fischer, G. W. Some evident synonymous relationships in certain graminicolous smut fungi. *Mycologia* **35**:610–19.

Fulling, E. H. Plant life and the law of man. IV. Barberry, currant and gooseberry, and cedar control. *Bot. Rev.* **9**:483–592.

McCallan, S. E. A. & Wellman, R. H. A greenhouse method of evaluating fungicides by means of tomato foliage diseases. *Contrib. Boyce Thompson Inst.* **13**:93–134.

Moore, W. C. The measurement of plant diseases in the field. Preliminary report of a sub-committee of the Society's Plant Pathology Committee. *Trans. Br. mycol. Soc.* **26**:28–35.

Wallace, T. *The diagnosis of mineral deficiencies in plants by visual symptoms. A colour atlas and guide.* London.

*Wehnelt, B. *Die Pflanzenpathologie der deutschen Romantik als Lehre vom kranken Leben und Bilden der Pflanze.* Bonn. [466 refs.]

1944   Dimock, A. W. Hot-water treatment for control of *Phytophthora* root rot of calla. *Phytopathology* **34**:979–81.

Garrett, S. D. *Root disease fungi.* Waltham, Mass.

Gram, E. & Weber, Anna. *Plantesygdomme handbog for frugtavlere, gartnere og haveejere.* Copenhagen. [Eng. trans. by R. W. G. Dennis, 'Plant diseases in orchard, nursery, and garden crops', 1952 (London).]

Heuberger, J. W. & Wolfenbarger, D. O. Zinc dimethyl dithiocarbamate and the control of early blight and anthracnose on tomatoes and of leaf hoppers and blight of potatoes. *Phytopathology* **34**:1003 (abst.).

Walker, J. C. Efficacy of fungicidal transplanting liquids for control of clubroot of cabbage. *Phytopathology* **34**:185–95.

1945   Gregory, P. H. The dispersal of airborne spores. *Trans. Br. mycol. Soc.* **28**:26–72.

Horsfall, J. G. *Fungicides and their action.* Waltham, Mass.

Large, E. C. Field trials of copper fungicides for the control of potato blight. I. Foliage protection and yield. *Ann. appl. Biol.* **32**:319–29.

May, K. R. The cascade impactor: an instrument for sampling coarse aerosols. *J. scientific Instr.* **22**:187–95.

Mitchell, K. J. Preliminary note on the use of ammonium molybdate to control whiptail in cauliflower and broccoli. *N.Z.J. Sci. Tech.* **A27**:287–93.

1946   Chester, K. S. *The nature and prevention of cereal rusts as exemplified in the leaf rust of wheat.* Waltham, Mass.

Gäumann, E. *Pflanzliche Infektionslehre.* Basel. [Eng. trans. W. B. Brierley (Editor), 'Principles of plant infection', 1950 (London).]

Johnson, J. Soil-steaming for disease control. *Soil Sci.* **61**:83–91.

Magrou, J. *Les maladies des végétaux.* Paris.

Meehan, Frances & Murphy, H. C. A new *Helminthosporium* blight of oats. *Science* **104**:413–14. [See also *Phytopathology* **36**:406, 407, 1946 (*H. victoriae* sp.n.).]

Wellman, R. H. & McCallan, S. E. A. Glyoxidine derivatives as foliage fungicides. I. Laboratory studies. *Contrib. Boyce Thompson Inst.* **14**:151–60.

Wolfenbarger, D. O. Dispersion of small organisms. Distance dispersion rates of bacteria, spores, seeds, pollen, and insects; incidence rates of diseases and injuries. *Am. Midl. Nat.* **35**(1):1–152.

Zentmyer, G. A., Horsfall, J. G. & Wallace, P. P. Dutch elm disease and its chemotherapy. *Bull. Conn. agric. Exp. Stn* **498**, 70 pp.

1947   Anon. The measurement of potato blight. *Trans. Br. mycol. Soc.* **31**:140–1.

Anderson, H. W. & Nienow, Inez. Effect of streptomycin on higher plants. *Phytopathology* **37**:1 (abst.).

Atkinson, J. D. Tomatoes injured by hormone weedkillers. *N.Z. J. Agric.* **75**:349–51.

Barratt, R. W. & Horsfall, J. G. Fungicidal action of metallic alkyl bisdithiocarbamates. *Bull. Conn. agric. Exp. Stn* **508**, 51 pp.

Beaumont, A. The dependence on the weather of the dates of outbreak of potato blight epidemics. *Trans. Br. mycol. Soc.* **31**:45–53.

Bouriquet, G. *Les maladies des plantes cultivées à Madagascar.* Paris.

*Buttress, F. A. & Dennis, R. W. G. The early history of seed treatment in England. *Agric. Hist.* **21**:93–103.

Marsh, R. W. Fruit spraying trials with certain recently-introduced fungicides. *J. Pomology* **23**:185–205.

Meehan, Frances & Murphy, H. C. Differential phytotoxicity of metabolic byproducts of *Helminthosporium victoriae*. *Science* **106**:270–1.

Waterston, J. M. The fungi of Bermuda. *Bull. Dep. Agric. Bermuda* **33**, 305 pp.

Yoshii, H. & Kawamura. *Kaibō Sokubutsa Byōrigaku* [Anatomical plant pathology. Jap.]. Tokyo.

1948   Bawden, F. C. *Plant diseases.* London & Edinburgh.

Bawden, F. C. & Roberts, F. M. Photosynthesis and predisposition of plants to infection by certain viruses. *Ann. appl. Biol.* **35**:418.

Hara, K. *Byogaichu-Hoten* [Manual of pests and diseases. Jap.]. Tokyo.

Morel, G. Recherches sur la culture associée de parasites obligatoires et de tissues végétaux. *Ann. Epiphyt.* NS **14**:123–234.

Weston, W. A. R. Dillon & Taylor, R. E. *The plant in health and disease.* London.

1949   Butler, E. J. & Jones, S. G. *Plant pathology.* London.

Dowson, W. J. *Manual of bacterial plant diseases.* London. [Edn 2 as *Plant diseases due to bacteria*, 1957 (Cambridge).]

Mundkur, B. B. *Fungi and plant disease.* London.

Porter, R. H. Recent developments in seed technology. *Bot. Rev.* **15**:221–344, 502 refs.

*Schlumberger, O. Wesen und Wirken der Biologischen Zentralanstalt für Land- und Forstwirtschaft 1898–1948. In *Fünfzig Jahre Deutsche Pflanzenschutzforschung*: 1–28. Berlin.

Viennot-Bourgin, G. *Les champignons parasites des plantes cultivées.* Paris.

1950   Bawden, F. C., Kassanis, B. & Nixon, H. L. The mechanical transmission and some properties of potato paracrinkle virus. *J. gen Microbiol.* **41**:viii.

Black, L. M. A plant virus multiplies in its insect vector. *Nature* **166**:852–3.

Garrett, S. D. Ecology of the root-inhabiting fungi. *Biol. Rev.* **25**:220–54.

Grainger, J. Forecasting outbreak of potato blight in West Scotland. *Trans. Br. mycol. Soc.* **33**:82–91.

Kassanis, B. Heat inactivation of leaf-roll virus in potato tubers. *Ann. appl. Biol.* **37**:339–41.

Owen, J. H., Walker, J. C. & Stahmann, M. A. Pungency, color, and moisture supply in relation to disease resistance in the onion. *Phytopathology* **40**:292–7.

Walker, J. C. *Plant pathology*. New York. [Edn 3, 1969.] [*Chap. 2, history.]

Westcott, Cynthia. *The plant doctor*. New York. [Edn 3, 1971.]

1951   *Beran, F. *50 Jahre Österreichischen Pflanzenschutz 1901–1951*. Vienna. [Geschichtlichen Rückblick, pp. 11–20.]

Bliss, D. E. The destruction of *Armillaria mellea* in citrus soils. *Phytopathology* **41**:665–83.

Brian, P. W., Wright, Joyce M., Stubbs, J. & Way, Audry, M. Uptake of antibiotic metabolites of soil micro-organisms by plants. *Nature* **167**:347–9.

Gregory, P. H. Deposition of air-borne *Lycopodium* spores on cylinders. *Ann. appl. Biol.* **38**:357–76.

Katznelson, H. & Sutton, M. D. A rapid phage count method for the detection of bacteria as applied to the demonstration of internally borne bacterial infection of seed. *J. Bact.* **61**:689–701.

Rishbeth, J. Observations on the biology of *Fomes annosus*, with particular reference to East Anglian pine plantations. III. Natural and experimental infection of pines, and some factors affecting severity of the disease. *Ann. Bot. Lond.* **15**:221–46.

Sarejanni, J. A. Essai sur le concept de la maladie en pathologie végétale. *Ann. Inst. Pasteur Benaki* **5**(2):88–127.

*Stevenson, J. A. A résumé of the activities of the mycological collections of the United States Department of Agriculture, with a phytopathological slant, 1885–1950. *Pl. Dis. Reptr*, suppl. **200**:21–9.

1951–4   Roger, L. *Phytopathologie des pays chauds*. 3 vols. Paris.

1952   Bant, J. H. & Storey, I. F. Hot-water treatment of celery seed in Lancashire. *Pl. Path.* **1**:81–3.

Black, W. A genetical basis for the classification of strains of *Phytophthora infestans*. *Proc. roy. Soc. Edinb.* B**65**:36–51.

Ciferri, R. The criteria for the definition of species in mycology. *Ann. mycol. Berlin* **30**:122–36.

Croxall, H. E., Glynne, D. C. & Jenkins, J. E. E. The rapid assessment of apple scab on fruit. *Pl. Path.* **1**:89–92.

Hirst, J. M. An automatic volumetric spore trap. *Ann. appl. Biol.* **39**:257–65.

Kittleson, A. R. A new class of organic fungicides. *Science* **115**:84–6.

Large, E. C. The interpretation of progress curves for potato blight and other plant diseases. *Pl. Path.* **1**:109–17.

Miller, P. A. & O'Brien, Muriel. Plant disease forecasting. *Bot. Rev.* **18**:547–601.

Morel, G. & Martin, C. Guérison de dalias atteints d'une maladie à virus. *C. R. Acad. Sci. Paris* **235**:1324–5.

Ordish, G. *Untaken harvest. Man's losses of crops from pests, weed and disease*. London.

Rogers III, A. D. *Erwin Frink Smith. A story of North American plant pathology*. Philadelphia.

Stevens, N. E. & Stevens, R. B. *Disease in plants. An introduction to agricultural phytopathology*. Waltham, Mass.

1953   Black, W., Mastenbroek, E., Mills, W. R. & Peterson, L. C. Proposal for an international nomenclature of races of *Phytophthora infestans* and of genes controlling immunity in *Solanum demissum* derivatives. *Euphytica* **2**:173–9.

Dimond, A. E. & Waggoner, P. E. On the nature and role of vivotoxins in plant disease. *Phytopathology* **43**:229–35.

Fischer, G. W. *Manual of the North American smut fungi*. New York.

Kamat, M. N. *Practical plant pathology.* Poona.

Large, E. C. Potato blight forecasting investigation in England and Wales, 1950–52. *Pl. Path.* **2**:1–15.

*Ling, Lee. International Plant Protection Convention: its history, objectives and present status. *FAO Pl. prot. Bull.* **1**:65–8.

Stefferud, A. (Editor). *Plant diseases: the yearbook of agriculture 1953.* Washington, DC (USDA).

1954   Bean, J., Brian, P. W. & Brooks, F. T. Physiologic races of the brown rust of brome grasses. *Ann. Bot. Lond.* NS **18**:129–42.

Chupp, C. *A monograph of the fungus genus Cercospora.* Ithaca, NY.

Gregory, P. H. The construction and use of a portable volumetric spore trap. *Trans. Br. mycol. Soc.* **37**:390–404.

Kassanis, B. Heat-therapy of virus-infected plants. *Ann. appl. Biol.* **41**:470–4.

Katznelson, H., Sutton, M. D. & Bailey, S. T. The use of bacteriophage of *Xanthomonas phaseoli* in detecting infection in beans, with observations on its growth and morphology. *Can. J. Microbiol.* 1:22–9.

McCubbin, W. A. *The plant quarantine problem.* Copenhagen.

*Stevenson, J. A. Plants, problems, and personalities: the genesis of the Bureau of Plant Industry. *Agric. Hist.* **28**:155–62.

Thirumalachar, M. J. Inactivation of potato leaf roll by high temperature storage of seed tubers in Indian plains. *Phytopath. Z.* **22**:429–36.

*Viennot-Bourgin, G. Pathologie végétale. In D. de Virville, *Histoire de la botanique en France,* 1954:289–300. Paris.

Wallin, J. R. & Polhemus, D. N. A dew recorder. *Science* **119**:294–5.

1955   Del Cañizo Gómez, J. & Gonzáles de Andrés, C. *Manual práctico de fitopatología y terapéutica agricola.* Madrid.

Grainger, J. The 'Auchincruive' potato blight forecast recorder. *Weather* **10**:213–22.

Hilton, R. N. South American leaf blight. A review ... *J. Rubb. Res. Inst. Malaya* **14**:287–337. (With an Appendix by R. A. Alston, pp. 338–54.)

Pemberton, C. E. Sugar-cane quarantine in Hawaii. *Commonw. phytopath. News* **1**(4):49–52.

*Rae, R. The work of the National Agricultural Advisory Service. *Ann. appl. Biol.* **42**:260–71.

*Richter, H. *et al.* 50 Jahre deutschen Pflanzenschutzdienst. *NachrBl. dtsch. PflSchutzdienst* **7**:65–99.

Ripper, W. E. Application methods for crop protection. *Ann. appl. Biol.* **42**:288–324.

Sheffield, Frances M. L. The East African Quarantine Station. *Commonw. phytopath. News* **1**(3):33–5. (*Ibid.* **10**(2):25–6, 1964, extension to station.)

Tarr, S. A. J. *The fungi and plant diseases of the Sudan.* Kew (CMI). [Supplement, *Mycol Pap.* **85**, 31 pp., 1963.]

1956   Garrett, S. D. *Biology of root-infecting fungi.* Cambridge.

Grainger, J. Host nutrition and attack by fungal parasites. *Phytopathology* **46**:445–56.

Hirst, J. M. & Stedman, O. J. The effect of height of observations in forecasting potato blight by Beaumont's method. *Pl. Path.* **5**:135–40.

Horsfall, J. G. *Principles of fungicidal action.* Waltham, Mass.

Kamat, M. N. *Introductory plant pathology.* Poona. [Edn 3, 1967.]

Müller, K. Einige einfache Versuchezum Nachweis von Phytoalexin. *Phytopath. Z.* **27**:237–54.

Padwick, G. Watts. Losses caused by plant diseases. *Phytopath. Pap.* **1**, 60 pp.

*du Plessis, S. J. Plant control services in South Africa. *Commonw. phytopath. News* **2**(2):17–19.

Smith, L. P. Potato blight forecasting by 90 per cent humidity criteria. *Pl. Path.* **5**:83–7.

Stolp, H. Bakteriophagenforschung und Phytopathologie. *Phytopath. Z.* **26**:171–2181.

Taylor, C. F. A device for recording the duration of dew deposits. *Pl. Dis. Reptr* **40**:1025–8.

Wilson, J. Plant quarantine for the Caribbean. *Commonw. phytopath. News* **2**(1):1–3.

1957   Baker, R. E. D. & Holliday, P. Witches' broom disease of cacao (*Marasmius perniciosus* Stahel). *Phytopath. Pap.* **2**, 42 pp.

Fischer, G. W. & Holton, C. S. *Biology and control of the smut fungi.* New York. [History, pp. 77–87.]

Hirst, J. M. A simplified surface-wetness recorder. *Pl. Path.* **6**:57–61.

Kassanis, B. The use of tissue cultures to produce virus-free clones from infected potato varieties. *Ann. appl. Biol.* **45**:422–7.

Matthews, R. E. F. *Plant virus serology.* Cambridge.

Miller, P. A. & O'Brien, Muriel. Prediction of plant disease epidemics. *Ann. Rev. Microbiol.* **11**:77–110.

Stakman, E. C. & Harrar, J. G. *Principles of plant pathology.* New York.

1957–8 Gaümann, E. A. Über Fusarinsäure als Welketoxin, *Phytopath. Z.* **29**:1–44, 1957; Fusaric acid as a wilt toxin, *Phytopathology*, **47**:342–57, 1957; The mechanisms of fusaric acid injury, *ibid.* **48**:670–86, 1958.

1958   *Grainger, T. H. *A guide to the history of bacteriology.* New York.

Hewitt, W. B., Raski, D. J. & Goheen, A. C. Nematode vector of fanleaf virus of grapevines. *Phytopathology* **48**:586–95.

Noble, Mary, de Tempe, J. & Neergaard, P. *An annotated list of seed-borne diseases.* 159 pp. CMI & Internat. Seed Testing Assn. [Edn 3, by M. J. Richardson, 320 pp., 1979. *Phytopath. Pap.* **23**.]

Riggenbach, A. A note on the chemical control of the white root disease of rubber. *Quart. Circ. Rub. Res. Inst. Ceylon* **34**:8–10.

Sheffield Frances, M. L. Requirements of a post-entry quarantine station. *FAO Pl. prot. Bull.* **6**:149–52.

Stapp, C. *Pflanzenpathogene Bakterien.* Berlin. [Eng. trans. by A. Schoenfeld, 'Bacterial plant pathogens', 1961 (Oxford).]

1958–62 Baudys, E. *et al.* (Editors). *Zemědělská fytopatologie.* [Czech.] Prague.

1959   Cammack, R. H. Studies on *Puccinia polysora* Underw. II, III. *Trans. Br. mycol. Soc.* **42**:27–32, 55–8.

Gaümann, E. *Die Rostpilze Mitteleuropas.* Berne (Beitr. Kryptogamenfl. Schweiz **12**(1)).

Holton *et al.* (Editors). *Plant pathology, problems and progress 1908–1958.* Madison, Wisc.

*Keitt, G. W. History of plant pathology. In Horsfall & Dimond (1959–60) **1**:61–97.

Khristov, A. *Spetsialna fitopathologiya.* Edn 2. Sofia.

*McCallan, S. E. A. The American Phytopathological Society – the first fifty years. In Holton *et al.* (1959):24–31.

*McNew, G. L. Landmarks during a century of progress in the use of chemicals to control plant disease. In Holton *et al.* (1959):42–54.

*Mayer, K. *4500 Jahre Pflanzenschutz.* Stuttgart.

Moore, W. C. *British parasitic fungi. A host-parasite index and guide to British literature on the fungus diseases of cultivated plants.* Cambridge.

Săvulescu, T. & Săvulescu, Olga. *Tratat de pathologie vegetală*. Bucharest.

*Stevenson, J. A. The beginnings of plant pathology in North America. In Holton *et al.* (1959):14–23.

Yerkes, W. D. & Shaw, C. G. Taxonomy of *Peronospora* species on Cruciferae and Chenopodiaceae. *Phytopathology* **49**:499–5.

1959–60 Horsfall, J. G. & Dimond, A. E. (Editors). *Plant pathology: an advanced treatise.* 1 (The diseased plant), 1959; 2 (The pathogen), 3 (The diseased population, epidemics and control), 1960. New York.

1959–63 Rishbeth, J. Stump protection against *Fomes annosus*. I, Treatment with creosote. II, Treatment with substances other than creosote. III, Inoculation with *Peniophora gigantea*. *Ann. appl. Biol.* **47**:519–28, 529–41, 1959; **52**:63–77, 1963.

1960 Baker, K. F. & Olsen, C. M. Aerated steam for soil treatment. *Phytopathology* **50**:82 (abst.).

Billing, Eve, Cross, J. E. & Garrett, Constance M. E. Laboratory diagnosis of fire blight and bacterial blossom blight of pear. *Pl. Path.* **9**:19–25.

Cox, A. E. & Large, E. C. Potato blight epidemics throughout the world. *Agric. Handb. USDA* **174**, 230 pp.

Hirst, J. M. & Stedman, O. J. The epidemiology of *Phytophthora infestans*. I. Climate, ecoclimate and the phenology of disease outbreak. II. The source of inoculum. *Ann. appl. Biol.* **48**:471–88, 489–517.

*Kausar, A. G. *Fifty years of investigation on plant diseases at Agricultural College and Research Institute, Lyallpur.* 59 pp. Dep. Agric., W. Pakistan.

*Moreau, C. La vocation phytopathologique du Laboratoire de Cryptogamie du Muséum National d'Histoire Naturelle. *Rev. Mycol. Paris* **25**:85–134.

Nakata, K. *Sakumotsu byōgai zuhen* [Illustrated manual of crop diseases. Jap.] 2nd revised edition by H. Yoshii *et al.* Tokyo. [Edn 1, 1934.]

Săvulescu, Alice, Hulea, Ana & Bucur, Elena (Editors). *Protectia plantelor in sprijinul zonării productiei agricole in R. P. R.* Bucharest.

Teakle, D. A. Association of *Olpidium brassicae* and tobacco necrosis virus. *Nature* **188**:431–2. [See also, *Virology* **18**:224–31, 1962.]

USDA. Index of plant diseases in the United States. *Agric. Handb. USDA* **165**, 531 pp.

Waggoner, P. E. Forecasting epidemics. In Horsfall & Dimond (1959–60), **3**:291–312.

1961 Crosse, J. E. & Garrett, Constance M. E. Relationship between phage type and host plant in *Pseudomonas mors-prunorum* Wormald. *Nature* **192**:379–80.

Dunin, M. S. [M. S. Woronin. Collected works.] Moscow. [Russ.]

Duran, R. & Fischer, G. W. *The genus* Tilletia. Pullman, Wash.

Foister, C. E. Economic diseases of Scotland. *Tech. Bull. Dep. Agric. Fish. Scotland* **1**, 209 pp.

Gregory, P. H. *The microbiology of the atmosphere*. London. [Edn 2, 1973.]

Scheffer, R. P. & Pringle, R. B. A selective toxin produced by *Periconia circinata*. *Nature* **191**:912–13.

Wilhelm, S., Storkan, R. C. & Sagen, J. E. Verticillium wilt of strawberry controlled by fumigation of soil with chloropicrin and chloropicrin-methyl bromide mixtures. *Phytopathology* **51**:744–8.

1961–70 Fox, R. A. White root disease of *Hevea brasiliensis*: recent developments in control techniques; the role of fungicides in control techniques. *Rep. 6th Commonw. mycol. Conf. 1960*: 41–8, 97–100, 1961. Also in Baker & Snyder (1965), 348–62 (role of biological eradication in root-disease control) and in Toussoun *et al.* (1970): 179–87 (dispersal, survival, and parasitism in some fungi causing root disease of tropical plantation crops).

1962   Carter, W. *Insects in relation to plant disease.* New York. [Edn 2, 1973.]
        Ramsfjell, T. & Fjelddalen, J. *Sjukdommer og skadedyr på jordbruskvekster.* Oslo.
        Stakman, E. C., Stewart, D. M. & Loegering, W. Q. Identification of
        physiologic races of *Puccinia graminis* var. *tritici. USDA agric. Res. Serv.* **E616**, 53
        pp.
        Stover, R. H. Fusarial wilt (Panama disease) of bananas and other *Musa* species.
        Phytopath. Pap. **4**, 117 pp.
        Thorn, G. D. & Ludwig, R. A. *The dithiocarbamates and related compounds.*
        Amsterdam.
        *Woodham-Smith, Cecil. *The great hunger. Ireland, 1845–9.* London.

1963   *Angus, A. Northern Rhodesia. An historical note. *Commonw. phytopath. News*
        9(2):31–2.
        Cadman, C. H. Biology of soil-borne viruses. *Ann. Rev. Phytopath.* **1**:143–72.
        Cruickshank, I. A. M. Phytoalexins. *Ann. Rev. Phytopath.* **1**:351–74.
        Garrett, S. D. *Soil fungi and soil fertility.* Oxford.
        Harrison, B. D., Peachy, J. E. & Winslow, R. D. The use of nematicides to
        control spread of arabis mosaic virus by *Xiphinema diversicaudatum* (Micol.). *Ann.
        appl. Biol.* **52**:243–55.
        Horsfall, J. G. & Dimond, A. E. A perspective on inoculum potential. *J. Indian
        bot. Soc.* **42A**:46–57.
        Saghún, B. de. *General history of the things of New Spain. Florentine Codex.* Book 11,
        Earthly things. Transl. by A. J. O. Anderson & C. E. Dibble. Salt Lake City
        (Univ. Utah Press. Monogr. Sch. Am. Res. & Mus. New Mexico no. 14, part
        xii).
        Sprague, G. F., McKinney, H. H. & Greeley, L. Virus as mutagenic agent in
        maize. *Science* **141**:1052–3.
        Vanderplank, J. E. *Plant diseases: epidemics and control.* New York.
        Wheeler, H. & Luke, H. H. Microbial toxins in plant disease. *Ann. Rev.
        Microbiol.* **17**:223–42.

1964   Corbett, M. K. & Sisler, H. D. (Editors). *Plant virology.* Gainsville, Fla.
        Garbowski, L. *Zarys fitopatologii ogólnej.* [Polish.] Warsaw.
        McCallan, S. E. A. The nature of the action of sulphur fungicides. *Agrochimica*
        9:15–44.
        Malone, J. P. & Muskett, A. E. Seed-borne fungi. Description of 77 fungus
        species. *Proc. internat. Seed Testing Assn* **29**(2):177–384.
        Noveroske, R. L., Williams, E. B. & Kuć, J. β-glucosidase and phenoloxidase in
        apple leaves and their possible relation to resistance to *Venturia inaequalis.*
        Phytopathology **54**:98–103. [See also *ibid.* **54**:92–7, 1964 (phloridzin and
        phloretin).]
        *Orlob, G. B. The concepts of etiology in the history of plant pathology.
        PflSch.-Nachr. Bayer **17**(4):1:85–268.
        Pringle, R. B. & Scheffer, R. P. Host specific plant toxins. *Ann. Rev. Phytopath.*
        2:133–56.
        Staron, T. & Allard, C. Propriétés antifongiques du 2-(4'-thiazolyl) benzidazole
        ou thiabendazole. *Phytiat.-Phytopharm.* **13**:163–8.

1965   Baker, K. F. & Snyder, W. C. (Editors). *Ecology of soil-borne plant pathogens.
        Prelude to biological control.* Berkeley, Calif.
        Bawden, F. C. & Kassanis, B. The potato variety King Edward VII and
        paracrinkle virus. *Rep. Rothamsted exp. Stn 1964*:282–90.
        Bouhot, D. & Mallamaire, A. *Les principales maladies des plantes cultivées au Sénégal.*
        2 vols Dakar.
        *Braun, H. *Geschichte der Phytomedizin.* Berlin. [Reprint of Sorauer, *Handb. d.*

Pflanzenkrankh., Aufl. 7, **1**(1):1–135, 1965.] [11 pp. refs.] Cf. Braun, 1923.

Fulton, R. H. Low-volume spraying. *Ann Rev. Phytopath.* **3**:175–96.

Gorlenko, M. V. *Bacterial diseases of plants.* Jerusalem. [Trans. from Russ.]

Ou, S. H. (Editor). *The rice blast disease.* Baltimore, Md.

Stolp, H., Starr, M. P. & Baigent, Nancy L. Problems in speciation of phytopathogenic pseudomonads and xanthomonads. *Ann. Rev. Phytopath.* **3**:231–64.

Yoshinaga, E., Uchida, T. & Iwakura, T. Studies on rice blast control. IV. Rice blast control of kitazin by root treatment. [Jap.] *Ann. phytopath. Soc. Japan* **30**:307 (abst.).

1966   *Bawden, F. C. Some reflexions on thirty years of research on plant viruses. *Ann. appl. Biol.* **58**:1–11.

Gibbs, A. J., Harrison, G. B., Watson, D. H. & Wildy, P. What's in a virus name? *Nature* **209**:450–4.

Mackenzie, D. R., Anderson, P. M. & Wernham, C. C. A mobile air blast inoculator for plot experiments with maize dwarf mosaic virus. *Pl. Dis. Reptr* **50**:363.

Netherlands journal of plant pathology **72**(2):33–168. [Issued to commemorate the 75th anniversary of the Netherl. Soc. Pl. Path. A useful series of contributions on the history of plant pathology in the Netherlands.]

*Rademacher, B. Vor 75 Jahren; Begründung der Zeitschrift für Pflanzenkrankheiten. *Z. PflKrankh. PflSchutz* **73**:1–3

Schmeling, B. von & Kulka, M. Systemic fungicidal activity of 1,4-oxathiin derivatives. *Science* **152**:659–60.

Schmid, K. *et al.* 6 Jahre Blauschimmelkrankheit des Tabaks in der Bundesrepublik Deutschland (1959–1964). *Mitteil. biol. Bundesanst. Land-Forstwirtsch. Berlin-Dahlem* Heft **120**, 117 pp.

1967   *Buchwald, N. F. Die Entwicklung der Pflanzenpathologie in Danemark. *Acta Phytopathologia* **2**:183–94.

Conners, I. B. An annotated index of plant diseases in Canada. *Canada Dep. Agric. Res. Br. Publ.* **1251**, 381 pp.

Cramer, H. H. *Plant protection and world crop production.* Leverkusen. [= *PflSchutz-Nachr. Bayer* **20**(1).]

Dawson, J. R., Kilby, A. A. T., Ebben, Marion H. & Last, F. T. The use of steam/air mixtures for partially sterilizing soils infested with cucumber root rot pathogens. *Ann. appl. Biol.* **60**:215–22.

Doi, Y., Teranaka, M., Yora, K. & Asuyama, H. Mycoplasma- or PLT group-like microorganisms found in the phloem elements of plants infected with mulberry dwarf, potato witches' broom, aster yellows, or Paulowia witches' broom. *Ann. phytopath. Soc. Japan* **33**:259–66.

Edgington, L. V. & Barron, G. L. Fungitoxic spectrum of oxathiin compounds. *Phytopathology* **57**:1256–7.

Ishiie, T., Doi, Y., Yora, K. & Asuyama, H. Suppressive effects of antibiotics of the tetracycline group on symptom development of mulberry dwarf disease. *Ann. phytopath. Soc. Japan* **33**:267–75.

*McCallan, S. E. A. History of fungicides. In Torgeson (1967–8), **1**:1–37.

Morton, D. J. & Dukes, P. D. Serological differentiation of *Pythium aphanidermatum* from *Phytophthora parasitica* var. *nicotiana* and *Ph. parasitica. Nature* **213**:923.

Ravel d'Esclapon, G. de. La régéneration des variétés d'oeillots Américains par thermothérapie. *Rev. Hort.* **139**:1,227–9

*Raychaudhuri, S. P. Development of mycological and plant pathological research, education, and extension work in India. *Rev. appl. Mycol.* **46**:577–83.

Scheffer, R. P. & Nelson, R. R. Geographical distribution and prevalence of *Helminthosporium victoriae. Pl. Dis. Reptr* **51**:110–11.

Wood, R. K. S. *Physiological plant pathology.* Oxford & Edinburgh.

1967–8 Torgeson, D. C. (Editor). *Fungicides. An advanced treatise.* **1**, Agricultural and industrial applications, 1967; **2**, Chemistry and physiology, 1968. New York.

1968 *Anon. Organization for plant pathology in the British Isles. *Rev. appl. Mycol.* **47**:321–7.

Brenchley, G. H. Aerial photography for the study of plant diseases. *Ann. Rev. Phytopath.* **6**:1–21.

CMI. *Plant pathologists' pocketbook.* Kew (CMI).

Delp, C. J. & Klöpping, H. L. Performance attributes of a new fungicide and mite ovicide candidate. *Pl. Dis. Reptr* **52**:95–9.

Elias, R. S., Shephard, M. C. & Stubbs, J. 5-*n*-Butyl-2-dimethylamino-4-hydroxy-6-methylpyrimidine: a systemic fungicide. *Nature* **219**:1160.

*Fuller, J. G. *The day of St Antony's fire.* New York.

*Parris, G. K. *A chronology of plant pathology.* Starkville, Miss.

Raa, J. Polyphenols and natural resistance of apple leaves against *Venturia inaequalis. Neth. J. Pl. Path.* **74**, suppl. 1: 37–45.

Shaw, Dorothy E. Coffee rust outbreaks in Papua from 1892 to 1965 and the 1965 eradication campaign. *Res. Bull. (Pl. Path. Ser.) Dep. Agric. Port. Moresby* **2**:20–52. [See also *ibid.* **13**:33–7, 1975; *Papua New Guinea Agric. J.* **22**:59–61, 1970.]

Sheffield, Frances M. L. Closed quarantine procedures. *Rev. appl. Mycol.* **47**:1–8.

Shepherd, R. J., Wakeman, R. J. & Romanko, R. R. DNA in cauliflower mosaic virus. *Virology* **36**:150–2.

Vanderplank, J. E. *Disease resistance in plants.* New York & London.

1968–71 Martyn, E. B. (Editor). Plant virus names. Supplement 1. *Phytopath. Pap.* **9**, 1968, 1971. 204 pp.; 41 pp.

1969 *Ainsworth, G. C. History of plant pathology in Great Britain. *Ann. Rev. Phytopath.* **7**:13–30.

*Carefoot, G. L. & Sprott, E. R. *Famine on the wind.* London.

*Chamberlain, E. E. Plant pathology in New Zealand. *Rev. appl. Mycol.* **48**:1–10.

Clemons, G. P. & Sisler, H. D. Formation of a fungitoxic derivative from Benlate. *Phytopathology* **59**:705–6.

Hirumi, H. & Maramorosch, K. Mycoplasma-like bodies in the salivary glands of insect vectors carrying the aster yellows virus. *J. Virology* **3**:82–4.

*Kerling, C. C. P. Phytopathologisch Laboratorium 'Wille Commelin Scholten' 18 December 1894 – 18 December 1969. *Meded. Phytopath. Lab. WCS, Baarn* **75**:1–44.

*McCallan, S. E. A. A perspective on plant pathology. *Ann. Rev. Phytopath.* **7**:1–12.

Nyland, G. & Goheen, A. C. Heat therapy of virus diseases of perennial plants. *Ann. Rev. Phytopath.* **7**:331–54.

Waggoner, P. E. & Horsfall, J. C. EPIDEM. A simulator of plant disease written for a computer. *Bull. Conn. agric. Exp. Stn* **698**, 80 pp.

Wheeler, B. E. J. *An introduction to plant disease.* London.

1969–73 Robinson, R. A. Disease resistance terminology, *Rev. appl. Mycol.* **48**:593–606, 1969; Vertical resistance, *Rev. Pl. Path.* **50**:233–9, 1971; Horizontal resistance, *ibid.* **52**:483–501, 1973.

1970 *Bock, K. R. Plant pathology in East Africa. *Rev. Pl. Path.* **49**:1–6.

Bourke, P. M. A. Use of weather information in the prediction of plant disease epiphytotics. *Ann. Rev. Phytopath.* **8**:345–70.

Bové, F. J. *The story of ergot.* Basel.

*Chiarappa, L. Phytopathological organizations of the world. *Ann. Rev. Phytopath.* **8**:419–39.

Dement'eva, M. I. *Fitopatologiya.* Moscow. [Russ.]

*Estey, R. H. History and perspective of phytopathology in Quebec. *Phytoprotection* **51**:124–33.

*Fish, S. The history of plant pathology in Australia. *Ann. Rev. Phytopath.* **8**:13–36.

Garrett, S. D. *Pathogenic root-infecting fungi.* Cambridge.

Hansen, H. P. *Contribution to the systematic virology.* Copenhagen (Royal Vet. & Agric. Univ.).

Holliday, P. South American leaf blight (*Microcyclus ulei*) of *Hevea brasiliensis. Phytopath. Pap.* **12**, 31 pp.

Klinkowski, M. Catastrophic plant diseases. *Ann. Rev. Phytopath.* **8**:37–60.

Meredith, D. S. Banana leaf spot disease (Sigatoka) caused by *Mycosphaerella musicola* Leach. *Phytopath. Pap.* **11**, 147 pp.

*Tietz, Helga. One centennium of soil fumigation: its first years. In Toussoun *et al.* (1970):203–7.

Toussoun, T. A., Bega, R. V. & Nelson, P. E. (Editors). *Root diseases and soil-borne pathogens.* Berkeley, Calif.

Wellman, F. L. Announcement. Rust of coffee in Brazil. (The rust *Hemileia vastatrix* now firmly established on coffee in Brazil.) *Pl. Dis. Reptr* **54**:355 (539–41).

1971   Aelbers, E. Thiophanate and thiophanate-methyl: two new fungicides with systemic action. *Meded. Fak. Landb. wetensch. Gent* **36**:126–34.

Baker, K. F. Fire blight of pome fruits: the genesis of the concept that bacteria can be pathogenic for plants. *Hilgardia* **40**:603–33.

Booth, C. *The genus Fusarium.* Kew (CMI).

Flor, H. H. Current status of the gene-for-gene concept. *Ann. Rev. Phytopath.* **9**:275–96.

Glasscock, H. H. Fireblight epidemic among Kentish apple orchards in 1969. *Ann. appl. Biol.* **69**:137–45.

Graham, K. M. *Plant diseases in Fiji.* London.

*Orlob, G. B. History of plant pathology in the Middle Ages. *Ann. Rev. Phytopath.* **9**:7–20.

*Stubbs, L. L. Plant pathology in Australia. *Rev. Pl. Path.* **50**:461–78.

1972   Baker, J. J. Report on diseases of cultivated plants in England & Wales for the years 1957–1968. *Tech. Bull. Ministr. Agric. Fish. Food* **25**, 322 pp.

*Conners, I. B. (Editor) *Plant pathology in Canada.* Winnipeg (Can. Phytopath. Soc.).

Diener, T. O. Viroids. *Adv. Virus Res.* **17**:295–313.

Gibson, I. A. Dothistroma blight of *Pinus radiata. Ann. Rev. Phytopath.* **10**:51–72.

*Kiraly, Z. Main trends in the development of plant pathology in Hungary. *Ann. Rev. Phytopath.* **10**:9–20.

Kuć, J. Phytoalexins. *Ann. Rev. Phytopath.* **10**:207–32.

Marsh, R. W. (Editor). *Systemic fungicides.* London. [Edn 2, 1977.]

*Raychaudhuri, S. P., Verma, J. P., Nariani, T. K. & Sen, Bineeta. The history of plant pathology in India. *Ann. Rev. Phytopath.* **10**:21–36.

Schieber, E. Economic impact of coffee rust in Latin America. *Ann. Rev. Phytopath.* **10**:491–510.

Tarr, S. A. J. *The principles of plant pathology.* London.

Ullstrup, A. J. The impact of the Southern corn leaf blight epidemics of 1970–1971. *Ann. Rev. Phytopath.* **10**:37–50.

Waggoner, P. E., Horsfall, J. G. & Lukens, R. J. EIPMAY: a simulator of Southern corn leaf blight. *Bull. Conn. agric. Exp. Stn.* **729**, 84 pp.

Waller, J. M. Water-borne dispersal in coffee berry disease and its relation to control. *Ann. appl. Biol.* **71**:1–18.

Wellman, F. L. *Tropical American plant disease.* Metuchen, N.J.

1973 Cheremisinov, N. A. *Obshchaya patologiya rastenii.* Moscow. [Russ.]

Federation of Brit. Pl. Pathologists. A guide to the use of terms in plant pathology. *Phytopath. Pap.* **17**, 55 pp.

*Padmanabhan, S. Y. The great Bengal famine. *Ann. Rev. Phytopath.* **11**:11–26.

Erwin, D. C. Systemic fungicides: disease control, translocation, and mode of action. *Ann. Rev. Phytopath.* **11**:389–422.

FAO crop assessment methods. Farnham Royal (CAB).

Hopkins, D. L., French, W. J. & Mollenhauser, H. H. Association of a rickettsia-like bacterium with phony peach disease. *Phytopathology* **63**:443 (abst.).

Hopkins, D. L., Mortensen, J. A. & Adlerz, W. C. Protection of grapevines from Pierce's disease with tetracycline antibiotics. *Phytopathology* **63**:443 (abst.).

*Lazzari, G. *Storia della micologia Italiana.* Trento.

Nyland, G., Goheen, A. C., Lowe, S. K. & Kirkpatrick, H. C. The ultrastructure of a rickettsia-like organism from a peach infected with phony peach disease. *Phytopathology* **63**:1275–8.

*Orlob, G. B. Ancient and medieval plant pathology. *PflSchutz-Nachr. Bayer* **26**:(2):63–294.

Rangaswami, G. & Rajagopalan, S. *Bacterial plant pathology.* Coimbatore.

1974 *Akai, S. History of plant pathology in Japan. *Ann. Rev. Phytopath.* **12**:13–26.

Day, P. R. *Genetics of host-parasite interaction.* San Francisco.

Jenkyn, J. F. A comparison of seasonal changes in deposition of spores of *Erysiphe graminis* on different trapping surfaces. *Ann. appl. Biol.* **76**:257–67.

Kranz, J. (Editor). *Epidemics of plant disease. Mathematic analysis and modeling.* Berlin.

*Thompson, J. R. The International Seed Testing Association 1924–74. *Seed Sci. & Tech.* **2**:267–83.

1975 Dye, D. W. *et al.* Proposal for a reappraisal of the status of the names of plant pathogenic *Pseudomonas* species. *Internat. J. syst. Bact.* **25**:252–7.

*Estey, R. H. A note on Casimir-Joseph Davaine 1812–1882. *Agric. Hist.* **49**:549–52.

Kaspers, H. *et al.* 1,2,4-Triazole derivatives, a new class of protective and systemic fungicides. *Proc. VIII Internat. Pl. Prot. Congr. Moscow* Sect. III:395–401.

Krause, R. A., Massie, L. B. & Hyre, R. A. Blitecast: a computerized forecast of potato blight. *Pl. Dis. Reptr* **59**:95–8.

Nelson, R. R. (Editor). *Breeding plants for disease resistance. Concepts and applications.* University Park, Penn.

Shrum, R. Simulation of wheat stripe rust (*Puccinia striiformis* West.) using EPIMEMIC, a flexible plant disease simulator. *Progr. Rep. Pa Sta. Univ.* **347**, 81 pp.

*Ting, W. P. Plant pathology in Peninsular Malaysia. *Rev. Pl. Path.* **54**:297–305.

Vanderplank, J. E. *Principles of plant infection.* New York.

*Walker, J. C. Some highlights in plant pathology in the United States. *Ann. Rev. Phytopath.* **13**:15–29.

1976 Agricultural Development & Advisory Service. *Manual of plant growth stages and disease assessment keys.* Harpenden, Herts.

*Ainsworth, G. C. *Introduction to the history of mycology.* Cambridge.

Beale, Helen P. *Bibliography of plant viruses and index to research.* New York; London (Columbia Univ. Press). [29 000 refs.]

Dekker, J. Acquired resistance to fungicides. *Ann. Rev. Phytopath.* **14**:405–28.

*Ellis, D. E. *Plant pathology in North Carolina 1776–1976.* Raleigh, N.C. (Sch. Agric., N.C. Sta. Univ.).

*Faber, W. (Editor). *Land- und Forstwirtschaftliche Forschung in Oesterreich.* Band 7, 284 pp. Vienna.

Gibbs, A. & Harrison, B. D. *Plant virology. The principles.* London.

Hardison, J. R. Fire and flame for plant disease control. *Ann. Rev. Phytopath.* 14:355–79.

*Mathys, G. *et al.* 25° Anniversaire de l'OEPP/25th Anniversary of *EPPO.* Paris (EPPO).

*Muskett, A. E. Mycology and plant pathology in Ireland. *Proc. roy. Irish Acad.* B76(27):393–472.

*Nolla, J. A. B. & Fernandez Valiela, M. V. Contribution to the history of plant pathology in South America, Central America, and Mexico. *Ann. Rev. Phytopath.* 14:11–29.

*Ordish, G. *The constant pest: a short history of pests and their control.* London.

Robinson, R. A. *Plant pathosystems.* Berlin.

Tomlinson, J. A., Faithfull, E. M. & Ward, C. M. Chemical suppression of two virus diseases. *Ann. appl. Biol.* 84:31–41.

*Wilkinson, Lise. The development of the virus concept as reflected in corpora of studies on individual pathogens. 3. Lessons of the plant viruses – tobacco mosaic virus. *Med. Hist.* 20:111–34.

*Zadoks, J. C. & Koster, L. M. A historical survey of botanical epidemiology. A sketch of the development of ideas in ecological phytopathology. *Meded. Landb. Sch. Wageningen* 76(12), 56 pp.

1977   *Brader, L. *et al.* 25 years of FAO contributions to plant protection. *FAO Pl. Prot. Bull.* 25:145–209.

Byrde, R. J. W. & Willetts, H. J. *The brown rot fungi of fruit.* London.

*Kelman, A. & Cook, R. J. Plant pathology in the People's Republic of China. *Ann. Rev. Phytopath.* 15:409–29.

*Maan, G. C. & Zadoks, J. C. Trends in research reported in the Netherlands Journal of Plant Pathology, 1895–1973. *Neth. J. Pl. Path.* 83:91–5.

*Markham, R. Landmarks in plant virology: genesis of concepts. *Ann. Rev. Phytopath.* 15:17–39.

Neergaard, P. *Seed pathology.* 2 vols. London.

Schwinn, F. J., Staub, T. & Urech, P. A. A new type of fungicide against diseases caused by Oomycetes. *Meded. Fac. Landb. Rijksuniv. Gent* 42:1181–8.

Turner, Maxine P. & Henry, Caroll E. Plant pathology in Jamaica. *Rev. Pl. Path.* 56:659–68.

*Ubrizsy, A. A növénykortan útja. *Termeszet Vilaga 1977*(4): 181–4. [Hung.]

1977–8 Horsfall, J. G. & Cowling, E. B. (Editors). *Plant disease: an advanced* treatise. 1, How disease is managed, 1977; 2, How disease develops in populations; 3, How plants suffer from disease, 1978. New York.

1978   Corke, A. T. K. Microbial antagonism affecting tree diseases. *Ann. appl. Biol.* 89:89–93.

*Bitancourt, A. A. Phytopathology in a developing country [Brazil]. *Ann. Rev. Phytopath.* 16:1–18.

*Firman, I. D. Plant pathology in the region served by the South Pacific Commission. *Rev. Pl. Path.* 57:85–90.

Madden, L., Pennypacker, S. P. & MacNab, A. A. FAST, a forecast system for *Alternaria solani* on tomato. *Phytopathology* 68:1354–8.

Nelson, R. R. Genetics of horizontal resistance to plant disease. *Ann. Rev. Phytopath.* 16:359–78.

*Peacock, F. C. (Editor). *Jealott's Hill. Fifty years of agricultural research*. Bracknell, Berks. (Imperial Chem. Industr., Pl. Prot. Divn.)

Russell, G. E. *Plant breeding for pest and disease resistance*. London.

Vanderplank, J. E. *Genetic and molecular basis of plant pathogenesis*. Berlin.

Van Regenmortel, M. H. V. Applications of plant virus serology. *Ann. Rev. Phytopath.* **16**:57–81.

*Waterson, A. P. & Wilkinson, Lise. *An introduction to the history of virology*. Cambridge.

Young, J. M. *et al.* A proposed nomenclature and classification for plant pathogenic bacteria. *N.Z. J. agric. Res.* **21**:153–77.

1979   *Ebbels, D. L. A historical review of certification schemes for vegetatively-propagated crops in England and Wales. *ADAS Quart. Rev.* **32**:21–58.

Ebbels, D. L. & King, J. E. (Editors). *Plant health: the scientific basis for administrative control of plant diseases and pests*. Oxford.

*Kennedy, B. W., Widin, K. D. & Baker, K. F. Bacteria as the cause of disease in plants: a historical perspective. *ASM News* **45**:1–5.

*Moore, F. Joan. Presidential address. A discipline's debt to a learned society: plant pathology and the B.M.S. *Trans. Br. Mycol. Soc.* **73**:1–7.

*Noble, Mary. Outline of the history of seed pathology. In J. T Yorinori *et al.* (Editors), *Seed pathology – problems and progress*: 3–17. Paraná (Proc. 1st Latin American Workshop on seed pathology, 1977).

SOME LANDMARKS IN THE HISTORY OF
PLANT PATHOLOGY

1665 First illustration of a plant microfungus (rose rust) by Robert Hooke (p. 17).

1755 Wheat bunt shown to be seed-borne by Tillet (p. 38).

1802 Lime sulphur introduced by Forsyth (Table 3, p. 110).

1807 Experimental proof of fungal pathogenicity for plants by Prévost (p. 30).

1845–9 Potato blight epidemic in Ireland (p. 172).

1858 *Die Krankheiten der Kulturgewächse* by Julius Kühn (p. 36).

1865–6 Heteroecism of *Puccinia graminis* demonstrated experimentally by de Bary (p. 45).

1868–82 Coffee rust epidemic in Sri Lanka (p. 172).

1874 *Handbuch für Pflanzenkrankheiten* by Paul Sorauer (p. 230).

1878–85 Vine downy mildew epidemic in France (p. 111).

1885 Bordeaux mixture introduced by Millardet (Table 4, p. 111). Experimental proof of the bacterial cause of fire blight by Arthur (p. 65).

1886–98 Recognition of tobacco mosaic as a virus disease (Mayer, 1886; Ivanovski, 1892; Beijerinck, 1898) (p. 77).

1889 Introduction of hot water seed treatment against loose smut of wheat by Jensen (p. 136).

1891 *Zeitschrift für Pflanzenkrankheiten* began (p. 235). Nederlandsche Phytopatologische Vereeniging founded (p. 217).

1892 *Rivista di patologia vegetali* began (p. 234).

1894 Formae speciales of cereal rusts demonstrated by Eriksson (p. 48).

1895 *Tijdschrift over Plantenziekten* began (p. 234).

1902 First chair of plant pathology established in Copenhagen (p. 221).

1904 Cereal resistance to rust shown to be inherited as a Mendelian character by Biffen (p. 163).

1907 First university department of plant pathology founded at Cornell University, New York (p. 226). Destructive Insects & Pests Act passed in England (p. 180).

1908 American Phytopathological Society founded (p. 217).

1911 *Phytopathology* began (p. 234).

1913 Recognition of physiologic races of *Puccinia graminis* by Stakman (p. 48). Organic fungicides introduced (Table 5, p. 114).

1914 International Phytopathological Convention of Rome [never ratified] (p. 183).

1916 Phytopathological Society of Japan founded (p. 218).

1920 Commonwealth Mycological Institute founded (as 'Imperial Bureau of Mycology') (p. 214).

Introduction to bacterial diseases of plants by E. F. Smith (p. 70).

1922 *Review of plant pathology* (as *Review of applied mycology*, 1922–69) began (p. 214).

1923 Boron demonstrated to be an essential trace element for plant growth by Kathleen Warington (p. 101).

1929 International Convention for the Protection of Plants (p. 184).

1937 *A text-book of plant virus diseases* by K. M. Smith (p. 279).

1943 *The diagnosis of mineral deficiencies in plants* by T. Wallace (p. 101).

1946 *Pflanzliche Infektionslehre* by Ernst Gäumann (p. 175).

1951 European and Mediterranean Plant Protection Organisation (EPPO) founded (p. 186).

FAO International Plant Protection Convention (IPPC) (p. 185).

1963 *Annual review of phytopathology* began (p. 236).

Plant diseases: epidemics and control by J. E. Vanderplank (p. 175).

1967 Recognition of plant pathogenic mycoplasma-like organisms in Japan (p. 97).

1968 First International Congress for Plant Pathology held in London (p. 218).

NAMES INDEX

Adanson, M., 21, 252, 262
Aderhold, R. F. T., 56, 199, 252, 267–8
Adlerz, W. C., 291
Aelbers, E., 120, 290
Agulhon, H., 101, 169
Ainsworth, G. C., 58, 250–1, 279, 289, 291
Akai, S., 53, 59, 291
Alben, A. O., 101, 277
Albertus Magnus, 41
Alcock, N. L., 219, 252
Allard, C., 120, 287
Allard, H. A., 78, 83, 88, 252, 271
Allen, J. M., 278
Alston, R. A., 160, 284
Anderson, H. W., 121, 252, 282
Anderson, J., 248
Anderson, P. J., 172, 271
Anderson, P. M., 89, 288
Angell, H. R., 62, 132, 276, 278
Angus, A., 287
Anslow, W. K., 118, 281
Appel, F. C. L. O., 83, 199, 219, 252, 270
d'Arbois de Jubainville, 264
Arnaud, G., 199
Arthur, J. C., 65–6, 75, 123, 226, 252, 265, 267, 269, 275
Ashby, S. F., 209, 252
Asuyama, H., 288
Atanasoff, D. A., 248, 273, 277
Atkinson, G. F., 252
Atkinson, J. D., 103, 282
Austen, R., 162, 252, 261

Bacon, C. W., 273
Bacon, F., 122

Baigent, N. L., 288
Bailey, M. A., 272
Bailey, S. T., 284
Baker, J. J., 65, 168, 172–3, 181, 232, 290
Baker, K. F., 64, 136, 157, 286–7, 290, 293
Baker, R., 156
Baker, R. E. D., 168, 173, 285
Bald, J. G., 86, 276
Balls, W. L., 213
Bancroft, C. K., 252, 269, 270
Banks, J., 43, 54, 252, 262
Bant, J. H., 139, 283
Barclay, A., 210
Barger, G., 18, 172, 276
Barratt, R. W., 114, 282
Barron, G. L., 288
Bartholomew, E. T., 100, 270
de Bary, A., 34–6, 45, 54–5, 57, 59, 179, 220–1, 229, 252, 263–5
Basil the Great, 18
Bassus, C., 14, 261
Batista, A. C., 209
Baudys, 285
Bauer, E., 79, 253, 268
Bauer, F. A., 34, 247, 262
Bawden, F. C., 78, 84, 87–9, 173, 218, 253, 278–9, 282, 287–8
Beach, S. A., 266
Beale, H. P. (see also Purdy), 82, 92, 276, 291
Bean, J., 284
Beaumont, A., 84, 151, 169, 277, 282
Bega, R. V., 290
Beijerinck, M. W., 77, 221, 253, 267
Beinhart, E. G., 136–272
Bennett, C. W., 90, 280

SUBJECT INDEX